CW01263248

Simulation of
Electrochemical Processes II

WITPRESS

WIT Press publishes leading books in Science and Technology.
Visit our website for new and current list of titles.
www.witpress.com

WITeLibrary

Home of the Transactions of the Wessex Institute.
Papers presented at ELECTROCOR 2007 are archived in the WIT eLibrary
in volume 54 of WIT Transactions on Engineering Sciences (ISSN 1743-3533).
The WIT eLibrary provides the international scientific community with immediate and
permanent access to individual papers presented at WIT conferences.
Visit the WIT eLibrary at www.witpress.com.

SECOND INTERNATIONAL CONFERENCE ON
SIMULATION OF ELECTROCHEMICAL PROCESSES

ELECTROCOR 2007

CONFERENCE CHAIRMEN

V. G. DeGiorgi
Naval Research Laboratory, USA

C. A. Brebbia
Wessex Institute of Technology, UK

R. A. Adey
Wessex Institute of Technology, UK

INTERNATIONAL SCIENTIFIC ADVISORY COMMITTEE

K. Amaya	E. J. Lemieux	D. A. Shifler
S. G. R. Brown	P. Mandin	A. Taleb
J. Deconinck	A. Peratta	J. Toribio
A. Jivkov		

Organised by
Wessex Institute of Technology, UK

Sponsored by
WIT Transactions on Engineering Sciences

WIT Transactions on Engineering Sciences

Transactions Editor
Carlos Brebbia
Wessex Institute of Technology
Ashurst Lodge, Ashurst
Southampton SO40 7AA, UK
Email: carlos@wessex.ac.uk

Editorial Board

B. Abersek
University of Maribor
Slovenia

B Alzahabi
Kettering University
USA

K S Al Jabri
Sultan Qaboos University
Oman

A G Atkins
University of Reading
UK

J A C Ambrosio
IDMEC
Portugal

A F M Azevedo
University of Porto
Portugal

H Azegami
Toyohashi University of Technology
Japan

R Belmans
Katholieke Universiteit Leuven
Belgium

G Belingardi
Politecnico di Torino
Italy

E Blums
Latvian Academy of Sciences
Latvia

S K Bhattacharyya
Indian Institute of Technology
India

F-G Buchholz
Universitat Gesanthochschule Paderborn
Germany

A R Bretones
University of Granada
Spain

W Cantwell
Liverpool University
UK

J Byrne
University of Portsmouth
UK

S K Chakrabarti
Offshore Structure Analysis
USA

D J Cartwright
Bucknell University
USA

H Choi
Kangnung National University
Korea

A Chakrabarti
Indian Institute of Science
India

L De Biase
University of Milan
Italy

J J Connor
Massachusetts Institute of Technology
USA

R de Borst
Delft University of Technology
Netherlands

L Debnath
University of Texas-Pan American
USA

G De Mey
Ghent State University
Belgium

S del Giudice
University of Udine
Italy

M Domaszewski
Universite de Technologie de Belfort-Montbeliard
France

I Doltsinis
University of Stuttgart
Germany

J Dominguez
University of Seville
Spain

J P du Plessis
University of Stellenbosch
South Africa

M E M El-Sayed
Kettering University
USA

M Faghri
University of Rhode Island
USA

C J Gantes
National Technical University of Athens
Greece

R Gomez Martin
University of Granada
Spain

R H J Grimshaw
Loughborough University
UK

R Grundmann
Technische Universitat Dresden
Germany

J M Hale
University of Newcastle
UK

L Haydock
Newage International Limited
UK

C Herman
John Hopkins University
USA

M Y Hussaini
Florida State University
USA

D B Ingham
The University of Leeds
UK

Y Jaluria
Rutgers University
USA

D R H Jones
University of Cambridge
UK

S Kim
University of Wisconsin-Madison
USA

A S Kobayashi
University of Washington
USA

S Kotake
University of Tokyo
Japan

W Dover
University College London
UK

K M Elawadly
Alexandria University
Egypt

F Erdogan
Lehigh University
USA

H J S Fernando
Arizona State University
USA

E E Gdoutos
Democritus University of Thrace
Greece

D Goulias
University of Maryland
USA

D Gross
Technische Hochschule Darmstadt
Germany

R C Gupta
National University of Singapore,
Singapore

K Hameyer
Katholieke Universiteit Leuven
Belgium

P J Heggs
UMIST
UK

D A Hills
University of Oxford
UK

T H Hyde
University of Nottingham
UK

N Ishikawa
National Defence Academy
Japan

N Jones
The University of Liverpool
UK

T Katayama
Doshisha University
Japan

E Kita
Nagoya University
Japan

A Konrad
University of Toronto
Canada

T Krauthammer
Penn State University
USA

F Lattarulo
Politecnico di Bari
Italy

M Langseth
Norwegian University of Science and Technology
Norway

S Lomov
Katholieke Universiteit Leuven
Belgium

G Manara
University of Pisa
Italy

H A Mang
Technische Universitat Wien
Austria

A C Mendes
Univ. de Beira Interior
Portugal

A Miyamoto
Yamaguchi University
Japan

G Molinari
University of Genoa
Italy

D B Murray
Trinity College Dublin
Ireland

S-I Nishida
Saga University
Japan

B Notaros
University of Massachusetts
USA

M Ohkusu
Kyushu University
Japan

P H Oosthuizen
Queens University
Canada

G Pelosi
University of Florence
Italy

H Pina
Instituto Superior Tecnico
Portugal

L P Pook
University College London
UK

D Prandle
Proudman Oceanographic Laboratory
UK

F Rachidi
EMC Group
Switzerland

K R Rajagopal
Texas A & M University
USA

D N Riahi
University of Illinios-Urbana
USA

Y-W Mai
University of Sydney
Australia

B N Mandal
Indian Statistical Institute
India

T Matsui
Nagoya University
Japan

R A W Mines
The University of Liverpool
UK

T Miyoshi
Kobe University
Japan

T B Moodie
University of Alberta
Canada

D Necsulescu
University of Ottawa
Canada

H Nisitani
Kyushu Sangyo University
Japan

P O'Donoghue
University College Dublin
Ireland

K Onishi
Ibaraki University
Japan

E Outa
Waseda University
Japan

W Perrie
Bedford Institute of Oceanography
Canada

D Poljak
University of Split
Croatia

H Power
University of Nottingham
UK

I S Putra
Institute of Technology Bandung
Indonesia

M Rahman
Dalhousie University
Canada

T Rang
Tallinn Technical University
Estonia

B Ribas
Spanish National Centre for Environmental Health
Spain

W Roetzel
Universitaet der Bundeswehr Hamburg
Germany

K Richter
Graz University of Technology
Austria

S Russenchuck
Magnet Group
Switzerland

V Roje
University of Split
Croatia

B Sarler
Nova Gorica Polytechnic
Slovenia

H Ryssel
Fraunhofer Institut Integrierte Schaltungen
Germany

R Schmidt
RWTH Aachen
Germany

A Savini
Universita de Pavia
Italy

A P S Selvadurai
McGill University
Canada

B Scholtes
Universitaet of Kassel
Germany

L C Simoes
University of Coimbra
Portugal

G C Sih
Lehigh University
USA

J Sladek
Slovak Academy of Sciences
Slovakia

P Skerget
University of Maribor
Slovenia

D B Spalding
CHAM
UK

A C M Sousa
University of New Brunswick
Canada

G E Swaters
University of Alberta
Canada

C-L Tan
Carleton University
Canada

J Szmyd
University of Mining and Metallurgy
Poland

A Terranova
Politecnico di Milano
Italy

S Tanimura
Aichi University of Technology
Japan

S Tkachenko
Otto-von-Guericke-University
Germany

A G Tijhuis
Technische Universiteit Eindhoven
Netherlands

E Van den Bulck
Katholieke Universiteit Leuven
Belgium

I Tsukrov
University of New Hampshire
USA

R Verhoeven
Ghent University
Belgium

P Vas
University of Aberdeen
UK

B Weiss
University of Vienna
Austria

S Walker
Imperial College
UK

T X Yu
Hong Kong University of Science & Technology
Hong Kong

S Yanniotis
Agricultural University of Athens
Greece

M Zamir
The University of Western Ontario
Canada

K Zakrzewski
Politechnika Lodzka
Poland

Simulation of Electrochemical Processes II

Editors

V. G. DeGiorgi
Naval Research Laboratory, USA

C. A. Brebbia
Wessex Institute of Technology, UK

R. A. Adey
Wessex Institute of Technology, UK

WITPRESS Southampton, Boston

Editors:

V. G. DeGiorgi
Naval Research Laboratory, USA

C. A. Brebbia
Wessex Institute of Technology, UK

R. A. Adey
Wessex Institute of Technology, UK

Published by

WIT Press
Ashurst Lodge, Ashurst, Southampton, SO40 7AA, UK
Tel: 44 (0) 238 029 3223; Fax: 44 (0) 238 029 2853
E-Mail: witpress@witpress.com
http://www.witpress.com

For USA, Canada and Mexico

Computational Mechanics Inc
25 Bridge Street, Billerica, MA 01821, USA
Tel: 978 667 5841; Fax: 978 667 7582
E-Mail: infousa@witpress.com
http://www.witpress.com

British Library Cataloguing-in-Publication Data

A Catalogue record for this book is available
from the British Library

ISBN: 978-1-84564-071-2
ISSN: 1746-4471 (print)
ISSN: 1743-3533 (on-line)

The texts of the papers in this volume were set individually by the authors or under their supervision. Only minor corrections to the text may have been carried out by the publisher.

No responsibility is assumed by the Publisher, the Editors and Authors for any injury and/or damage to persons or property as a matter of products liability, negligence or otherwise, or from any use or operation of any methods, products, instructions or ideas contained in the material herein.

© WIT Press 2007

Printed in Great Britain by Athenaeum Press Ltd.

All rights reserved. No part of this publication may be reproduced, stored in a retrieval system, or transmitted in any form or by any means, electronic, mechanical, photocopying, recording, or otherwise, without the prior written permission of the Publisher.

Preface

This book contains the majority of papers presented at the Second International Conference on the Simulation of Electrochemical Processes held in Myrtle Beach, USA in May 2007. The meeting was organized by the Wessex Institute of Technology.

The first ELECTROCOR conference held in 2005 was successful and brought about meaningful interchanges between researchers. As in the first conference, the purpose of the meeting was to bring together researchers from a variety of fields that have a common focus on computational modeling of electrochemical processes. Electrochemical processes are the basis of diverse areas of interest ranging from corrosion to fabrication processes. While computational modeling is stressed, the need to understand the underlying physical phenomena is not forgotten. Also, the linkage between computational and experimental methods is highlighted in several papers in this volume.

The papers in this book are divided into the following main topics:
- Modelling methodologies
- Cathodic protection systems
- Experimental measurements and computer results
- Interference and signature control
- Stress corrosion, cracking and corrosion fatigue
- Detection and monitoring of corrosion

The Editors are indebted to the members of the International Scientific Advisory Committee for their help in reviewing abstracts and papers.

The Editors
Myrtle Beach, 2007

Contents

Section 1: Modelling methodologies

Micro-scale computer modelling of the relationship between
metallurgical microstructure and localized corrosion effects
S. G. R. Brown & N. C. Barnard 3

Mesoscopic modeling of corrosion processes:
pitting morphology evolution
A. Taleb, C. Vautrin-Ul, H. Mendy, J. Stafiej & A. Chausse 13

Dual reciprocity boundary element method for iron corrosion in
acidic solution
A. Peratta 23

An analytical modeling method for calculating the maximum
cathode current deliverable by a circular cathode under
atmospheric exposure
Z. Y. Chen & R. G. Kelly 33

Phase field modeling of phase boundary motion due to
transport-limited electrochemical reactions
A. Powell & W. Pongsaksawad 43

Kinetic Monte Carlo modelling the leaching of Raney Ni-Al alloys
N. C. Barnard & S. G. R. Brown 53

Dynamic simulation of a deposition process
J. P. Caire & A. Javidi 63

Two-phase electrolysis process modelling: from the bubble to the
electrochemical cell scale
P. Mandin, H. Roustan, R. Wüthrich, J. Hamburger & G. Picard 73

Simulating electro-coating of automotive body parts using BEM
J. M. W. Baynham, R. Adey, V. Murugaian & D. Williams 89

Section 2: Cathodic protection systems

3D cathodic protection design of ship hulls
L. Bortels, B. Van den Bossche, M. Purcar, A. Dorochenko & J. Deconinck 103

Validation plan for boundary element method modeling of impressed current cathodic protection system design and control response
E. A. Hogan, J. E. McElman, E. J. Lemieux, M. S. Krupa, V. G. DeGiorgi & A. L. LeDoux 113

Predicting the performance of cathodic protection systems with large scale interference
R. Adey, J. M. W. Baynham & T. Curtin 123

Cathodic protection application possibilities in activity building of overhead lines: protection of armature tower foundations
A. Muharemovic, I. Turkovic & S. Bisanovic 133

Section 3: Experimental measurements and computer results

Dipole modelling and sensor design
S. A. Wimmer, E. A. Hogan & V. G. DeGiorgi 143

Transport phenomena in an electrochemical rotating cylinder reactor
F. Tomasoni, J. F. Thomas, D. Yildiz, J. van Beeck & J. Deconinck 153

Simulation of the electrical transient during the porous anodizing of pure aluminium substrate
F. Le Coz, L. Arurault & R. S. Bes 163

Extraction of a quantitative reaction mechanism from linear sweep voltammograms obtained on a rotating disk electrode
E. Tourwé, T. Breugelmans, R. Pintelon & A. Hubin 173

IRDE and RDE electrochemical cells evaluation: comparison of electron and mass transfer
H. Van Parys, E. Tourwé, M. Depauw, T. Breugelmans, J. Deconinck & A. Hubin 183

Experimental study and modelling of heat transfer during anodizing
in a wall-jet set-up
T. Aerts, G. Nelissen, J. Deconinck, I. De Graeve & H. Terryn 193

Scanning Vibrating Electrode Technique as an application for
measuring corrosion activity of carbon steel welded pipelines
A. Abdurrahim & R. Akid 203

Section 4: Interference and signature control

Predicting corrosion related signatures
R. Adey & J. M. W. Baynham 213

Fast solution techniques for corrosion and signatures modelling
A. J. Keddie, M. D. Pocock & V. G. DeGiorgi 225

Corrosion related electromagnetic signatures measurements and
modelling on a $1:40^{th}$ scaled model
*L. Demilier, C. Durand, C. Rannou, E. Hogan, M. Krupa,
A. M. Grolleau, J. Blanc & A. Guibert* 235

Section 5: Stress corrosion, cracking and corrosion fatigue

Modelling environmentally assisted cracking in pipeline steels
A. Plumtree & S. B. Lambert 247

On the SCC behaviour of austenitic stainless steels in boiling
saturated magnesium chloride solution
O. M. Alyousif & R. Nishimura 257

Electrochemical behaviour and corrosion sensitivity of prestressed
steel in cement grout
*E. Blactot, C. Brunet-Vogel, F. Farcas, L. Gaillet, I. Mabille,
T. Chaussadent & E. Sutter* 267

Sub-Modelling Boundary Element approach for stress
concentration in samples exposed to pitting corrosion
A. Peratta & C. Brebbia 277

Corrosion of reinforcing steel in reinforced and prestressed
concrete bridges
A.-K. M. Hussain & A. M. Rifai 287

Section 6: Detection and monitoring of corrosion

Corrosion detection by multi-step genetic algorithm
K. Amaya, M. Ridha & S. Aoki 299

A reactive transport model for evaluating the long-term performance of stainless steels in concrete
M. Boulfiza 309

Characteristic evaluation of corrosion of aluminum-based rigid conductor line
S. Mohri & Y. Sato 319

Special weathering steel, a contribution to environmental protection
E. S. Ayllón 329

Author Index 339

Section 1
Modelling methodologies

Micro-scale computer modelling of the relationship between metallurgical microstructure and localized corrosion effects

S. G. R. Brown & N. C. Barnard
Materials Research Centre, School of Engineering,
Swansea University, Singleton Park, Swansea, UK

Abstract

An attempt is made to model the corrosion effects experienced by zinc and zinc–aluminium coating alloys in contact with conductive, chloride containing aqueous electrolytes. 3D predictions of localized corrosion of zinc alloy coatings are made using a control-volume finite difference approach, where an irregular structured mesh is employed. The model deals with the movement of multiple evolving conserved concentration fields. Perturbations in concentration are used to determine the conditions at an exposed surface, in addition to influencing any processes operating. The effect of coupled-materials and electrolyte present on the progression of localized corrosion processes is calculated using established relationships and experimentally derived data.

Quantitative predictions of the corrosion performance at the exposed surface and cut-edge to electrolyte (used in the accelerated corrosion testing of these coating alloys) are made. Comparison is made to the predicted form of the current density fields around corroding surfaces and those observed experimentally. In addition, absolute metal losses calculated across exposed areas of the simulated alloys are compared with those estimated during experimental investigations.

Keywords: zinc alloy coatings, localized dissolution, finite difference method.

1 Introduction

As the cost of corrosion processes accounts for *ca* 5% GDP of the developed nations, it is no wonder that a great deal of interest exists in protecting the world's infrastructure and durable goods. The use of steel in many applications means that the protection of these goods/structures against the deleterious effects of corrosion processes is of considerable concern. For in excess of a century, zinc has played a major role in the efforts of corrosion scientists and engineers to protect steel against corrosive attack. The use of zinc and its alloys in this respect derives from its ability to provide both barrier and sacrificial protection to the electrically-connected steel.

Zinc may be used in a completely sacrificial capacity, although it is more commonplace to coat a substrate (predominantly strip steel) either via electroplating or more commonly via hot-dip methods. The hot-dipping process consists of the continuous-feeding of a steel strip through a molten zinc alloy. The composition of this alloy, the processing conditions [1] and the complex bath chemistry taking-place ultimately lead to the formation of widely varying microstructures observed for the different Zn alloys produced. For example, a common addition the hot-dip bath is Al that can be added at low levels during the production of ordinary galvanized steels and an increased level (eutectic composition) of ~4.5wt.% Al for a Galfan-type alloy. An increase to 55wt.% Al produces a Galvalume ZnAl alloy, all of which exhibit different microstructures owing to the processes taking place prior to, during and post hot-dipping, e.g. reaction-diffusion at the solid-liquid interface. Other factors such as texture [2] play a role in the corrosion performance, although these are not considered here. Further additions to the bath modify the microstructure further, leading to a wide variety of Zn alloy steel coatings with characteristic corrosion properties.

There is significant interest in the corrosion performance Zn-Al alloy coated steel in the service environment. To this end many attempts to assess the degradation experienced in aqueous and atmospheric [3] environments over different timescales are made. Scanning electrochemical techniques [4] and impedance spectroscopy [5] are often employed in order to predict the degradation of a Zn alloy coating in an aggressive chloride-containing electrolyte in order to elucidate the corrosion behaviour in the service environment. The focus of this paper is an attempt to model the corrosion behaviour of Zn/ZnAl coatings in 5% NaCl solutions: typically used in accelerated testing using the scanning vibrating electrode technique (SVET). This permits validation of the computer model using experimentally determined data.

This paper outlines an attempt to model the localized dissolution of zinc from zinc and zinc–aluminium alloy coatings used in the protection of steel. Developed from a metallurgical standpoint, the aim is to quantitatively predict the ionic evolution of multiple species from the coating layers as well as the temporal changes in electrical potential in the electrolyte brought about by metallic dissolution. In order to capture the dominant processes taking place during the dissolution from the steel coatings and the effects of the microstructures present, the model has been developed at the micro-mesoscale. A 3D irregular orthogonal grid is used to represent the solid and liquid phases present. Finite difference calculations of potential and concentration are performed on this structured mesh, thus predicting the localized dissolution from the coating layers of interest using established electrochemical relationships.

2 Dissolution modelling in aqueous NaCl

The algorithm presented is an extension of a first order model developed previously, whereby the evolving ionic concentration was approximated by the changes in effective free proton concentration [6,7]. All simulations are performed at pH neutral conditions and are isothermal at 298K.

2.1 Computational domain

As previously stated, the model is designed to operate at the micro-mesoscale. A 3D irregular structured orthogonal mesh is used to represent steel and coatings in contact with an electrolyte using a typical area of exposure of about 1mm^2. A usual setup for the grid would be the allocation of half of the cells present as solid. These cells are labelled as being zinc, steel or zinc–aluminium sites. A solid site contains information about the composition, volume fraction that is solid (f), and electrical potential (ϕ). Each liquid site possesses an electrical potential (ϕ), volume fraction of corrosion product (p), current density vector (i_x, i_y, i_z) and concentrations (c) of the conserved fields; [M^{n+}], [O$_2$], [OH$^-$] and [H$^+$]. Variable grid spacings ($\Delta x, \Delta y, \Delta z$) for each cell in the mesh are also stored.

The representation of the microstructure is entirely synthetic and a near eutectic composition of Zn – 4.5wt.% Al is used, since the intermetallic formation in this hot-dip coating is limited. The microstructure of this Galfan-type coating is one of primary zinc crystals within a ZnAl eutectic that contains approximately 5wt.% Al.

2.2 Metallic dissolution

The basis of the model is the removal of metallic constituents from the coating layers via equation 1. This then brings about concentration perturbations, altering the surface/bulk electrolyte conditions and thereby the cathodic protection offered to the steel.

$$M_{(s)} \rightarrow M^{n+}_{(aq)} + ne^- \tag{1}$$

$$O_{2(g)} + 2H_2O_{(l)} + 4e^- \rightarrow 4OH^-_{(aq)} \tag{2}$$

The assumption is made that any dissolution taking place is fully supported cathodically by oxygen reduction (equation 2). Anodic and cathodic sites are determined at the solid/liquid interface according to the phases in contact with the liquid. At the beginning of each iteration, the exposed areas of each phase are assessed in order to determine any galvanic effects present, in addition to determining whether the interfacial areas are acting in an anodic or cathodic manner.

2.3 Electrode potential

An electrode potential is calculated for all liquid sites adjacent to solid phases using the Nernst equation.

$$E = E_0 + \frac{RT}{nF} \ln\left(\frac{[Ox]}{[Red]}\right) \qquad (3)$$

where E_0 is the standard electrode potential (V), R is the universal gas constant (Jmol^{-1}K^{-1}), F is Faraday's Constant (Cmol^{-1}), T is the temperature (K) and [Ox] and [Red] are the concentrations (mol.dm^{-3}) of the oxidized and reduced species respectively. If the site resides next to an anodic site (zinc rich phase) then the standard electrode potential for that metal would be used and the oxidized species would be the concentration of metal cations present. Conversely, if the site is adjacent to a material predicted to behave cathodically, then the standard electrode potential for oxygen reduction in neutral conditions is used and the relative amounts of OH$^-$ and O_2 is taken into account. Should a liquid site be bounded by both anodic and cathodic material then the electrode potential is taken as the weighted geometric mean of all the faces in contact with conductive solid.

2.4 Interface potential and corrosion rate

The electrical potential is calculated throughout the metallic solid present using a cell-centred control-volume finite difference approach to solve the Laplace equation to describe the steady-state voltage fields.

$$\nabla(-\kappa\nabla\phi) = 0 \qquad (4)$$

Where κ is the electrical conductivity (Ω^{-1}m^{-1}). The electrode potential is used as a fixed boundary condition at the exposed faces of the simulated solid and insulation conditions are applied to all other sides. A harmonic mean is used at the solid-liquid interface to determine the conductivity at these sites,

$$\kappa = \frac{2.\kappa_S \kappa_L}{\kappa_S + \kappa_L} \qquad (5)$$

Next, the voltages of the liquid sites at the solid-liquid interface are determined using the voltages in neighbouring solid sites, weighting each value according to interfacial area. These voltages are then used as Dirichlet boundary conditions in determining the steady-state electrical potential field throughout the electrolyte using the Laplace equation (4), assuming uniform conductivity. Again, insulation conditions are applied at all other sides of the mesh.

Having determined those phases that will act anodically and the potentials experienced at the interface, a decision is made as to what amount of material may be removed from each anodic site. This is calculated from experimentally derived data using the rotating disc electrode technique, which relates potential to current density. The local potential in the Neumann neighbourhood of an anodic site is calculated, giving a current density that is used to estimate the dissolution from the site using the area of contact. The cumulative total of electrons to be transferred during dissolution, M_A, is given by,

$$M_A = \sum_{i=1}^{l} \frac{n(\delta f_i \Delta x_i \Delta y_i \Delta z_i)\rho_i}{M_r} \quad (6)$$

where l is the number of sites corroding, f_i is the volume fraction of the cell that is permitted to transform at that potential, n is the number of electrons to be transferred in the oxidation of a phase, ρ_i is the density (kg.m^{-3}) and M_r is the molecular mass (kg.mol^{-1}) of the phase.

2.5 Cathodic capacity

Assuming that the anodic dissolution taking place is fully supported cathodically by the reduction of oxygen, equation 2, it is necessary to calculate the amount of oxygen that is available to the cathodic sites. This is the amount of oxygen that is within the diffusion length from the solid liquid interface. In our case the oxygen that is considered to be available for the cathodic processes occupies a position in the y direction from the surface that satisfies equation 7.

$$y_j - y_{surf} \leq \sqrt{4D\Delta t} \quad (7)$$

where D is the diffusivity of O_2 (m^{2}s^{-1}) and Δt is the time increment (s). The maximum amount of dissolution that can be supported can thus be calculated, which may result in a scaling down of the anodic processes taking place. Once the amount of material removed is determined, the volume fraction of the cells is updated, along with concentration increases made to these cells in terms of cation and OH⁻ at anodic and cathodic sites respectively. Decreases in dissolved oxygen levels at the cathode surface are considered later in the algorithm.

2.6 Corrosion product deposition

Corrosion products are deposited at the surface of the simulated solid within arbitrary voltage tolerances and to an equivalent of 5% of the volume corroded within a time increment.

$$Zn^{2+}_{(aq)} + H_2O_{(l)} \rightarrow Zn(OH)^+_{(aq)} + H^+_{(aq)} \quad (8)$$

$$Zn(OH)^+_{(aq)} + H_2O_{(l)} \rightarrow Zn(OH)_{2(s)} + H^+_{(aq)} \quad (9)$$

This is done to follow the processes outlined in equations 8 and 9, whereby the metal cations, predominantly Zn^{2+}, emanating from dissolution events taking place on the coating layers rapidly undergo hydrolysis reactions to produce a hydroxide film in solution. This is characterized in the model by a decrease in $[M^{n+}]$, an increase in $[H^+]$ and the assignment of a value for volume fraction of a liquid cell occupied by corrosion product (p). This is distributed evenly amongst those cells that satisfy the tolerances arbitrarily set out in the model.

2.7 Mass transport

The evolution of all of the concentration fields in the 5% NaCl solution is carried out in the model using equation 10, describing the ordinary diffusion (left-hand term) and migration effects (right-hand term) experienced by a charged particle in an electrical field.

$$\frac{\partial c_i}{\partial t} = -\nabla(D_i \nabla c_i) + z_i F \nabla (u_i c_i \nabla \phi) \qquad (10)$$

D_i is the diffusion coefficient (m²s⁻¹) for species i, z_i is the unit charge and u_i is the ionic mobility (m²/(Vs)) calculated using the Nernst-Einstein equation:

$$u_i = D_i \frac{z_i F}{RT} \qquad (11)$$

The diffusion coefficient of each species is altered at electrolyte sites containing corrosion products by,

$$D_i^* = D_i (1-p)^2 \qquad (12)$$

Clearly, if p reaches unity then this cell becomes a complete barrier to diffusion. Convection is assumed to be negligible throughout the electrolyte and movement of dissolved oxygen is considered only with respect to Fickian diffusion effects.

The concentration fields are calculated iteratively using a cell-centred finite-difference method [8]. During this procedure the dissolved oxygen content of the cells adjacent to cathodic sites is continually reduced until the amount necessary to facilitate the preceding anodic dissolution is reached. After this is achieved, ordinary diffusion proceeds without the removal of any O_2 concentration. During the diffusion calculation insulation boundary conditions are applied at the solid-liquid interface and sides of the mesh, the top of the mesh is fixed at the initial concentration.

In order to maintain the equilibrium of the ionization of water, a procedure is undertaken post-diffusion and/or migration. This is to ensure that the corresponding [H⁺] and [OH⁻] fields are such that the ionic product remains 10^{-14}, This procedure is carried out as to return the concentration fields to equilibrium, although still allowing fluctuations in local pH.

3 Results and discussion

The results presented here are representative of those obtained from the model when simulating the destruction of synthetic microstructures in 5% NaCl solutions. Concentration fields for evolving species characteristic of the anodic and cathodic processes are depicted. The normalized form of current density fields shown, taken from the steady-state voltage fields, agree with numerical solutions [9].

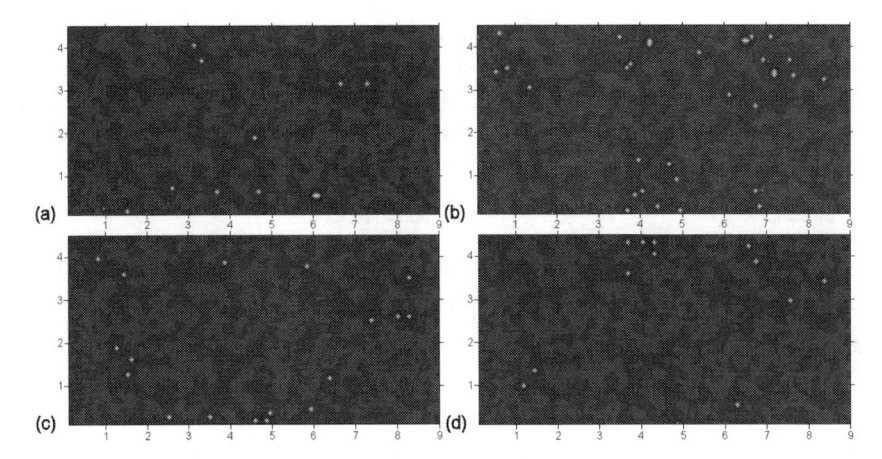

Figure 1: Vertical current density maps 100μm above an EZ coating after (a)12, (b)14, (c)17, and (d)19 hours, showing anodic (light) and cathodic (dark) areas. (Dimensions in mm).

3.1 Simulation of corrosion on pure zinc

In contrast to hot-dipping, electroplating a steel substrate (EZ) yields a virtually pure layer of zinc, effectively absent of spatial variations in microstructure. Figure 1 shows the movement of anodic current densities across the surface of EZ-coated steel, which is proportional to the corrosion taking place at the surface of the simulated solid. Since the time between sequential maps is relatively small, this indicates that the location of anodic dissolution in this case is determined by the instantaneous local fluctuations in cation concentration. In addition to this the location of the cathodic activity is influenced by the differential aeration experienced across the coating.

3.2 Influence of Hot-Dip Zn – 4.5wt.% Microstructure

The microstructure of a Galfan coating consists of primary zinc dendritic regions within ZnAl eutectic. In constructing this microstructure in the model the anodic and cathodic sites can be seen to occur at preferential sites across the solid-liquid interface, shown in figure 2. The anodic actions operating across the coating layer remain in the same position throughout the simulation whereas the cathodic potentials appear to be the subject of some flux, highlighting the availability of oxygen plays a major role in the predictions made by the model. The assumption can be made that within the model the anodes (primary zinc phase) will continue to be the subject of anodic attack until either these dendrites have been removed or the ingress of electrolyte is prevented by corrosion product formation.

The simulation of the cut-edge corrosion of a Galfan-coated-steel – steel substrate with 25μm coating on either side – can be seen in figure 3. In this case the steel substrate can be seen to become the site dominant for cathodic oxygen reduction, thus a depletion of the oxygen field can be seen over this area in figure

3 (right). The primary zinc phase can still be seen to be the area of the most intense anodic activity (shown by local cation releases shown in left-hand side of figure 3). In cut-edge corrosion simulations, the eutectic phase is also subject to some dissolution due to calculated galvanic effects in the model, albeit to a lesser degree.

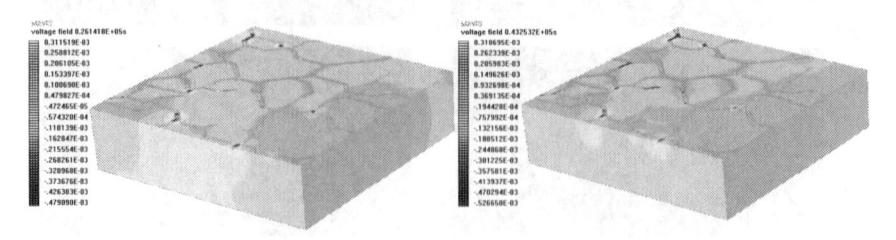

Figure 2: Electrical potential throughout the 5% NaCl solution in contact with Zn – 4.5wt.% Al coating after 7 (left) and 12 (right) hours.

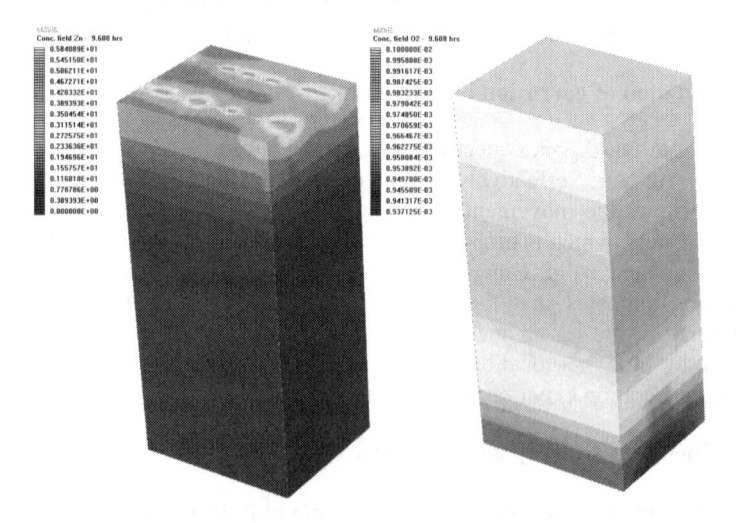

Figure 3: Evolved cation (left, inverted) and dissolved oxygen (right) concentration fields (mol.dm^{-3}) after 9.6 hours above a Galfan-coated steel cut-edge. (1mm cut-edge length).

3.3 Effect of magnesium addition

For an Al composition of 4.5wt.%, small Mg additions increase the volume fraction of primary zinc phase. (A 0.04% Mg addition yields a primary volume fraction equivalent to 18%, compared to 6% without Mg). Figure 4 shows the current density fields 100μm above the coating surface in aqueous NaCl, which suggests that the coating layer with the higher primary Zn fraction experiences a higher frequency of anodic attack at a higher intensity. This is not unexpected since the greater area of primary zinc exposed would give a higher corrosion rate.

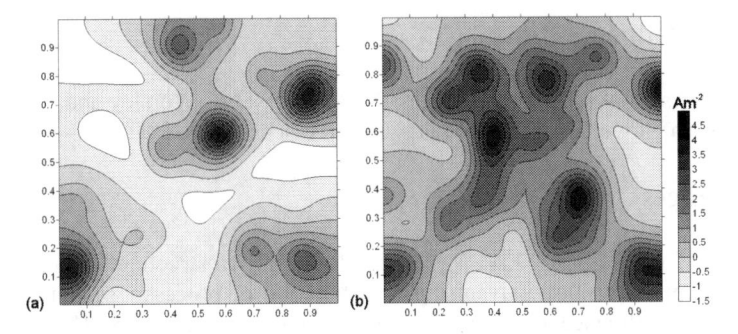

Figure 4: Current density (vertical component, i_y) maps for surfaces of Galfan coatings exposed to 5% NaCl; containing (a) 6% and (b) 18% primary phase after 7 hours simulated exposure. (Dimensions in mm).

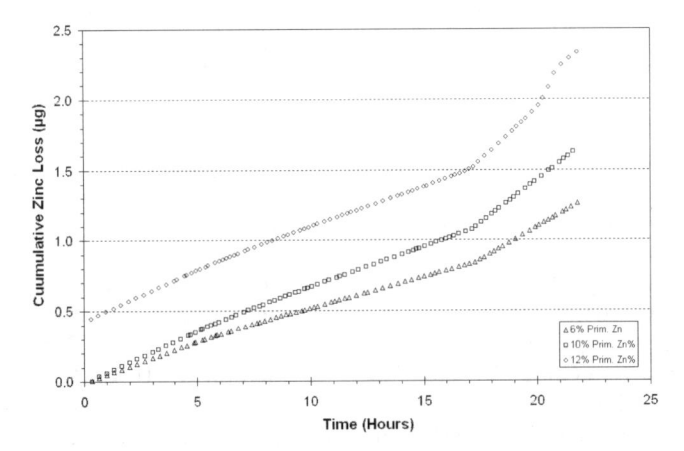

Figure 5: Zn losses from the Zn – 4.5wt.% Al coatings with varying volume fractions of primary Zinc phase. (1mm^2 area exposed).

Figure 5 shows varying zinc losses from Galfan coatings of different volume fractions of primary phase with different Mg additions. As expected, a proportionally higher corrosion rate is experienced from the coatings containing higher volume fractions of primary Zn.

4 Conclusions

The localized corrosion effects experienced by pure zinc and Galfan-coated steel have been simulated. 3D conserved concentration fields are used to predict the electrical conditions in 5% NaCl solution, which are related to experimentally derived data to predict material losses from Galfan coating layers in this aqueous environment. The form of the current density fields, along with predicted zinc losses from coating layers compares favourably with short-term experimental testing.

Acknowledgements

The authors would like to thank H.N. McMurray, D.A. Worsley and J.A. Spittle at Swansea University (UK) for many useful discussions.

References

[1] Elvins, J., Spittle, J.A., & Worsley D.A., "Microstructural changes in zinc aluminium alloy galvanising as a function of processing parameters and their influence on corrosion." *Corrosion Science*, 2005. **47**: pp. 2740-2759.

[2] Seré, P.R., Culcasi, J.D., Elsner, C.I., & Di Sarli, A.R., "Relationship between texture and corrosion resistance in hot-dip galvanized steel sheets." *Surface & Coatings Technology*, 1999. **122**: pp. 143-149.

[3] Panossian, Z., Mariaca, L., Morcillo, M., Flores, S., Rocha, J., Peña, J.J., Herrera, F., Corvo, F., Sanchez, M., Rincon, O.T., Pridybailo, G., & Simancas, J., "Steel cathodic protection afforded by zinc, aluminium and zinc/aluminium alloy coatings in the atmosphere." *Surface & Coatings Technology*, 2004. **190**(2-3): pp. 244-248.

[4] Böhm, S., McMurray, H.N., Powell, S.M., & Worsley, D.A., "Photoelectrochemical investigation of corrosion using scanning electrochemical techniques." *Electrochimica Acta*, 2000. **45**: pp. 2465-2174.

[5] McMurray, H.N., Parry, G., & Jeffs, B.D., "Corrosion resistance of Zn-Al alloy coated steels investigated using electrochemical impedance spectroscopy." *Iron and Steelmaking*, 1998. **25**(3): pp. 210-215.

[6] Brown, S.G.R., Barnard, N.C., & McMurray, H.N., "3-dimensional modelling of localized corrosion effects in zinc and zinc alloy steel coatings." *CIMTEC* 2004. Acireale(CT), Italy: Techna Group.

[7] Brown, S.G.R., & Barnard, N.C., "3D computer simulation of the influence of microstructure on the cut edge corrosion behaviour of a zinc aluminium alloy galvanized steel." *Corrosion Science*, 2006. **48**(8): pp. 2291-2303.

[8] Patankar, S.V., *Numerical Heat Transfer and Fluid Flow*. Computational methods in mechanics and thermal sciences. 1980: Hemisphere. 214.

[9] Livingstone-Bridge, D., Myland, J.C., & Oldham, K.B., "A model of ionic current densities in the vicinity of a corroding disk-shaped region." *Electrochemistry Communications*, 2001. **3**: pp. 384-389.

Mesoscopic modeling of corrosion processes: pitting morphology evolution

A. Taleb[1], C. Vautrin-Ul[2], H. Mendy[2], J. Stafiej[3] & A. Chausse[2]
[1]*Laboratoire d'Electrochimie et de Chimie Analytique, Université Pierre. et Marie Curie, CNRS UMR 7575, ENSCP, Paris, France*
[2]*Laboratoire Analyses et Modelisations pour la Biologie et l'Environnement, Universite d'Evry Val d'Essonne, Evry, France*
[3]*Institute of Physical Chemistry, Polish Academy of Sciences, Warsaw, Poland*

Abstract

We use cellular automata simulations at a mesoscopic scale to describe the phenomenon governing the morphology of pitting corrosion. We show that spontaneous separation of anodic and cathodic reaction zones plays a major role in determining the shape of the corroded space. The location of these zones versus the corroded surface is a stochastic event that influences strongly the direction of the corrosion propagation.
Keywords: cellular automata, corrosion, pitting morphology.

1 Introduction

The corrosion phenomena lead to a strong modification of the metal surface properties. Both physical and chemical aspects of the surface are affected by this change. Depending on the corrosive environment properties in terms of pH, temperature, aggressive ion concentration and potential, the metal surface is covered by an oxide film or not. In certain conditions, the presence of the film leads to a passivation of the surface that is separation of the corrosive media from the metal greatly reducing the rate of corrosion. The local breakdown of this protecting layer induces local corrosion named pitting corrosion or crevice corrosion if the site is occluded.

Pitting corrosion has been studied for several decades by many authors and all of them state that this kind of local corrosion takes place in several stages: pit nucleation or initiation, pit propagation and for certain conditions pit healing [1-4].

In the present study, we are interested in the pit morphology during the propagation stage. This morphology is documented experimentally and established to be connected to the material and the aggressive solution properties. Different authors have reported relations between the pitting corrosion and crystallographic properties of the corroded metal [5-7]. This relation was explained mainly by the metal atom coordination. In fact, metal at crystal defects have less bonds than in the rest of material and thus is more reactive [5-7]. Some other authors show that for strong salt solutions pits grow hemispherically but change to more saucer-like shape in a later stage which means that the ratio pit depth and pit width is time dependent [8-11]. They show also that this ratio is temperature dependent [8-11]. Theses results can be understood in term of diffusion processes. In fact, the pit depth increases according to $t^{1/2}$ as expected for a diffusion controlled process [11]. The pit grows faster without cover when only the pit depth provides the diffusion barrier [11].

On the theoretical side local corrosion is a very complicated interfacial growth problem with transport, diffusion, bulk and heterogeneous chemical reactions coupled with each other and leading to specific surface phenomena such as passivation, depassivation and dissolution. The first principle atomic scale approaches to this phenomenon seem to be a difficult task. To understand and reproduce the rich phenomenology associated with corrosion system different approaches going from a mesoscopic to a macroscopic level were developed.

In our modelling we remain within the mesoscopic approaches which combine the macroscopic phenomenology with the stochastic character of the processes originating from the microscopic scale phenomena. Working at the mesoscopic scale we avoid the difficulties of the microscopic level and retain only the essential features of the studied phenomena.

The present work is a theoretical attempt to understand some characteristic of pitting morphology. We show using a simple model that we can reproduce and understand qualitatively some pitting morphologies. Hereafter we introduce anodic and cathodic reactions that can be separated in space. As a result of this separation heterogeneities in the distribution of H^+ and OH^- ions can be formed [12]. Diffusion of these ions and their possible neutralization counteract the formation of these heterogeneities which determines the pit morphology. Since the pitting corrosion is a stochastic process in our model we use a stochastic cellular automaton (CA) as done in [12-17].

The paper is organized as follows. In section 2 we summarize the physicochemical model and its cellular automata representation. In section 3 we describe our simulations results with the corresponding discussion. In the final section we give some conclusions.

2 The model

The model has been already presented elsewhere [12]. We recall here the points which are important for understanding of the new results presented in this work.

We consider a piece of metal material with a flat surface in contact with a corrosive environment. The surface is covered by an insulating layer (painted). The corrosion process starts from a single punctual damage of this layer. Then the role of the layer is to impose open circuit conditions with strict anodic and cathodic reaction balance inside the developing corrosion cavity. The metal corrosion is governed by a suite of surface local chemical and electrochemical reactions followed by bulk reactions. The metal anodic dissolution is followed by metal cation hydrolysis in acidic or neutral medium, with corrosion products detached from the surface.

$$Me + H_2O \rightarrow MeOH_{aq} + e^- + H^+ \quad (1)$$

Metal anodic oxidation in basic medium leads to the formation of corrosion products on the metal surface.

$$Me + OH^- \rightarrow MeOH_{solid} + e^- \quad (2)$$

Anodic reactions require simultaneous cathodic reactions. These are reduction of the hydrogen ion or water depending on the acidity of the environment.

$$H^+ + e^- \rightarrow 1/2\, H_2 \quad (3)$$

$$H_2O + e^- \rightarrow 1/2\, H_2 + OH^- \quad (4)$$

When the oxidation and reduction are spatially separated, thereafter named SSE reactions, anodic reactions (cathodic reactions) acidify (basify) the environment leading to pH inhomogeneity of the solution.

We study the effect of prevailing SSE reactions on the corrosion process. We take an implicit account of the presence of aggressive ions assuming that they may induce dissolution of $MeOH_{solid}$

$$MeOH_{solid} \rightarrow MeOH_{aq} \quad (5)$$

especially when the medium is acidic. We consider $MeOH_{aq}$ as a part of the environment solution.

The ions H^+ and OH^- generated in the solution by SSE reactions diffuse and may neutralize one another:

$$H^+ + OH^- \rightarrow H_2O \quad (6)$$

To represent the above physicochemical model in the framework of a cellular automat approach we have to use discrete lattice. The lattice sites are labelled in our model as M, R, P, E, A, C and W. E stands for the solution sites having the original pH level. A (C) stands for the solution sites with increased acidity (basicity). M, R and P stand for bulk metal, surface reactive and passive metal sites. Bulk metal sites cannot have E, A and C sites as nearest neighbours. Surface metal sites are those that have at least one solution site as a nearest neighbour. Reactive sites, R, can undergo anodic dissolution while the passive sites, P, cannot as explained further. Finally W are the isolating layer sites unchanged during the system evolution.

The transformation rules and probabilities associated with a given transformation path in our model depend on the local chemistry of the site. To characterize this local chemistry we use a simplified acidity scale – the difference of the numbers of acidic and basic sites, $N^{exc} = N_A - N_C$, in the nearest neighbourhood. Consequently we speak about acidic, neutral and basic environment for $N^{exc} > 0$, $N^{exc} = 0$ and $N^{exc} < 0$, respectively.

The anodic reactions (1) or (2) are represented by:

$$R \rightarrow A \qquad (7)$$

$$\text{or } R+C(nn) \rightarrow P+E(nn) \qquad (8)$$

where reaction (7) occurs when $N^{exc} \geq 0$ and reaction (8) otherwise. The "(nn)" indicates the nearest neighbour of the site transformed. The anodic reaction is possible if the site R is connected with another surface site S (S = R, P) by a path of nearest neighbours of types M, R or P so that the cathodic counterpart of the reaction is possible according to one of the schemes corresponding to reactions (3) and (4) respectively

$$S+A(nn) \rightarrow S+E(nn) \qquad (9)$$

$$S+E(nn) \rightarrow S+C(nn) \qquad (10)$$

where reaction (9) occurs when $N^{exc} \geq 0$ and reaction (10) otherwise. If all neighbour sites are of type C the cathodic reaction cannot occur because we do not admit multiple C species on the same site and limit placing of the C site created in the nearest neighbour shell.

The reactions (7) – (9) correspond to SSE reactions. In our model we assume that these reactions occur with a probability $p_{sse} = 1$. It can be lower due to the blocking of cathodic reaction discussed above.

We can mimic the dissolution of MeOH$_{solid}$ (reaction 5) as

$$P \rightarrow E \qquad (11)$$

The probability of this reaction is taken as $p_{PE} = 0.25 \, N^{exc}$ if $N^{exc} \geq 0$ and $p_{PE} = 0$ otherwise. When an R or P site disappears, the M sites in contact with the solution become R sites.

To account for diffusion C and A sites perform a random walk. The target site is selected at random from the nearest neighbours of the walker. If the target site is E the walker goes to it. If the walker is C and the target site is occupied by the walker A or vice versa both disappear to mimic reaction (6),

$$A+C \rightarrow E+E \qquad (12)$$

Otherwise the random walker stays in its initial position. We set the diffusion rate with respect to the corrosion rate with N_{dif} being an integer parameter indicating that we perform N_{dif} steps of random walk per one corrosion step. N_{dif} plays an important role in the corrosion evolution as diffusion counteracts the pH inhomogeneities created by SSE reactions.

We select to work with a square lattice in a 2D system for the computational efficiency reasons. The square lattice has 4 connectivity for random walk and metal site update rules. For the front connectivity we modify the rules. Two front sites are connected if they are von Neumann nearest neighbours or they have an M site as a common von Neumann neighbour. This prevents that a connected M area has a border composed of disconnected pieces.

The rules of evolution of the system are applied starting from an initial state on. The system is a square of M sites surrounded by inert W sites imitating isolating protective layer. Two W sites in the middle of the topmost line are turned into two E sites and the M two sites below become R sites to imitate the damage of the covering layer.

The simulation box sizes are from 1000 x 1000 to 1500 x 1500. Simulation runs up to several thousands steps are collected.

3 Results and discussion

The SSE reactions described above induce local pH inhomogeneities which in turn modify the corrosion process. The diffusion and neutralization of H^+ and OH^- ions re-establishes the solution homogeneity.

3.1 Snapshots

In figure 1(a) and 1(b) we present the snapshots obtained for various simulation time steps. We distinguish two different behaviours:

First, at the early stage of simulation (figure 1(a)) the solution is locally neutral and the corrosion front uniformly rough. The corrosion front contains the same number of R and P sites.

Later (figure 1(b)) the separation of the solution (corrosion front) into acidic (smooth surface) and basic zones (rough surface) occurs with an intermediate neutral zone. Simultaneously the corroded space changes from hemispherical shape to complex asymmetrical one.

The transition from a homogeneous state to an inhomogeneous one for both the solution and the corrosion front is seen for N_{dif} ranging from 100 to 10000. For small N_{dif} this transition appears earlier. This suggests that for each N_{dif}

value there is a critical size of the corrosion cavity. Below this critical size the diffusion of A and C species is fast enough to neutralize of all of them. Above this critical size the diffusion is too slow for a complete neutralization.

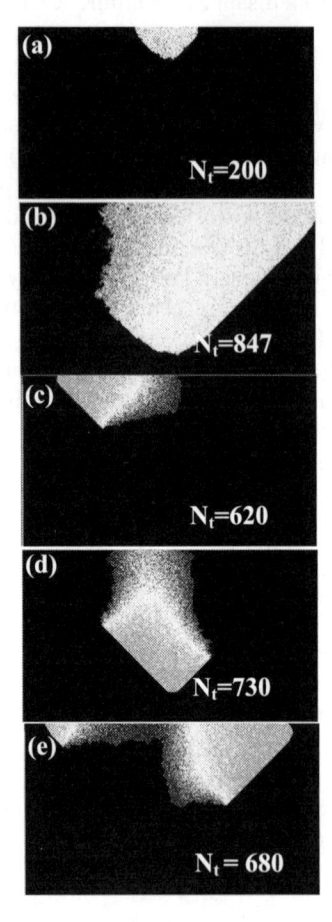

Figure 1: Snapshots corresponding to indicated simulation time steps (N_t). (a) and (b) corresponds to the number of diffusion step $N_{dif} = 6000$ whereas (c), (d) and (e) to $N_{dif} = 100$. The green, blue and white colours correspond respectively to Acidic, basic and neural zones.

The reproducibility of the simulation results has been tested on 10 independent runs with the same set of simulation parameters for each N_{dif}.

The snapshots of figure 1 represent typical configurations of anodic and cathodic zones. In figure 1(b) and 1(c), the single anodic zone develops the corrosion either to the right side or to the left side respectively depending on a randomly created initial asymmetry. In figure 1(d) an in depth cavity develops downwards in contrast to the sideways cavities of figures 1(c) and 1(e). In figure 1(e) two anodic zones are separated by one cathodic zone.

The difference observed in the cavity shapes of figures 1(b)-(e) outlines the stochastic character of the phenomena occurring during the corrosion process.

3.2 Equivalent radius evolution

In figure 2 we present the equivalent radius as a function of simulation time steps for several values of N_{dif}. The equivalent radius is defined assuming that the total number of M species corroded, N_{corr} can be placed in a semicircle:

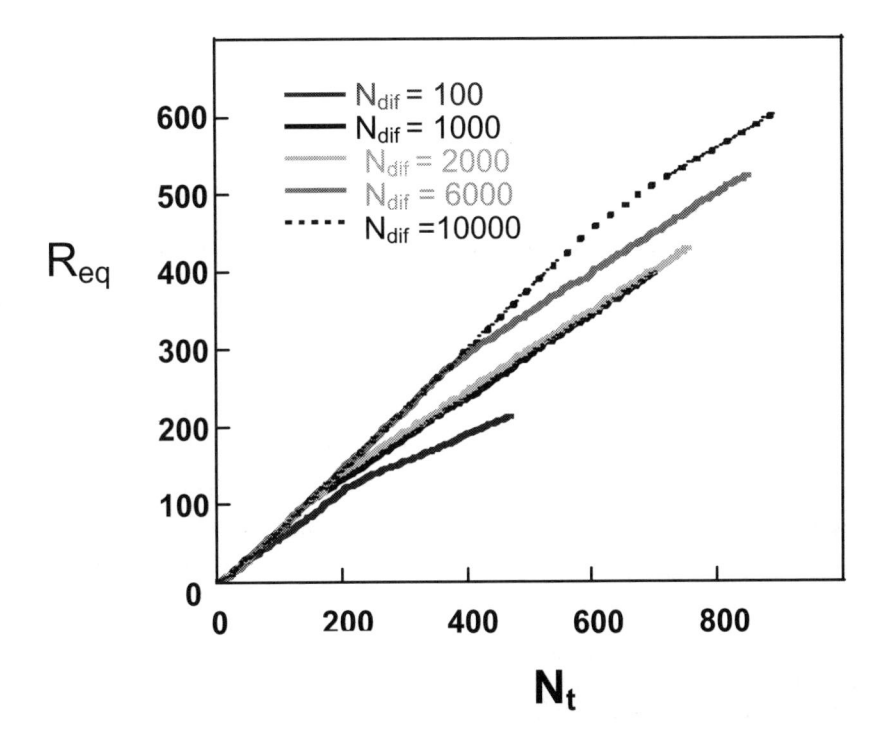

Figure 2: Evolution of the equivalent radius versus the simulation time steps, N_t, for indicated number of diffusion step N_{dif}.

$R_{eq} = \sqrt{2N_{corr}/\pi}$ in the a length units. We use this equivalent radius because it depends linearly on time in the case of homogeneous development of the cavity as can be observed in early stage of simulation (figure 2).

In the initial stage of the corrosion the curves coincide and form a straight line for all N_{dif} value. After that the smaller N_{dif}, the earlier the curve deviates from this line decreasing the slope. It coincides to zone separation. We call the first regime stationary regime (initial stage of the corrosion) and the second diffusion limited regime (later stage of the corrosion).

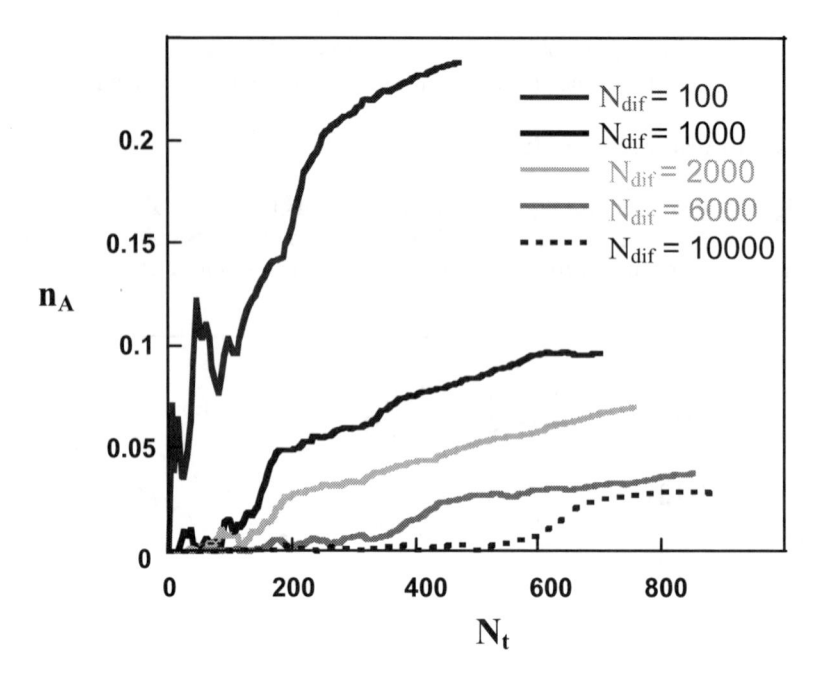

Figure 3: The evolution of acidic sites fraction, n_A, versus simulation time steps, N_t, for indicated number of diffusion step N_{dif}.

In figure 3 we present the fraction of the number of A sites with respect to all the solution sites as a function of the time steps. A and C sites are created and disappear in pairs. Therefore the numbers of A and C sites are equal. The fraction of A sites is practically negligible until a certain time step where it begins to grow. It occurs later for larger N_{dif}. The increase of the fraction of A sites corresponds to the deviation of the R_{eq} vs. time step dependence from the stationary regime (figure 2). For the cavity larger than the critical size the diffusion is too slow for a complete neutralization of A and C sites. A and C sites survive in acidic and basic zones of the globally neutral solution. The fraction of A sites is smaller in the case of large N_{dif} because the critical cavity size is larger and contains already more E sites.

The transition from what we call stationary regime to diffusion limited regime is known in another model and observed in experimental characteristics of corrosion systems [18]. Here in the diffusion limited regime the cavity shapes are governed by an initial disposition of the zones resulting from a stochastic fluctuation.

4 Conclusion

We show that a simple mesoscopic stochastic model reproduces the main features observed in real experiments. The simulation results establish a clear connection between the geometrically complicated form of cavities and the solution inhomogeneity. Two types of separated zones appear with different pH. The location of theses zones is a stochastic phenomenon influencing the morphology of the corroded area. We can see connection between the inhomogeneity of the corrosion process and the surface roughness.

Acknowledgements

The collaboration of our laboratories has been financially supported within PAN-CNRS project no.14470.

References

[1]   Z. Szklarska-Smialowska, Pitting and Crevice Corrosion, (NACE International, 2005).
[2]   D. Landoldt, Traité des Matériaux. Corrosion et Chimie de Surfaces de Matériaux. Presses Polytechniques et Universitaires Romandes, 1993, Lausanne.
[3]   G. S. Frankel, J. Electrochem. Soc. 145, 2186 (1998).
[4]   N. J. Laycock, S. P. White, J. Electrochem. Soc. 148 (7), B264 (2001).
[5]   H. Kaesche, Werkstoffe Korr. 39, 152 (1988).
[6]   M. Baumgartner and Kaesche, Corr. Sci. 29, 363 (1989).
[7]   Lillard, R. S. Electrochem. Solid-state letters, 6(8), B29, (2003).
[8]   J. Mankowski, Z. Szklarska-Smialowska, Corros. Sci. 15, 493, (1975).
[9]   R. C. Alkire, K. P. Wong, Corros. Sci. 28, 411, (1988)
[10] W. Schwenk, Corrosion, 20, 129, (1964).
[11] P. Ernst, R. C. Newman Corro. Scie. 927, (2002).
[12] C. Vautrin-Ul, A. Chaussé, J. Stafiej and J. P. Badiali, Polish Journal of chemistry 78, 1795 (2004)
[13] P. Cordoba-Torres, R. P.Nogueira, L. De Miranda, L. Brenig, J. Wallenborn, V. Fairén, Electrochemica Acta, 46, 2975.
[14] A. Taleb, A. Chauss\'e, M. Dymitrowska, J. Stafiej, and J. P. Badiali, J. Phys. Chem. B 108, 952 (2004).
[15] J. Saunier, A. Chausse, J. Stafiej, and J. P. Badiali, J. Electroanal. Chem. 563, 239 (2004).
[16] J. Saunier, M. Dymitrowska, A. Chauss\'e, J. Stafiej, and J. P. Badiali, J. Electroanal. Chem. 582, 267 (2005).
[17] M. Malki, B. Baroux, Corro. Scie. 47, 171 (2005).
[18] L. Balazs, Phys. Rev B, 54, 1183, (1996).

Dual reciprocity boundary element method for iron corrosion in acidic solution

A. Peratta
Wessex Institute of Technology, Southampton, UK

Abstract

This paper presents a hybrid two dimensional Dual Reciprocity Boundary Element Method (DRBEM) for modelling the primary reactions of iron corrosion in an acidic solution. The main focus is on the establishment of the DRBEM approach for this kind of problems in which the mathematical formulation is based on the Planck-Nerst law, the conservation of charge and the conservation of mass for the species participating in the process. The model is capable to describe the transport process of Fe^{+2} and H^+ ions in the aqueous electrolyte driven by the instantaneous electric field that results from the existent charge distribution. The results of this work compare qualitatively well with previously established references.

1 Introduction

Numerical modelling of corrosion cells is necessary in order to understand the basic principles involved in the process. Even the simplest system with primary chemical reactions can represent a challenging problem from both the theoretical and numerical points of view. The high complexity of the phenomenon is mainly due to the non-linear coupling between the transport and electric field equations in the electrolyte, and more specifically in the region close to the anodic and cathodic nests. The understanding of the basic principles may lead to novel techniques for corrosion control, and may help to avoid expensive experimental setups.

The goal of this work is to establish a Dual Reciprocity Boundary Element Method (DRBEM) for solving iron corrosion problems in which the anodic and cathodic surfaces are well identified in advance. The conceptual model discussed in this work is sketched in Fig. 1.

In particular, this paper focuses in the primary reactions occurring in the anode and cathode, and defers the secondary reactions (i.e. rust formation) for

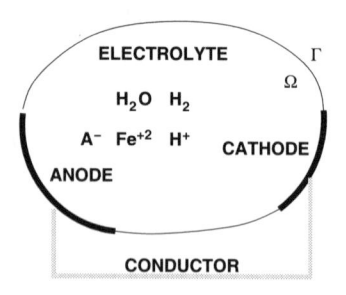

Figure 1: Conceptual model of the corrosion model to solve.

other models. The participating species are Iron ion (Fe^{+2}), Hydrogen ion (H^+) and a general ion A^- coming from an additional dissolved acid. The problem consists in solving the time dependent transport equation for Fe^{+2}, H^+ and A^-, in the aqueous acidic electrolyte, coupled with the electric field equations due to the instant ionic distribution. The computational approach is the Dual Reciprocity Boundary Element Method (DRBEM). Numerical modelling with Boundary Element Methods (BEM) [1, 2] is very attractive in the sense that it avoids domain discretisation and at the same time it uses the fundamental solution of the leading differential operator in the equation to solve. In addition, BEM applied to the time dependent advection-diffusion equation (A-DE) has been widely developed in the last decades and a large variety of efficient formulations were established [3, 4]. Usually, the DRBEM yields a system of equations whose condition number grows with a certain power law of the number of degrees of freedom [5], thus imposing an upper practical limit to the size of the model to solve, this problem for DRBEM has not been completely sorted out so far, and the practical limitations of this formulation need to be explored in more detail.

2 Governing equations

The mathematical model is based on the work done by V. Botte et al [6], where a one-dimensional finite difference approach is employed to solve the iron corrosion problem in acidic aqueous solution. This section summarises the main theoretical aspects developed in the cited reference. By assuming low solute concentration and neglecting gravitational type convective motions in the electrolyte, the isothermal mass flux $\boldsymbol{J}_k$ of species k in a non-dense aqueous electrolyte can be described by the Planck-Nernst law [7] according to:

$$\boldsymbol{J}_k = -D_k \boldsymbol{\nabla} c_k - \frac{z_k F}{R_0 T} c_k \boldsymbol{\nabla} \varphi \qquad (1)$$

where D_k, c_k and z_k are the diffusion coefficient, the concentration, and the charge number of k-th species in the aqueous solution, $F \sim 96500 \; sA \; mol^{-1}$ is the Faraday constant, $R_0 \sim 8.314 \; JK^{-1}mol^{-1}$ the universal gas constant, T is the temperature of the system and φ is the electric potential.

The mass conservation equation can be expressed as:

$$\frac{\partial c_k}{\partial t} + \nabla \cdot \boldsymbol{J}_k = \rho_k \tag{2}$$

where t is time and ρ_k is a source term which represents the creation/annihilation of species k due to secondary chemical reactions (i.e. rust production). The mass conservation equation leads to the time dependent A-DE with reaction term, see ref. [6], which describes the transport of species k in the electrolyte:

$$\frac{\partial c_k}{\partial t} + \nabla \cdot \left[-\left(\frac{z_k F D_k}{R_0 T} \nabla \varphi\right) c_k \right] - D_k \nabla^2 c_k = \rho_k. \tag{3}$$

The instantaneous charge conservation equation which implies: $\nabla \cdot (\sum z_k \boldsymbol{J}_k) = 0$, leads to

$$\nabla \cdot \left[\left(\sum_k \frac{z_k F D_k}{R_0 T} c_k \right) \nabla \varphi \right] = -\sum_k D_k \nabla^2 c_k \tag{4}$$

The boundary conditions for the electric problem are imposed by means of the dominant anode and cathode reactions: $Fe \rightarrow Fe^{+2} + 2e^-$ and $2H^+ + 2e^- \rightarrow H_2$, respectively. Then the normal current in the anode can be modelled by means of a simplified Butler-Volmer equation with transfer coefficient equal to $1/2$:

$$i_A = 2i_0 \sinh\left[\frac{z_{Fe^{+2}} F}{R_0 T} \eta_A\right] \tag{5}$$

where i_A is the current density of the anode (used as boundary condition for the electrical problem), i_0 is obtained experimentally, and

$$\eta_A = \delta\varphi_A - E_A \tag{6}$$

is the anode over-voltage, where $\delta\varphi$ is the difference of electric potential between the anode and the adjacent electrolyte film, and E_A is the electrode potential at zero current, computed as:

$$E_A = E_0 + \frac{R_0 T}{z_{Fe^{+2}}} \ln[c_{Fe^{+2}}] \tag{7}$$

Once the current density at the anode is found in terms of the over-voltage and iron ion concentration, the mass flux given by eq. (2) can be obtained, thus yielding the necessary boundary conditions for the transport problem.

When an acid is added to the solution, it dissociates according to $HA \rightarrow H^+ + A^-$, thus reducing the pH in the electrolyte [6]. Then, $c_{A^-} = c_{Fe^{+2}} + c_{H^+}$.

3 Dual Reciprocity Boundary Element Method

The governing equations for the electric potential and the solute transport for each species can all be casted into Poisson-like equations of the form:

$$\nabla^2 c_k(\mathbf{x}, t) = b(\mathbf{x}, t) \qquad \mathbf{x} \in \Omega, \qquad (8)$$

defined in the integration domain Ω with boundary $\Gamma = \partial\Omega$, where the right hand side (RHS) term $b(\boldsymbol{x}, t)$ can be regarded as a generic source term to be defined later for each problem. Then, the DRBEM [3, 8] is applied in order to convert this equation into an algebraic system of linear equations which require boundary only discretisation. Following the well established boundary integral approach, the weighted boundary integral expression of eq. (8) with function G yields [2]:

$$\mu_i c_k(\boldsymbol{x}_i) + \int_\Gamma \left(\frac{\partial G(\boldsymbol{x}_i, \boldsymbol{x})}{\partial \hat{n}} c_k(\boldsymbol{x}) - G(\boldsymbol{x}_i, \boldsymbol{x}) \frac{\partial c_k(\boldsymbol{x})}{\partial \hat{n}} \right) d\Gamma =$$
$$\int_\Omega G(\boldsymbol{x}_i, \boldsymbol{x}) b(\boldsymbol{x}, t) d\Omega, \qquad (9)$$

where $\mu_i = 1/2$ for smooth boundaries, and the weight function G is the fundamental solution of Laplace equation in 2D: $\nabla^2 G(\boldsymbol{x}_i, \boldsymbol{x}_j) + \delta(x_i, x_j) = 0$. The domain integral appearing in the RHS of eq. (9) is treated with the Dual Reciprocity Method (DRM) [8]. In this work we have employed a set of N_r radial basis functions (RBF) $f_i(\boldsymbol{x}_j) = \{1, r_{ij}^2 \ln(r_{ij}), x_j, y_j\}$ known as Augmented Thin Plate Splines (ATPS) in order to approximate b as:

$$b(\boldsymbol{x}, t) \approx \sum_{i=1}^{N_r} a_i(t) f_i(\boldsymbol{x}), \qquad (10)$$

where r_{ij} is the distance between the source and field points denoted by indices i and j, respectively, and $a_j(t)$ are the interpolating coefficients to be adjusted in order to minimise the error between b and its approximation. The DRM is based on the fact that the RBFs can be considered as source terms of potential functions $\hat{u}$ defined by: $\nabla^2 \hat{u}_j(x_i) = f_{ij}$. Henceforth, the RHS of eq. (9) can be integrated by parts in order to obtain a boundary-only integral equation of the form:

$$\mu_i c_k + \int_\Gamma \left(\frac{\partial G}{\partial \hat{n}} c_k - G \frac{\partial c_k}{\partial \hat{n}} \right) d\Gamma = \sum_{j=1}^{N_r} a_j \left[\mu_\xi \hat{u} + \int_\Gamma \left(\frac{\partial G}{\partial \hat{n}} \hat{u} - G \frac{\partial \hat{u}}{\partial \hat{n}} \right) d\Gamma \right]. \qquad (11)$$

Details of this approach can be found in refs. [5, 8]. In the standard BEM matrix notation, eq. (11) can be expressed as:

$$(\mathbf{H}\{c_k\} - \mathbf{G}\, \partial_n\{c_k\}) = \left(\mathbf{H}\hat{U} - \mathbf{G}\,\hat{Q} \right) \{a\}, \qquad (12)$$

where $\boldsymbol{H}$, $\boldsymbol{G}$, $\hat{U}$ and $\hat{Q}$ are the usual DRBEM collocation matrices [9, 8, 2], $\{c_k\} = c_k(\boldsymbol{x}_i, t)$ is a 1-column array of c_k evaluated at the collocation nodes, and the coefficients $\{a\} = a_j$ can be determined inverting the linear eq. (10) provided that $\{b_i\}$ is known at the collocation nodes:

$$\{a\} = F^{-1}\{b\}. \tag{13}$$

Then, eq. (12) becomes: $H c_k - G \partial_n c_k = Sb$, where $S := \left(H\hat{U} - G\hat{Q}\right) F^{-1}$, $S \in R^{M \times M}$, and $F = f_{ij}, \in R^{(M+3) \times (M+3)}$ is the 2D ATPS RBFF matrix. Then, $\hat{U} = \hat{u}_{ij}, \in R^{M \times (M+3)}$ and $\hat{Q} = \partial \hat{u}_{ij}/\partial \hat{n}_i \in R^{N \times (M+3)}$ are the usual DRM matrices. N is the number of freedom nodes used to discretise Γ, $M = N + L$, and L is the number of DRM nodes in Ω. Next, we express the 1-column array $\{b\}$ representing the RHS term evaluated at the collocation nodes in a way that results suitable for the DRM treatment. The time dependent term is approximated as:

$$\frac{\partial c_k}{\partial t} \approx \frac{c_k^{m+1} - c_k^m}{\delta t}; \qquad c_k^m = \theta c_k(t + \delta t) + (1 - \theta) c_k(t) \tag{14}$$

where index m represents the time level and θ can be adjusted between 0 and 1 in order to tune the time integration scheme between fully explicit ($\theta = 0$) and implicit ($\theta = 1$). The convective term is split into the following two terms: $\nabla \cdot (V c_k) = (V \cdot \nabla) c_k + (\nabla \cdot V) c_k$. The first one on the RHS represents advection in incompressible media, i.e. $\nabla \cdot V = 0$, while the second one can be regarded as a first order reaction term, provided that $\nabla \cdot V$ is known in advance.

4 Computational implementation

This section outlines the computational implementation of the simple iron corrosion process in a two dimensional cell. The problem is solved with an iterative two-stage sequential approach. The time integration scheme involves a finite difference approach with two time levels. First, the electric field is computed with the most updated values of species concentration, second the transport problem for each species is solved. Then, a new time level can be solved once the electric and transport problems are consistent with each other at present time level.

The integration domain is discretised with a mixed unstructured mesh, where some regions may be decomposed into many sub-domains and some others may be discretised only in their boundary. The former is known as multi-domain region while the latter is identified with single domain region. The assembly of multi and single-domain regions into the same problem provides a suitable pre-processing flexibility which allows the treatment of complicated geometries. In this work, the characteristic size of a cell in the multi-domain approach is $\sim 0.01m$.

4.1 Electric problem

The electric problem at time level m can be expressed as:

$$\nabla^2 \varphi^m = b_{\text{elec}}. \tag{15}$$

The source term is expressed in terms of the concentrations found in the last time level according to:

$$b_{\text{elec}} = -\frac{\sum_k D_k \nabla^2 c_k^{m-1}}{\sum_k \alpha_k c_k^{m-1}}. \tag{16}$$

For simplicity, the DRBEM is implemented throughout a multi-domain approach in which instant diffusion coefficients can be regarded as piecewise constant isotropic and homogeneous in each sub-domain. After applying collocation DRBEM, the discrete version of the electric problem becomes:

$$\boldsymbol{H}\{\varphi\} + \boldsymbol{G}\{q\} = \boldsymbol{S}\{b_{\text{elec}}\}. \tag{17}$$

The denominator in eq. (16) represents an apparently instant diffusion coefficient D_e for the electric problem, and $q = -D_e \nabla \varphi \cdot \hat{n}$ can be regarded as a current density, to be conserved throughout the interface between two sub-domains.

4.2 Transport problem

The second stage solves the transport problem for each chemical species involved. The transport problem can be formulated by casting the transport equation eq. (3) into a Poisson-like equation as follows:

$$\nabla^2 c_k(\mathbf{x}, \mathbf{t}) = b_{\text{tran}} \tag{18}$$

where the source term b_{tran} given by:

$$b_{\text{tran}} = \frac{1}{D_k} \left[\frac{\partial c_k}{\partial t} - \nabla \cdot (\boldsymbol{V} c_k) - \rho_k \right], \tag{19}$$

and the apparent convective velocity is defined as: $\boldsymbol{V} = \frac{z_k F D_k}{R_0 T} \boldsymbol{E}$, where $\boldsymbol{E} = \nabla \varphi$ is the electric field in the electrolyte. After applying the collocation DRBEM technique, the integral matrix equation per sub-domain becomes:

$$\boldsymbol{H}\{\mathbf{c_k}\} - \boldsymbol{G}\{\partial_n \mathbf{c_k}\} = \frac{\boldsymbol{S}}{D_f} \left[\frac{\partial}{\partial t} + \sum_{p=1}^{2} (\boldsymbol{V}_p \cdot \boldsymbol{T}_p) + \nabla \cdot \boldsymbol{V} \right] \{\mathbf{c_k}\}, \tag{20}$$

where the following matrices were employed [5, 9]: $\boldsymbol{V}_p = \text{diag}\{v_p\}$ and $\boldsymbol{T}_p = \nabla_p \boldsymbol{F} \cdot \boldsymbol{F}^{-1}$; where $\boldsymbol{V} \in R^{M \times M \times 3}$ and $\boldsymbol{T} \in R^{M \times M \times 3}$, see ref. [9] for details. By applying same procedure as in [9] a close linear system of equations of the form

$$\boldsymbol{A}_{\text{TRAN}} \boldsymbol{x} = \boldsymbol{B} \tag{21}$$

can be obtained, in which the solution represents fluxes and potentials at the boundary of the integration domain.

4.3 Coupling strategy

The coupling strategy in each time level can then be summarised as follows:
1. Define initial conditions for c_k obtained from the previous time level
2. Compute b_{elec} with the most updated c_k field
3. Update boundary conditions with Butler-Volmer equation (5)
4. Solve electric problem given by eq. (15)
5. Use the electric field in order to compute effective velocity $\mathbf{V}$
6. Compute b_{tran} with the most updated φ and $\mathbf{E}$
7. Solve A-DE system given by eq. (21) for each species
8. If c_k is consistent with φ stop, otherwise repeat step 2

The consistency is checked if the relative error of c_k between two subsequent iterations L and $L-1$ is smaller than a small arbitrary factor, i.e. $\sum |c_k^L - c_k^{(L-1)}| < \varepsilon c_k^L$, and $\varepsilon \sim 0.1$.

5 Results

This section presents the numerical results obtained for two cases: in the first one, the cathode and anode are separated by a homogeneous column of electrolyte. In this example, symmetrical lateral boundary conditions, have been imposed in order to retrieve one-dimensional results. This is done in order to compare results with the work of Botte et al. [6].

The second example, has non-symmetrical conditions and describes qualitatively the behaviour of a 2D corrosion cell with two adjacent electrodes.

5.1 One-dimensional example

This example consists of a rectangular domain of $0.1m \times 0.025m$, with its larger dimension oriented in x direction. The anode is placed on the left boundary ($x = 0$) while the cathode is on the right ($x = 0.1m$) as shown in Figure 2.

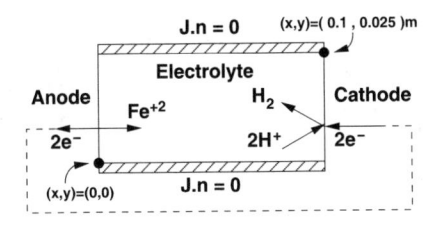

Figure 2: Conceptual model of the corrosion cell in example 1.

Impermeable conditions, i.e. $\mathbf{J} \cdot \hat{n} = 0$, were imposed on both top and bottom boundaries. The simulation parameters are summarised in table 1. Figure 3 shows the concentration profiles of Fe^{+2} and H^+ ions in the electrolyte at different time levels. The concentration of Fe^{+2} near the anode increases with time and

Table 1: Simulation parameters.

Description	Symbol	Value	
Concentration of Fe^{+2} at $t=0$	$c_{Fe^{+2}}(x, t=0)$	1×10^{-14}	mol/m^3
Concentration of H^+ at $t=0$	$c_{H^+}(x, t=0)$	1×10^{-3}	mol/m^3
Time step	δt	2×10^1	s
Exchange anode current density	i_0	1×10^1	A/m^2
Anode reaction potential	E_0	1.44	V
Diffusion coefficient Fe^{+2}	$D_{Fe^{+2}}$	5×10^{-6}	m^2/s
Diffusion coefficient H^+	D_{H^+}	1×10^{-6}	m^2/s
Temperature	T	300	K

propagates towards the bulk of the solution as time passes by. On the other hand, the concentration of H^+ decreases with time and the maximum of its distribution moves towards the cathode at a very slow rate, where it is consumed in the cathode. The slow motion of the distribution of H^+ towards the cathode seems to be driven by the instant electric field. Figure 4 shows the concentration profiles of species A^- at different time levels. Figure 4 (right) shows the electric potential profile along the electrolyte at different times $3h$, $6h$, $9h$ $12h$ and $15h$. it can be observed that the solution in the bulk of the electrolyte is quite close to the solution of the Laplacian equation, i.e. the gradient of concentration does not play a significant role in the electric problem far from the electrodes.

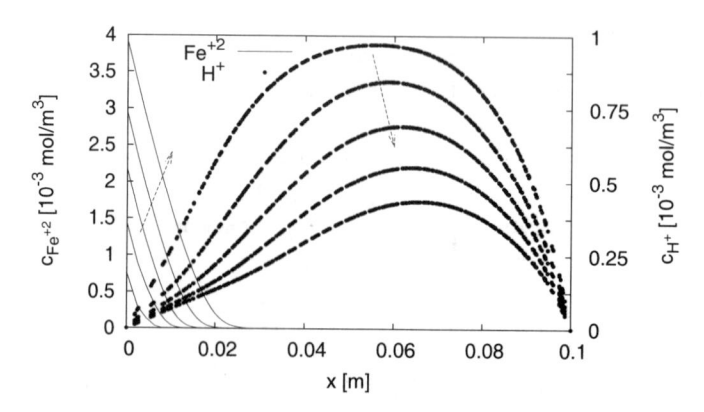

Figure 3: Concentration profiles of Fe^{+2} ($c_{Fe^{+2}}$ - left axis) and H^+ (c_{H^+}) in $10^{-3} mol/m^3$ at times $3h$, $6h$, $9h$ $12h$ and $15h$. The arrows indicate the time direction.

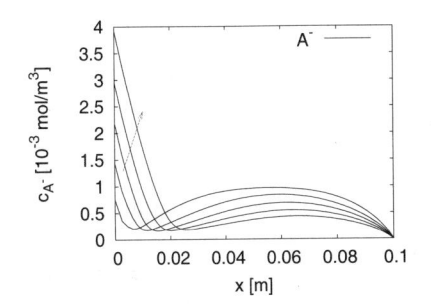

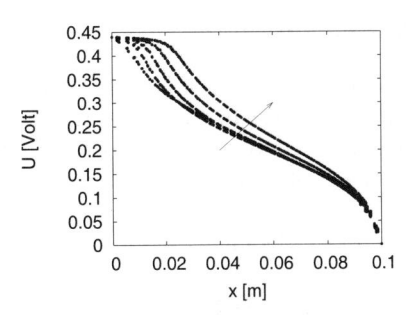

Simulation of Electrochemical Processes II   31

Figure 4: Concentration profiles of A^- ion (c_{A^-}) (left), and electric potential profiles (right) at times $3h$, $6h$, $9h$ $12h$ and $15h$. The arrows indicate positive time direction.

5.2 Two dimensional example

The integration domain in this example is shown in Figure 5(a). It consists of a sample of $10cm \times 5cm$. The initial conditions were assumed in the same way as in the previous example. The boundary conditions for the anode were given according to the Butler-Volmer equation eq. (5) and the over-voltage was updated at each time step. The cathodic current density was the same as the anodic, in view of the global charge conservation equation. This can be done because all other boundary conditions apart from the anodic and cathodic surfaces have assigned zero normal current. Fig. 5(a) shows the concentration distribution of Fe^{+2} at time $t = 1h$. Fig. 5(b) shows the corresponding electric potential, the numbers associated with the iso-lines represent concentration in $[mol/m^3]$. The numbers inside the figure indicate potential in [V].

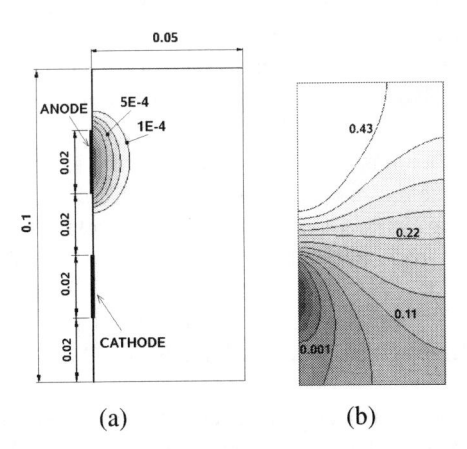

(a)   (b)

Figure 5: (a) Concentration of Fe^{+2}, and (b) electric potential at time $t = 1\,h$.

6 Conclusions

An iterative approach based on the multi-domain DRBEM for solving an iron corrosion model in acidic aqueous electrolyte has been established and successfully implemented for two-dimensional problems.

The results obtained compare qualitatively well with previous publications [6], although more study is required in order to represent properly the thin layer of electrolyte close to the electrodes.

Further extensions of this work will involve the solution of the secondary reactions that yield rust and the study of the influence of external electric fields. This is in order to assess the impact of environmental electric pollution on simple corrosion cells.

References

[1] C.A. Brebbia and J. Dominguez. *Boundary Elements, An Introductory Course. Second Edition*. Computational Mechanics Publications, McGraw-Hill, New York Colorado San Francisco Springs Mexico Montreal Oklahoma City San Juan Toronto, 1992.

[2] C.A. Brebbia, J. C. Telles, and L. C. Wrobel. *Boundary Elements Techniques*. Springer-Verlag, Berlin, Heidelberg New York and Tokio, 1984.

[3] L.C. Wrobel and D. B.De Figueiredo. A dual reciprocity boundary element formulation for convection-diffusion problems with variable velocity fields. *Engng Analysis with Boundary Elements*, 8(6):312–319, 1991.

[4] Z. H. Qiu, L. C. Wrobel, and H. Power. An evaluation of boundary element schemes for convection-diffusion problems. In C. A. Brebbia and J.J. Rencis, editors, *Boundary Elements XV, Transactions on Modelling and Simulation*, volume 1. Wessex Institute of Technology, UK and Worcester Polytechnic Institute, USA, WIT Press, 1993.

[5] A. Peratta. *BEM applied to Flow and Transport in Fractured Porous Media*. PhD thesis, University of Wales - Wessex Institute of Technology, Southampton, UK, December 2004.

[6] V. Botte, D. Mansutti, and A.Pascarelli. Numerical modelling of iron corrosion due to an acidic aqueous solution. *Applied Numerical Mathematics*, 55:253–263, June 2005.

[7] S. Furini, F. Zerbetto, and S. Cavalcanti. A numerical solver of 3D poisson nernst planck equations for functional studies of ion channels. In M. Ursino, editor, *Modelling in Medicine and Biology*, volume 2 of *Advances in Bioengineering, Vol 2*, pages 111–120, University of Bologna, Italy, 2005. WIT Press.

[8] P.W. Partridge, C.A. Brebbia, and L.C. Wrobel. *The Dual Recoprocity Boundary Elements Method*. Computational Mechanics Publications, Southampton Boston, 1992.

[9] A. Peratta and V. Popov. A new scheme for numerical modelling of flow and transport processes in 3d fractured porous media. *Advances in Water Resources*, 29:42–61, 2006.

An analytical modeling method for calculating the maximum cathode current deliverable by a circular cathode under atmospheric exposure

Z. Y. Chen & R. G. Kelly
Center for Electrochemical Science and Engineering,
Department of Materials Science and Engineering,
University of Virginia, Charlottesville, USA

Abstract

An analytical method for evaluating the stability of pitting corrosion of corrosion-resistant alloys under thin-layer (or atmospheric) conditions is presented. The method uses input data that are either thermodynamic in nature or easily obtained experimentally. The maximum cathode current available (referred to as the cathode capacity) depends on the cathode geometry, temperature, relative humidity, deposition density of salt (i.e., mass of salt per unit area of cathode), and interfacial electrochemical kinetics. The anode demand depends on the pit geometry and the pit stability product. By coupling these two approaches, the stability of a pit can be determined for a given environmental scenario. The method has been applied to the atmospheric pitting corrosion of Type 316L stainless steel, leading to a quantitative description of limiting stable pit sizes.
Keywords: galvanic corrosion, relative humidity, NaCl particles, localized atmospheric corrosion, stainless steel, pitting.

1 Introduction

Corrosion resistant materials exposed to marine atmospheric conditions can suffer from localized corrosion (pitting, crevice corrosion, stress-corrosion cracking). The stability of such a localized corrosion site requires that the site (anode) must dissolve at a sufficiently high rate to maintain its critical chemistry [1] and a wetted surrounding area (cathode) must provide a matching cathodic current. The objective of this study was to computationally characterize the stability of such a local corrosion site and explore the effects of physiochemical parameters and electrochemistry of this stability. Specifically, this work considers the stability of a pit on Type 316L stainless steel exposed to atmospheric conditions after the deposition of a known amount of sodium chloride.

1.1 Background

The conservation of charge requires that the sum of the cathodic reaction rates be equal to the sum of the anodic reaction rates for a corrosion system under open circuit conditions. A stable localized corrosion system (depassivated anode, dominant external cathode) requires that the two regions be compatible in terms of the current demanded by the anode and the current that could be supplied by the cathode.

In localized corrosion, it is generally accepted that the stability of the anode site requires that its dissolution rate be sufficient to maintain an aggressive solution against the diluting effects of diffusion. The aggressiveness required is a function of the alloy composition, bulk environment, and temperature. Galvele [2–4] was the first to address this issue analytically, and considered the diffusion and migration relevant to the pit, assuming that there was sufficient cathodic reaction to maintain electroneutrality. Galvele demonstrated that for a one-dimensional pit with an active base and passive walls, the stability parameter could be put in terms of the dissolution rate at the base of the pit, i, and the depth of the pit, x, in the form of a pit stability product, $i \cdot x$. For a hemispherical pit, it has been shown that the stability product takes the form I/r where I is the anodic current from the pit, and r is its radius [5,6]. Several researchers have experimentally determined the so-called *pit stability product* for different materials, and values are generally close to 10^{-2} A/cm for stainless steel [5,7,8].

Although the pit stability product has been successfully used to assess the minimum current required for a pit to be stable, little work has been done considering what controls the current that a cathode can supply, particularly under atmospheric exposure conditions.

For most localized corrosion systems, the vast majority of the cathodic current is supplied by the surface outside the localized corrosion site. For atmospheric exposures, only thin layers of electrolyte will be present. These layers represent a substantial ohmic resistance between the localized corrosion site and more remote areas on the cathode. This resulting ohmic drop distributed along the cathode will limit the total cathodic current that it can provide in support of the localized corrosion site.

Recently, Cui *et al.* [9] and Kelly *et al.* [10] presented an analytical solution for a one-dimensional cathode which allowed a solution for the cathode current capacity at a constant temperature as a function of interfacial kinetics, RH, deposition density, and the deliquescence behavior of the salt deposited. For stability considerations, knowledge of only the total net current from the cathode, I_{net}, is required; its distribution along the cathode is unimportant. Thus, an *equivalent cathode* can be constructed which provides the same current as the actual cathode, but has a configuration that allows a straightforward calculation of I_{net}. They validated the approach by comparisons of analytic calculations to the results of finite element analysis methods. In addition, they showed that a bounding solution can be found, termed $I_{net,max}$, which describes the maximum current available from a given cathode irrespective of its size. This maximum, referred to as the *cathode current capacity*, results from the fact that once an area of the cathode is sufficiently far from the anode that the potential drop leads to an interfacial potential of the open circuit corrosion potential of the cathode, it cannot provide any *net* cathodic current to the anode as it is all being used locally to support the passive dissolution.

Extending the approach of Cui *et al.* [9] and Kelly *et al.* [10] to a circular cathode is required to address the less restrictive mass transport conditions of relevance to a pit surrounded by a wetted area. Pitting is a very common form of corrosion damage for stainless steels exposed to marine atmospheres.

2 Computational approach

We follow the approach of Kelly *et al.* [10] in defining an equivalent cathode as shown in Figure 1. Figure 1 shows a one-dimensional cathode, but the tenets are applicable for two-dimensional cathodes as well. As mentioned above, the issue of localized corrosion stability requires knowledge of the total net cathodic current, not its distribution along the cathode. Thus, we can more easily calculate the I_{net} by defining an equivalent current density, i_{eq}, and an equivalent length (in 1D), L_{eq}, that results in the same I_{net} as the actual cathode.

$$I_{net} = i_{eq} \cdot L_{eq} \cdot W \tag{1}$$

The equivalent current density is defined by:

$$i_{eq} = \frac{\int_{E_L}^{E_{rp}} (i_c - i_p) \cdot dE}{E_L - E_{rp}} \tag{2}$$

where $(i_c - i_p)$ is the cathodic current less the passive current at a given potential (i.e., the electrochemical kinetics of the cathode), E_{rp} is the repassivation potential which is assumed to be that at the mouth of the pit, E_L is the potential at the position on the cathode furthest from the pit, and W is the width of the cathode.

If we consider only ohmic drop along the cathode, then the difference in potential between the anode and the farthest point on the cathode would be related to the current and the nominal resistance:

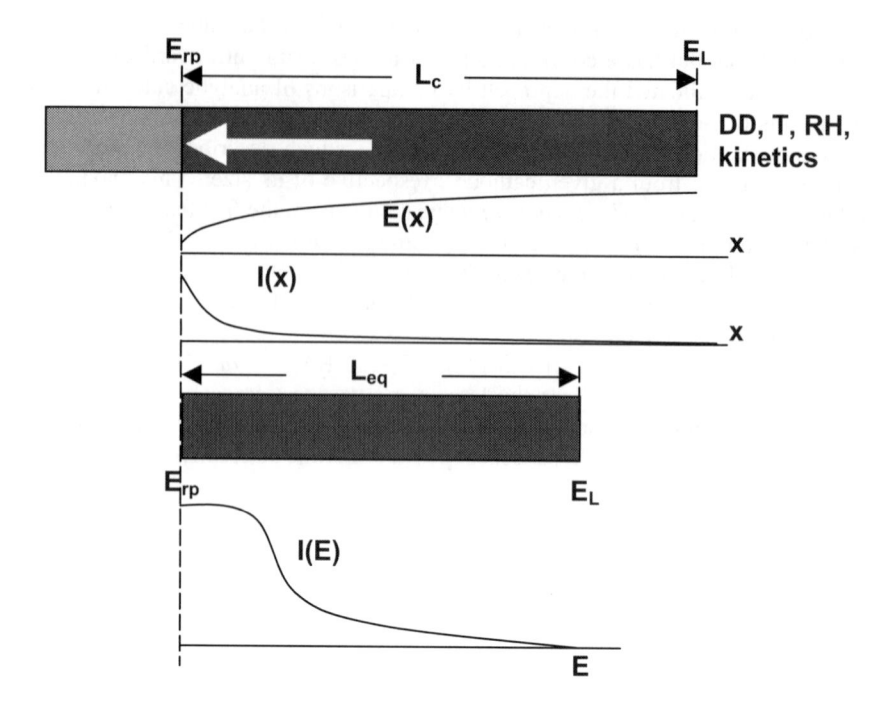

Figure 1: Schematic showing the connection between a real 1D cathode and an equivalent 1D cathode.

$$\Delta E = I \cdot R = \frac{\int_{E_L}^{E_{rp}} (i_c - i_p) \cdot dE}{E_L - E_{rp}} \cdot \frac{L_{eq}^2}{2 \cdot \kappa \cdot WL} \tag{3}$$

Combing eqns (1) through (3) results in the expression for I_{net}:

$$I_{net} = W \cdot \sqrt{2 \cdot \kappa \cdot WL \cdot \int_{E_L}^{E_{rp}} (i_c - i_p) \cdot dE} \tag{4}$$

where κ is the conductivity of the electrolyte layer and WL is the thickness of the electrolyte layer.

The accuracy of eqn (4) has been validated by independent FEM calculations of the current distributions under the same conditions [10]. Replacing E_L with the open circuit potential of the cathode, E_{corr}, leads to an expression for the maximum net cathodic current that a cathode can deliver under the conditions considered.

Extending this approach to a circular geometry requires a new definition of the anode and cathode dimensions as shown in Figure 2. The central anode (the pit) has a radius r_a and the equivalent cathode has a radius r_{eq}. The I_{net} is defined in an identical way to the 1D case. Consideration of ohmic drop results in an expression for I_{net}:

$$\ln I_{net} = \frac{4\pi \cdot \kappa \cdot WL \cdot (E_L - E_{rp})}{I_{net}} + \ln\left(\frac{\pi \cdot e \cdot r_a^2 \cdot \int_{E_L}^{E_{rp}} (i_c - i_p) \cdot dE}{E_L - E_{rp}}\right) \quad (5)$$

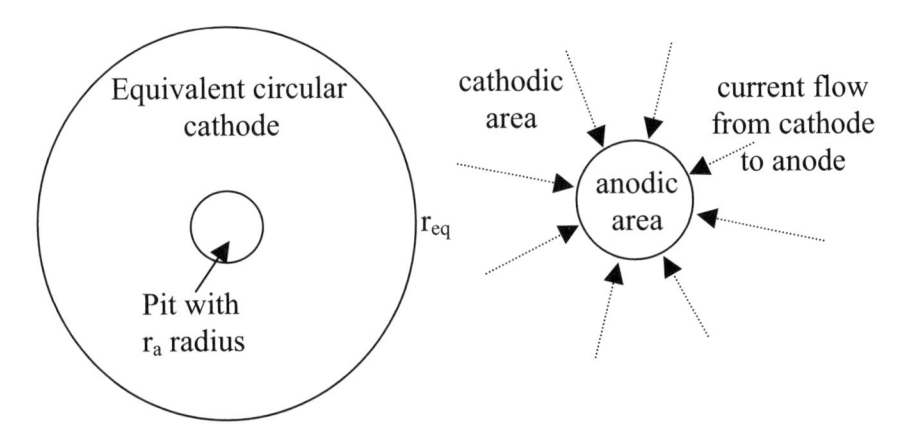

Figure 2: Schematic showing the circular cathode around a pit and the radial current distribution.

Again, replacement of E_L with E_{corr} leads to an expression for $I_{net,max}$.

The WL and κ are difficult to control or measure in either the laboratory or the field. However, under equilibrium conditions, their values can be calculated knowing the relative humidity (RH), the amount of salt deposited per unit area (hereafter referred to as the deposition density, DD), and quantitative information about the deliquescence behavior of the deposited salt [11]. Deliquescence data are available for a wide range of salts and mixtures of salts as a function of temperature.

3 Results

Figure 3 shows a comparison of the results of the analytical expression (eqn (5)) with the results of FEM analyses of the same conditions. In the case of the FEM analyses, the current distributions were determined, then integrated over the length of the cathode. The agreement is excellent over more than two orders of magnitude of current.

One means for using eqn (5) is to assess the effect of RH and DD on the $I_{net,max}$. Figure 4 shows an example for Type 316L stainless steel. The cathode kinetics were taken from the literature, with the diffusion limited current density adjusted to be equal to the water layer thickness. For $WL < 200$ µm, this has been demonstrated experimentally to be the case [12].

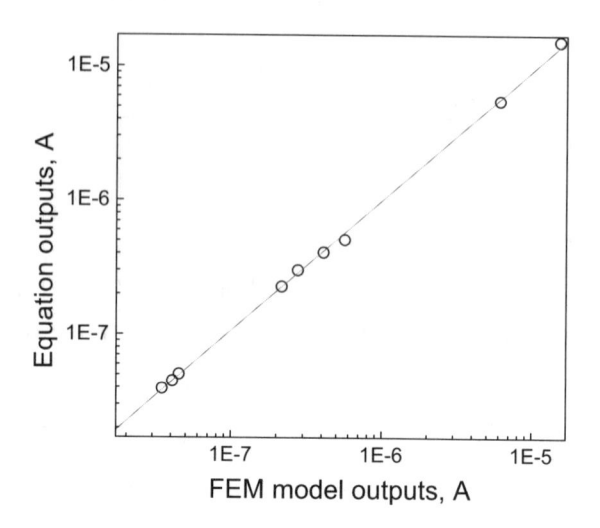

Figure 3: Comparison of the net cathodic current results obtained from finite element model calculations and from eqn (5).

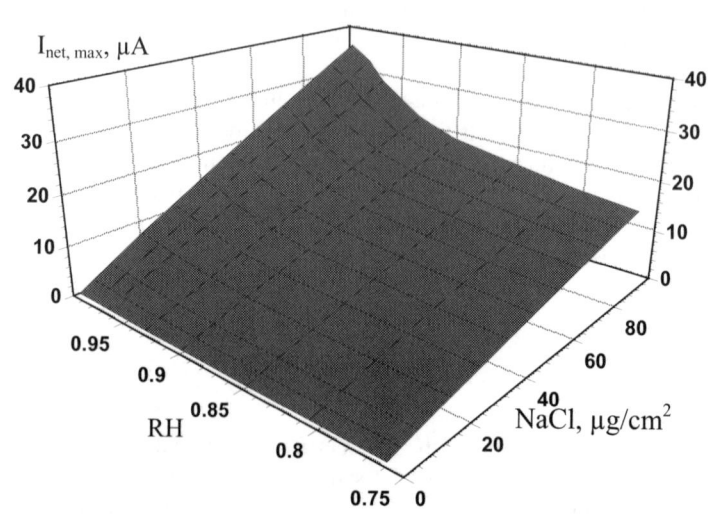

Figure 4: The effect of relative humidity and amount of deposited NaCl particles on the cathode capacity of SS 316L under atmospheric exposure. The cathode is circular around a pit with radius of 10 µm. Temperature is 25 °C.

4 Discussion

An important application of the $I_{net,max}$ is its coupling to the pit stability product. Pits in stainless steels often grow on the edge of stability even under potentiostatic conditions [5]. In atmospheric exposures, the limitations on cathode current available can become important in determining the stability of the localized corrosion system.

In considering such an analysis, the bounding nature of $I_{net,max}$ must be justified. As constituted, the $I_{net,max}$ expression assumes that the chemistry of the surface solution is constant, there are no inert particulate on the surface, and the solution perfectly wets the a flat surface. Each of these represents a practical upper bound. In reality, the solution chemistry will evolve with time, typically increasing in pH due to the production of hydroxyl on the cathode surface via reduction of oxygen and/or water. Higher pH solutions lead to slower cathodic kinetics, thus leading to a decrease in the term within the integral of eqn (5), and lower I_{net}. In service, inert particulate are usually present on surfaces. These particulate lower the effective conductivity of the surface, which leads directly to a decrease in the I_{net}. Finally, surface tension considerations as well as surface curvature, result in physical limitations on the cathode size which would lower I_{net}.

Figure 5 illustrates the coupling of the $I_{net,max}$ expression and the pit stability product for Type 316L stainless steel exposed to a marine environment. It considers two deposition densities (10 and 100 µg/cm^2) and two RH (85 and 98%) at 25°C. A pit stability product of 10^{-2} A/cm is also assumed [5,7,8]. As is obvious from the definition of the pit stability product, as the pit grows, the minimum current it requires increases linearly with r_a. As noted in Figure 5, $I_{net,max}$ also increases with increasing r_a, but at a less than linear rate. The intersection of the $I_{net,max}$ and I_{LC} curves represents the maximum pit size that the cathode can support. A series of critical pit radii are shown for the different DD/RH combinations, ranging from 1.5 µm (for 85% RH, 10 µg/cm^2) to 39.1 µm (for 98% RH, 100 µg/cm^2).

Pits smaller than the critical pit radius can grow, whereas once the critical pit radius is reached, the pit must either repassivate, or have only a portion of its surface remain active. The latter choice is an example of a type of stifling of localized corrosion.

Note that this type of analysis does not guarantee that pits will reach the maximum size. As noted above, the $I_{net,max}$ equation represents an upper bound. If conditions are such that the actual I_{net} is less than $I_{net,max}$, then the maximum supportable pit size will be smaller.

5 Conclusions

A bounding analytical solution for the maximum current that a circular cathode can deliver ($I_{net,max}$) to a pit exposed to a marine atmosphere has been derived and validated numerically. The solution shows that there is a linear dependence of $I_{net,max}$ on deposition density. Coupling of the analytical expression for the $I_{net,max}$

for the cathode with the Galvele pit stability product sets a defined limit on the rate at which a semicircular pit can grow as well as a maximum size for a given exposure scenario (T, deposition density, RH, cathode interfacial kinetics, anode critical pit stability product).

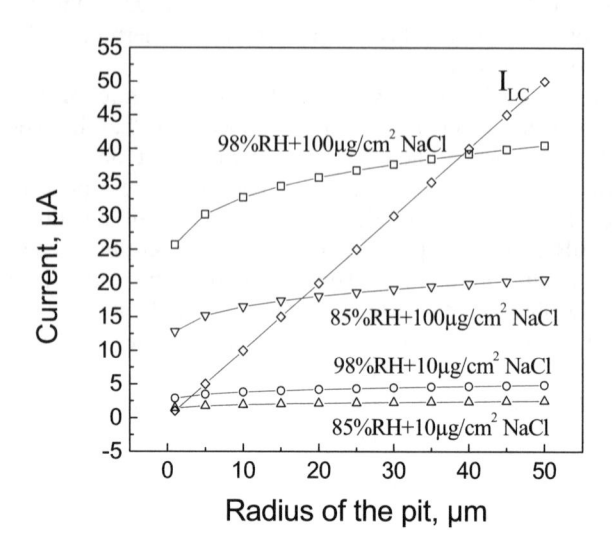

Figure 5: The effect of the radius of the pit on the circular cathode capacity, $I_{net,\,max}$, and anodic current demand, I_{LC}.

Acknowledgements

Support by the Office of Science and Technology and International of the U.S. Department of Energy (DOE), Office of Civilian Radioactive Waste Management is gratefully acknowledged. The work was performed under the Corrosion and Materials Performance Cooperative, DOE Cooperative Agreement Number: DE-FC28-04RW12252.

References

[1] Galvele, J. R., Tafel's law in pitting corrosion and crevice corrosion susceptibility. *Corros. Sci.*, **47(12)**, pp. 3053-3067, 2005.
[2] Galvele, J. R., Transport processes and the mechanism of pitting of metals. *J. Electrochem. Soc.* **123(4)**, pp. 464-474, 1976.
[3] Galvele, J. R., Transport processes in passivity breakdown—II. Full hydrolysis of the metal ions. *Corros. Sci.* **21(8)**, pp. 551-579, 1981.
[4] Gravano S. M. & Galvele, J. R., Transport processes in passivity breakdown—III. Full hydrolysis plus ion migration plus buffers. *Corros. Sci.*, **24(6)**, pp. 517-534, 1984.

[5] Williams, D.E., Stewart, J. & Balwill, P. H., "Nucleation, growth, and stability of micropits in stainless steels," in *Critical Factors in Localized Corrosion,* Frankel, G. S. and Newman, R. C., Editors, PV 92-9, The Electrochemical Society Proceedings Series, Pennington, NJ, pp. 36-64, 1992.

[6] Pride, S. T., Scully, J. R. & Hudson, J. L., Metastable pitting of aluminum and criteria for the transition to stable pit growth. *J. Electrochem. Soc.,* **141(11)**, pp. 3028-3040, 1994.

[7] Burstein, G. T., Pistorius, P. C. & Mattin, S. P., The nucleation and growth of corrosion pits on stainless steel. *Corros. Sci.,* **35(1-4)**, pp. 57~62, 1993.

[8] Moayed, M. H., & Newman, R. C., The relationship between pit chemistry and pit geometry near the critical pitting temperature. *J. Electrochem. Soc.,* **153(8)**, pp. B330-B335, 2006.

[9] Cui, F., Preseul-Moreno, F. J. & Kelly, R. G., Computational modeling of cathodic limitations on localized corrosion of wetted SS 316L at room temperature. *Corros. Sci.,* **47(12)**, pp. 2987-3005, 2005.

[10] Kelly, R. G., Cui, F. & Preseul-Moreno, F. J., Computational modeling of the stability of crevice corrosion of wetted SS316L. *ECS Trans.*, **1(16)** pp. 17-36 (2006).

[11] Chen, Z. Y., Cui, F. & Kelly, R. G., "An Analytical Modeling Method for Calculating the Current Delivery Capacity of a Thin-Film Cathode and the Stability of Localized Corrosion under Atmospheric Environments," accepted by *ECS Trans*, November, 2006.

[12] Nishikata, A., Ichihara, V., Hayashi, V. & Tsuru, T., Influence of electrolyte layer thickness and pH on the initial stage of the atmospheric corrosion of iron. *J. Electrochem. Soc.*, **144(4)**, pp. 1244-1252, 1997.

Phase field modeling of phase boundary motion due to transport-limited electrochemical reactions

A. Powell[1] & W. Pongsaksawad[2]
[1] *Veryst Engineering LLC, Needham, Massachusetts, USA*
[2] *National Metal and Materials Technology Center, Khlong Luang, Pathumthani, Thailand*

Abstract

A new phase field model is presented for simulation of phase boundary motion due to transport-limited electrochemical reactions. The model consists of Cahn-Hilliard diffusion based on a statement of free energy with an electrostatic energy term, and conservation of charge. It is shown that under assumptions of negligible charge transfer resistance (mass transfer dominance) and rapid charge redistribution, the conservation of charge equation reduces to zero divergence of current density. When simulating electrolytic metal oxidation/reduction with an unsupported electrolyte, the model reproduces analytical models of cathode interface stability. It can also simulate electronically mediated reactions at separate interfaces, such as those occurring in metallothermic reduction processes. Results are presented for both unsupported and supported electrolytes, and both solid-state transformations and those involving fluid flow, including fluid-structure interactions using the Mixed Stress model for diffuse interface fluid-structure modeling.
Keywords: electrochemistry, mathematical modeling, phase field, fluid flow, stability analysis, dendrite, streamer, titanium, steelmaking.

1 Introduction

Electrolysis enjoys widespread use for extraction of metals from their ores or aqueous solutions. However, metal electrodeposition very often results in a rough surface or dendrites due to a Mullins-Sekerka instability at the cathode/electrolyte interface. Designing around this problem involves *e.g.* using additives in aqueous processes or limiting the current in high-temperature systems.

The few tools available to the designer for modeling the onset and development of such instabilities can be roughly categorized into: linear stability theory, particle models such as Monte Carlo simulations, and continuum models which track the interface such as finite element models with Lagrangian interface descriptions and level set methods.

Continuum models which explicitly track the interface such as that of Cao *et al.* [1] work sufficiently well under some circumstances but break down when the interface topology changes. The level set method [2] tracks interface propagation over time based on the curvature-dependent speed. This method has been used to model roughness evolution during superconformal electrodeposition with additives and catalysts in the electrolyte [3, 4]. Unfortunately, it requires that the interface dynamics be understood *a priori*, making it difficult to extend to complex systems with many components.

The phase field methodology derives its governing equations from a thermodynamic equation for free energy. Guyer *et al.* have developed a general formulation for phase field modeling of electrodeposition and electrodissolution [5]. Although their approach includes the detail of charge distribution at the interface, it is currently limited in two ways: it requires discretization of the electrochemical double layer, limiting its use to nanometer-scale systems, and it is a numerically challenging technique, such that it has to date only been solved in one dimension.

When limited by mass transfer, the effect of charge transfer resistance at the interfacial double layer is negligible, removing the need to discretize the double layer, and permitting simulation of larger systems on scales up to millimeters and beyond. The resulting phase field method, described by the authors in detail elsewhere [6], considerably simpler than that of Guyer, but limited to those situations where charge transfer resistance can be neglected. This paper outlines that method as applied to a liquid-liquid iron/iron oxide system, and a ternary titanium/magnesium/chlorine system without flow.

2 Binary iron/iron oxide model

Electric Field-Enhanced Smelting and Refining (EFESR) was invented by Uday Pal [7] to drive the reaction between FeO in steelmaking slag and carbon in hot metal. This process operates by two half-cell reactions at the cathode (above) and anode (hot metal-slag interface) respectively:

$$Fe^{2+} + 2e^- \Rightarrow Fe \qquad (1)$$

$$C + O^{2-} \Rightarrow CO + 2e^- \qquad (2)$$

This has several beneficial effects: it increases the yield of steelmaking (about 2% of the iron is lost to the slag), reduces foaming, and removes carbon from hot metal, possibly reducing the need for subsequent vacuum degassing. For stainless steel, it also removes chromium oxide from the slag, bringing both economic and environmental benefits.

The kinetics of the overall process are limited by mass transfer of ferrous ions to the cathode. But the cathode reaction in equation 1 forms liquid iron, which grows as the liquid equivalent of dendrites, called *streamers* in the aluminum literature, which dart out into the fresh slag and enhance the mass transfer coefficient considerably. Streamers then break up into droplets due to the Rayleigh instability of a liquid tube; these droplets then sink down to join the liquid hot metal below.

Because the kinetics of the process are determined by the formation and breakup of streamers, a model which describes them would aid in the design of a solid cathode, flow conditions, magnetic fields, etc., to optimize mass transfer.

2.1 Thermodynamics

A phase field formulation for coupled fluid flow, diffusion, electromigration, and transport-limited electrochemical reactions was developed by Dussault [8], and expanded by Pongsaksawad and Powell [6]. This model begins with a dimensionless concentration variable C which is zero in the FeO slag and one in the iron metal, and uses a very simplistic polynomial function for the homogeneous free energy with minima at zero and one:

$$\Psi(C) = C^2(1-C)^2. \tag{3}$$

Following the Cahn-Hilliard formulation [9], as extended by the Guyer electrochemistry model [5, 10], the total chemical free energy adds terms for the concentration gradient and electrical potential:

$$F = \int \left(\beta\Psi(C) + \frac{\alpha}{2}|\nabla C|^2 + F\Phi \sum z_i c_i \right) dV, \tag{4}$$

where Φ is the electrical potential, F is the Faraday constant, and z_i and c_i are the charge and concentration (in moles/volume) of each species. Like the chemical potential in the Cahn-Hilliard model, the total electrochemical potential in the slag is the variational derivative of the free energy functional:

$$\mu_{\text{slag}} = \frac{\delta F}{\delta C} = -\alpha\nabla^2 C + \beta\Psi'(C) + 4F\rho_M \Phi. \tag{5}$$

Charge neutrality requires that the iron charge be given by:

$$z_{\text{Fe}}(C) = \frac{2(1-C)}{(1+C)}. \tag{6}$$

Because iron electromigration flux is proportional to its charge z_{Fe}, the electrostatic term in electrochemical potential can be multiplied by $(1-C)/(1+C)$ in order to make it applicable in both electrolyte and metal:

$$\mu = -\alpha\nabla^2 C + \beta\Psi'(C) + 4F\rho_M \frac{1-C}{1+C}\Phi. \tag{7}$$

This electrochemical potential then determines the flux.

2.2 Kinetics

The transport equation in a fluid with velocity field $\vec{u}$ is:

$$\frac{DC}{Dt} = \frac{\partial C}{\partial t} + \vec{\nabla} \cdot (\vec{u}C) = -\vec{\nabla} \cdot \vec{J}, \qquad (8)$$

where where $\frac{D}{Dt}$ is the substantial derivative which represents the rate of change in the frame of reference moving with velocity $\vec{u}$, and $\vec{J}$ is the flux. Flux in turn is given by the gradient in electrochemical potential times mobility κ:

$$\vec{J} = -\kappa \vec{\nabla} \mu = -\kappa \vec{\nabla} \left(-\alpha \nabla^2 C + \beta \Psi'(C) + 4F\rho_M \frac{1-C}{1+C} \Phi \right). \qquad (9)$$

For iron and ferrous oxide, the molar density is roughly constant, so this term can be taken outside the gradient. The complete conservation equation is given by:

$$\frac{DC}{Dt} = \vec{\nabla} \cdot \left(\kappa \vec{\nabla} \left(-\alpha \nabla^2 C + \beta \Psi'(C) \right) + 4\kappa F\rho_M \frac{1-C}{1+C} \vec{\nabla} \Phi \right) \qquad (10)$$

2.3 Conservation of charge

The equation for conservation charge is given by:

$$\frac{D\rho_f}{Dt} = -\vec{\nabla} \cdot \left(F\rho_M \sum z_i \vec{J}_i - \sigma_e(C) \vec{\nabla} \Phi \right), \qquad (11)$$

where ρ_f is the charge density and $\sigma_e(C)$ is the electronic conductivity. Electronic conductivity in turn is estimated by interpolation between that of the metal and zero in the slag. By comparing the timescales of diffusion and electromigration, of convective charge transport and current, and timescales of diffusion vs. charge buildup at interfaces, in a mass transfer-limited situation with rapid charge redistribution this can be reduced to:

$$0 = \vec{\nabla} \cdot (\sigma_{eff}(C) \vec{\nabla} \Phi), \qquad (12)$$

where the effective conductivity σ_{eff} is given by:

$$\sigma_{eff} = 4\kappa F^2 \rho_M^2 \frac{1-C}{1+C} + \sigma_e(C). \qquad (13)$$

In the ternary system, transport of charge due to chemical potential-driven diffusion must also be included.

2.4 Hydrodynamics

In the EFESR process, the reaction occurs between liquid phases, and it is necessary to include the effect of fluid flow. The Navier-Stokes equations describe conservation of mass and momentum in a fluid system. Using u, v and ω for x- and y-velocities and vorticity, the velocity-vorticity form is given by:

$$\nabla^2 u + \frac{\partial \omega}{\partial y} = 0 \tag{14}$$

$$\nabla^2 v - \frac{\partial \omega}{\partial x} = 0 \tag{15}$$

$$\frac{\partial \omega}{\partial t} + \vec{u} \cdot \nabla \omega = \nu \nabla^2 \omega - N \nabla \times (C \nabla \mu), \tag{16}$$

where ν is the kinematic viscosity and N is a Weber number $\frac{\gamma D}{\rho u^2 \epsilon}$. The vorticity transport equation includes the curl of the interfacial tension forcing term $(-C\nabla\mu)$ given by Jacqmin [11], which in turn describes driving force due to curvature of a diffuse interface. We assume uniform fluid properties, and neglect gravity.

2.5 Simulating streamer formation

The above equations are implemented in the open source RheoPlast tool [12]. Simulations begin with the three-layer metal-oxide-metal condition shown in figure 1. Periodic boundary conditions are applied in the x-direction, and symmetry planes used for C in the y direction. Voltage is set to 0 at the top and 1 at the bottom. Physical parameters are: $\rho_M = 1.31 \times 10^5$ mol/m^3, $\sigma = 12.0$ Ω^{-1} m^{-1} in the slag, $\sigma_e = 10^3$ Ω^{-1} m^{-1} in the metal.

Figures 1a and 1b show simulated cathode shape evolution under unstable and stable conditions respectively. The instability in the former case leads to not only shape change, but also topology change as droplets break free from the streamers.

2.6 Stability analysis benchmark

The Peclet number predicts stability of sinusoidal perturbation of wavelength λ:

$$\text{Pe} = \frac{u_{int} \lambda}{D} = \frac{\sigma_i \Delta \Phi \lambda}{zF\rho_M LD} = \frac{\Delta \Phi z F \rho_M \varepsilon \lambda}{2\sqrt{18} \cdot 3.1 \gamma L}. \tag{17}$$

That is, large Peclet number corresponds to fast plating due to large electric field and conductivity, with small diffusivity and interfacial tension, leading to unstable growth. Barkey's linear stability analysis [13] gives the criterion for stable growth in an unsupported electrolyte:

$$\frac{\Delta \Phi}{L} < \frac{8\pi^2 \gamma}{zF\lambda^2 \rho_M}; \text{ i.e. Pe} < \frac{8\pi^2}{6.2\sqrt{18}} \frac{\varepsilon}{\lambda} = 3.00 \frac{\varepsilon}{\lambda}. \tag{18}$$

This holds at the limit of a sharp interface, perfect cathode conductivity, infinite electrode separation, and solid electrode *i.e.* infinite viscosity.

Simulations show that the critical quantity $\text{Pe}_{\text{crit}} \lambda / \epsilon$ approaches 3.0 in the limits of sharp interface, perfect cathode conductivity, infinite electrode separation,

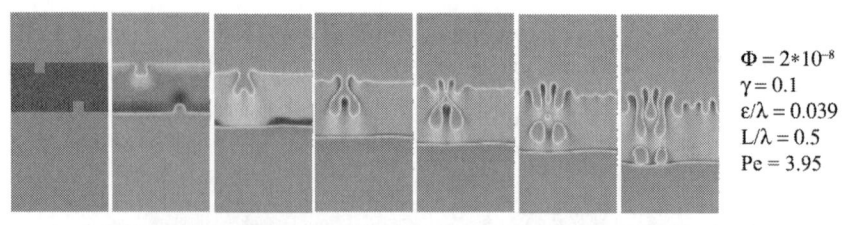

(a) Unstable simulation with high electric field on a 150×300 grid.

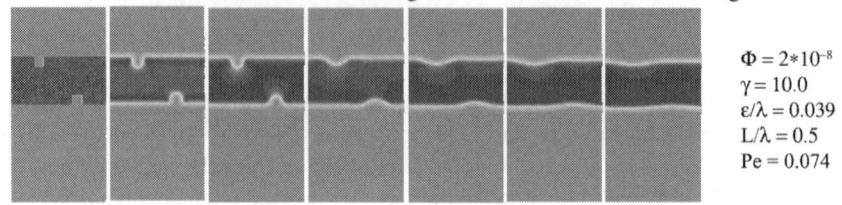

$\Phi = 2*10^{-8}$
$\gamma = 0.1$
$\varepsilon/\lambda = 0.039$
$L/\lambda = 0.5$
$Pe = 3.95$

(b) Stable simulation with high interfacial tension on an 80×160 grid.

$\Phi = 2*10^{-8}$
$\gamma = 10.0$
$\varepsilon/\lambda = 0.039$
$L/\lambda = 0.5$
$Pe = 0.074$

Figure 1: Cathode shape evolution without flow under conditions leading to (a) unstable and (b) stable behavior.

and solid electrode or infinite viscosity. That limiting value does not change appreciably as interface thickness approaches $\lambda/10$, nor for electrode separation as small as λ.

With liquid electrode and electrolyte, stability is determined by the Schmidt number, which is the ratio ν/D. When the Schmidt number is large, high viscosity prevents flow, and stability behavior approaches the case without flow. When it is small, the interface becomes more stable, as described in further detail in ref [6].

2.7 Relating the model to experiments

In the EFESR experiments of Pal *et al.*, liquid FeO has the following properties: $\mu = 0.20$ Pa·s [14], $D = 4.20 \times 10^{-7}$ m^2/s [15], $\rho = 3.50 \times 10^3$ kg/m^3, $\gamma = 6.45 \times 10^{-1}$ N/m [16]. Sc is thus 1.40×10^2, and the critical Peclet number is approximately 2.35×10^2. Experiments produce an electric field of 20 V/m, resulting in critical wavelength $\lambda_{crit} = 0.4$ mm.

One experiment quenched the slag-metal system at the cathode, and the streamers measured approximately $\frac{1}{4}$ mm across, corresponding to a critical wavelength approximately twice that. The results of this binary model can thus be considered in good qualitative agreement with experimental observations.

3 Ternary titanium/magnesium/chlorine model

The Kroll process, which has dominated titanium production for the past fifty years, is conducted in an iron crucible, resulting in iron contamination of the sponge product near its walls, reducing the process yield. If the crucible were made instead of titanium, none of the product would be discarded, however TiCl$_4$ would react with it to form TiCl$_3$ and TiCl$_2$, compromising the integrity of the walls. The subhalide reduction process of Okabe and Takeda [17] starts with magnesium and TiCl$_2$, eliminating the reaction with a titanium crucible. The overall reaction is:

$$\text{TiCl}_2 + \text{Mg} \Rightarrow \text{Ti} + \text{MgCl}_2. \tag{19}$$

Here phase field simulates sponge structure formation, which involves electronically-mediated reactions (EMR) between Ti^{2+} and Mg which are physically separate but connected by electronic conductors. The Ti-Mg-Cl ternary system is most easily represented by two dimensionless concentration parameters C_2 and C_3: C_2 is zero in Mg and MgCl$_2$ and one in Ti and TiCl$_2$, and C_3 is zero in the chlorides and one in the metals. The mole fractions of each element are then written:

$$X_{\text{Ti}} = C_2 \left(\frac{1}{3} + \frac{2}{3} C_3 \right) \quad X_{\text{Mg}} = (1 - C_2) \left(\frac{1}{3} + \frac{2}{3} C_3 \right) \quad X_{\text{Cl}} = \frac{2}{3}(1 - C_3). \tag{20}$$

3.1 Thermodynamics and kinetics

As before, free energy includes homogeneous, gradient, and electrostatic terms:

$$F = \int \left(\Psi(C_2, C_3) + K_{22} |\nabla C_2|^2 + K_{33} |\nabla C_3|^2 + F\Phi \sum z_i c_i \right) dV, \tag{21}$$

where again c_i is the molar concentration of species i, and we omit cross gradient terms. Homogeneous energy is unknown, and is constructed based on its binaries:

$$\Psi = C_2 \ln C_2 + (1 - C_2) \ln(1 - C_2) + \Omega_{12} C_3 C_2 (1 - C_2) + \tag{22}$$
$$C_3 \ln C_3 + (1 - C_3) \ln(1 - C_3) + \Omega_{13} C_3 (1 - C_3) -$$
$$(C_2 - 0.5)(C_3 - 0.5).$$

The first three terms form an ideal solution between chlorides ($C_3 = 0$), and a regular solution with a miscibility gap between metals ($C_3 = 1$), $\Omega_{12} = 3.3$. The next three form a regular solution with miscibility gap between metals and chlorides, $\Omega_{13} = 2.5$. The last term drives the reaction in equation 19.

The chemical potential for each concentration parameter μ_i is again the variation of the free energy functional with gradient coefficients $K_{22} = K_{33} = 5 \times 10^{-4}$:

$$\mu_2 = -K_{22} \nabla^2 C_2 + \frac{\partial \Psi}{\partial C_2} \tag{23}$$

$$\mu_3 = -K_{33} \nabla^2 C_3 + \frac{\partial \Psi}{\partial C_3} + 2F\rho_M \frac{1 - C_3}{1 + C_3} \Phi. \tag{24}$$

Conservation equations for each parameter can be written as:

$$\frac{DC_2}{Dt} = \frac{\partial C_2}{\partial t} + \vec{u} \cdot \nabla C_2 = M_{22} \nabla^2 \mu_2 + M_{23} \nabla^2 \mu_3 \tag{25}$$

$$\frac{DC_3}{Dt} = M_{32}\nabla^2\mu_2 + M_{33}\nabla^2\left(-K_{33}\nabla^2 C_3 + \frac{\partial\Psi}{\partial C_3}\right) + B\nabla^2\Phi, \quad (26)$$

where B is an electromigration constant set to one, cross mobilities M_{23} and M_{32} are set to zero, and diagonal mobilities M_{22} and M_{33} are both set to 2×10^{-2}.

3.2 Conservation of charge

As mentioned in section 2.3, conservation of charge in the ternary case must include a flux due to chemical potential gradient, and is thus given by:

$$0 = \vec{\nabla} \cdot \left[\tilde{z}M_{33}F\vec{\nabla}\left(-K_{33}\nabla^2 C_3 + \frac{\partial\Psi}{\partial C_3}\right)\right] + \vec{\nabla} \cdot (\sigma_{eff}\vec{\nabla}\Phi), \quad (27)$$

where $\tilde{z}$ is a constant describing charge transfer due to diffusion and the effective conductivity and electronic conductivity are written such that σ_e is zero in the chlorides, and that of titanium is half of that of magnesium:

$$\sigma_{eff} = \frac{1-C_3}{1+C_3} + \frac{\sigma_e}{\sigma_i}, \quad \sigma_e = \sigma_{e0}(3C_3^2 - 2C_3^3)\left(1 - \frac{1}{2}C_2\right). \quad (28)$$

3.3 Simulations

Two two-dimensional simulations are presented here. The first, shown in figure 2, begins with metal phases below, and chloride phases above. As expected, the titanium grows and magnesium shrinks at the Ti-Mg-chlorides triple line, and chlorides interdiffuse but overall become richer in $MgCl_2$. The titanium phase also grows upward into what is originally $TiCl_2$, and the magnesium phase shrinks away from the triple line, indicating electronically-mediated reactions (EMR). Electric potential is low (red) where electrons are generated at the Mg-$MgCl_2$ interface and high (blue) where they are absorbed at the Ti-$TiCl_2$ interface. Thus electrons move from low to high potential through metal phases, allowing the reaction to proceed toward completion.

A second simulation of unstable cathode growth without flow is shown in figure 3. This figure shows the formation of a $TiCl_2$-poor layer above the growing titanium which is considerably thinner in front of the "dendrite" tips, resulting in faster diffusion and enhancing the instability due to electric field concentration.

4 Summary

The phase field electrochemistry formulation presented here is a potentially powerful tool for simulation of electrochemical reactions. It is able to model such phenomena as roughness formation during plating from a supported or unsupported electrolyte, and electronically-mediated reactions between interfaces in electronic contact. And it shows good quantitative agreement with linear stability theory in the solid case, and develops a new stability theory for reduction

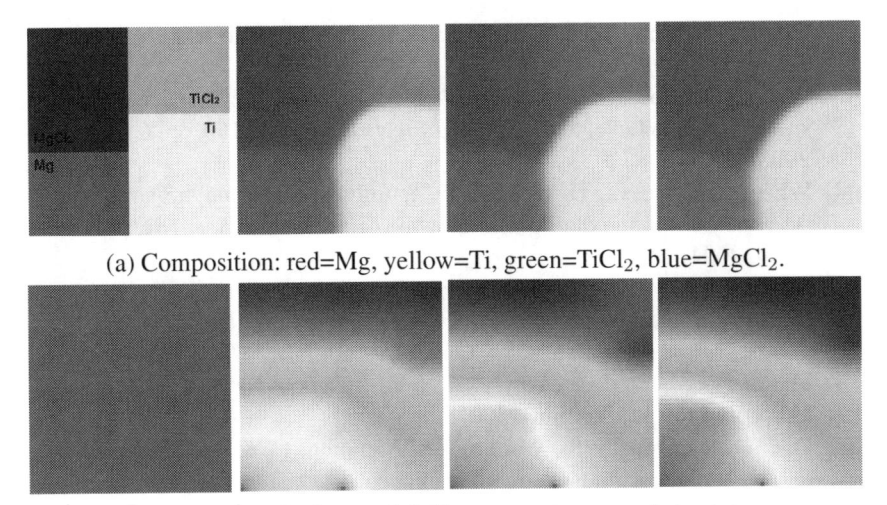

(a) Composition: red=Mg, yellow=Ti, green=TiCl$_2$, blue=MgCl$_2$.

(b) Relative electrical potential: blue at maximum, red at minimum.

Figure 2: Ternary simulation of electronically-mediated reaction (EMR).

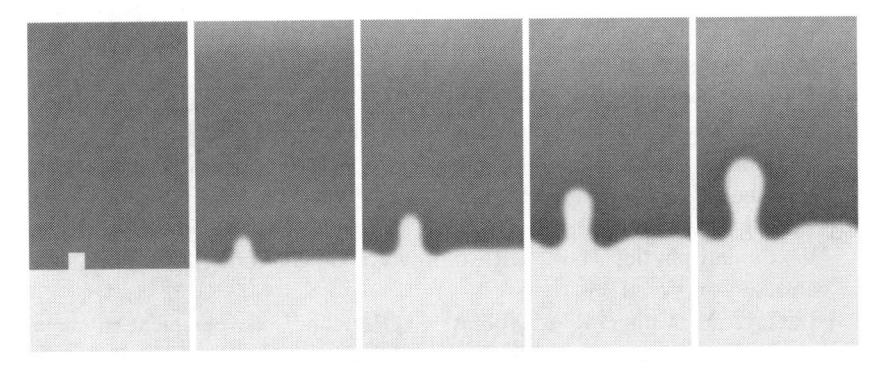

Figure 3: Simulation of titanium deposition, same composition scale as figure 2(a).

of liquid metal from liquid electrolyte which agrees well with an experimental measurement.

The model's assumption of transport-limited systems with negligible charge transfer resistance limits its use, but for high temperature electrometallurgy processes such as the two described here, this model provides a very useful tool.

References

[1] Cao, Y., Taephaisitphongse, P., Chalupa, R. & West, A., Linear stability analysis of unsteady galvanostatic electrodeposition in the two-dimensional diffusion-limited regime. *J Electrochem Soc*, **148**, pp. C466–C472, 2001.

[2] Osher, S. & Sethian, J.A., Fronts propagating with curvature-dependent

speed: Algorithms based on Hamilton-Jacobi formulations. *J Comput Phys*, **79**, pp. 12–19, 1988.

[3] Wheeler, D., Josell, D. & Moffat, T., Modeling superconformal electrodeposition using the Level Set Method. *J Electrochem Soc*, **150**, pp. C302–C310, 2003.

[4] Wheeler, D., Josell, D. & Moffat, T., Influence of catalytic surfactant of roughness evolution during film growth. *J Electrochem Soc*, **151**, pp. C538–C544, 2004.

[5] Guyer, J.E., Boettinger, W.J., Warren, J.A. & McFadden, G.B., Phase field modeling of electrochemistry. II. Kinetics. *Phys Rev E*, **69(021604)**, 2004.

[6] Pongsaksawad, W., Powell, A.C. & Dussault, D., Phase field modeling of transport-limited electrolysis in solid and liquid states. *J Echem Soc*, **(Accepted for publication)**, 2007.

[7] Pal, U., MacDonald, S., Woolley, D., Manning, C. & Powell, A., Results demonstrating techniques for enhancing electrochemical reactions involving iron oxide in slags and C in liquid iron. *Metall Mater Trans*, **36B**, pp. 209–218, 2005.

[8] Dussault, D. & Powell, A., Phase field modeling of electrolysis in a slag or molten salt. *Proc. Mills Symp.*, The Institute of Materials, London, UK, pp. 359–371, 2002.

[9] Cahn, J. & Hilliard, J. *J Chem Phys*, **28**, p. 258, 1958.

[10] Guyer, J., Boettinger, W., Warren, J. & McFadden, G., Phase field modeling of electrochemistry. I. Equilibrium. *Phys Rev E*, **69(021603)**, 2004.

[11] Jacqmin, D., Calculation of two-phase Navier-Stokes flows using phase-field modeling. *J Comp Phys*, **155**, pp. 96–127, 1999.

[12] Powell, A.C., Zhou, B., Vieyra, J. & Pongsaksawad, W., RheoPlast phase field multi-physics code. URL http://lyre.mit.edu/~powell/rheoplast.html, currently version 0.8.9.

[13] Barkey, D., Muller, R. & Tobias, C., Roughness development in metal electrodeposition: II. Stability theory. *J Electrochem Soc*, **136**, pp. 2207–2214, 1989.

[14] Urbain, G. & Boiret, M., *Ironmaking and steelmaking*. National Institute of Standards and Technology, 1990.

[15] Li, Y., Lucas, J., Fruehan, R. & Belton, G., The chemical diffusivity of oxygen in liquid iron oxide and a calcium ferrite. *Metall Mater Trans B*, **31B**, pp. 1059–1068, 2000.

[16] Mills, K. & Keene, B., Physical properties of BOS slags. *Int Mater Rev*, **32**, pp. 1–120, 1987.

[17] Okabe, T. & Takeda, O., A new high speed titanium production by subhalide reduction process. *Proc. Light Metals, San Francisco, CA, TMS Annual Meeting*, TMS, pp. 1139–1144, 2005.

Kinetic Monte Carlo modelling the leaching of Raney Ni-Al alloys

N. C. Barnard & S. G. R. Brown
Materials Research Centre, School of Engineering,
Swansea University, Singleton Park, Swansea, UK

Abstract

Raney nickel-aluminium alloys are widely used in powder form as catalysts for hydrogenation reactions. While the catalytic powder may be produced via several processing routes in this paper we are concerned with powders manufactured via spray atomization of liquid metal. Before use, the powdered nickel-aluminium alloy undergoes a leaching process carried out in sodium hydroxide solution which dissolves aluminium from the solid and leaves a nanoscopically porous nickel rich product. This product is sometimes referred to as 'spongy nickel'. The kinetics of the process is governed by both the rate of dissolution of aluminium into solution and the rate of surface diffusion of the nickel (which promotes the formation of the nanoscopically porous material). However, individual powder particles may contain more than one NiAl phase. Of particular interest to this work are the $NiAl_3$ and Ni_2Al_3 phases. This paper describes a preliminary attempt to model the leaching behaviour of these two phases using a kinetic Monte Carlo simulation.

Keywords: Raney-Ni, leaching, spray-atomization, kinetic Monte Carlo.

1 Introduction

Raney-type nickel catalysts have been used extensively in hydrogenation reactions, owing to their high catalytic activity in organic reactions. Raney-type Nickel catalysts are obtained from Ni-Al alloys, whereby a large proportion of the aluminium present is removed via caustic action: immersion in concentrated NaOH solution. The starting composition of the Ni-Al alloy and the leaching time/condition has been shown to have an impact on the structure and catalytic performance of the Raney-Ni product [1-5]. Traditionally, the composition used for the precursor alloy is Ni – 50wt.% Al, consisting of NiAl$_3$, Ni$_2$Al$_3$ and Al-NiAl$_3$ phases. The precursor alloy is often doped with other elements and in an effort to increase the catalytic activity/selectivity of the activated Raney-Ni catalyst, although the effect of these promoters in the leaching process is not considered here.

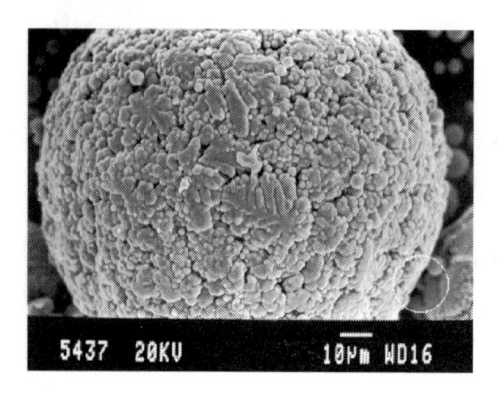

Figure 1: Magnified image of an atomised Ni – 50wt.% Al powder.

Particles of the Ni-Al alloy may be produced via cast-and-crush or spray atomisation routes. The particles produced range from a few, to in excess of 200 microns diameter. Concentrated aqueous NaOH is used to 'leach' away a large proportion of the aluminium whilst leaving the nickel. In addition to significantly increasing the surface area of the powder particles, leaching produces hydrogen, which activates the Ni catalyst:

$$2(\text{Ni-Al})_{(s)} + 2\text{OH}^-_{(aq)} + 6\text{H}_2\text{O}_{(l)} \rightarrow 2\text{Ni}_{\text{Raney}} + 2\text{Al(OH)}^-_{4(aq)} + 3\text{H}_{2(g)} \quad (1)$$

During the leaching process 10-20wt.% NaOH is used in order to prevent the formation of bayerite ($\text{Al}_2\text{O}_3.3\text{H}_2\text{O}$) which is produced when the aluminate formed in solution is precipitated at lower pH.

$$2\text{AlO}_2^- + 4\text{H}_2\text{O} \rightarrow \text{Al}_2\text{O}_3.3\text{H}_2\text{O} + 2\text{OH}^- \quad (2)$$

The leaching process is carried out at elevated temperatures (~50°C) and produces a highly pyrophoric catalyst that requires storage in water or alcohol.

Two dominant processes that occur during this process are the dissolution/diffusion of the aluminium and the solid-state diffusion of the nickel present in the phases. Although it has been stated that the Ni – 50wt.% Al alloy contains the Al-NiAl$_3$ phase, it is the NiAl$_3$ and Ni$_2$Al$_3$ phases that are the concern of this paper as observation shows they are by far the most dominant phases present. In order to model the differences in the leaching behaviour of these phases, containing nickel and aluminium in differing ratios, the kinetic Monte Carlo (kMC) technique has been employed. In a seminal paper Erlebacher *et al* [6] used this technique to model the leaching of silver from a Au – 50at.% Ag alloy, examining the evolution of nanoporosity.

An attempt is made here to model the structural transformations during the leaching of the precursor alloy to produce Raney-Ni catalysts using a kMC bond-breaking approach.

2 Kinetic Monte Carlo (kMC)

2.1 Rates of diffusion and dissolution

In order to simulate the time evolution of the leaching of a Raney-Ni structure a kinetic Monte Carlo method is used. As such, it is a requirement that the processes taking place, i.e. adatom diffusion (Ni or Al) and aluminium dissolution, occur at known rates that serve as the inputs of the model. In the case of the diffusion and dissolution of atoms, the processes are assumed to proceed at a rate described by the following Arrhenius relationships respectively:

$$k_n^{diff} = v_{diff} \exp\left(-\frac{nE_b}{k_bT}\right) \quad (3a)$$

$$k_n^{diss} = v_{diss} \exp\left(-\frac{nE_b}{k_bT}\right) \quad (3b)$$

where k_n^{diff} is the rate constant for diffusion of an atom (Ni or Al), k_n^{diss} is the rate constant for dissolution of an atom (Al only), the prefactors v_{diff} and v_{diss} are attempt frequencies, n is the number of bonds, k_b is the Boltzmann constant, T is the absolute temperature (K) and the bond energy is taken as $E_b = 0.15\text{eV}$.

The diffusion of both elements present means that the final nanoscopic spongy Nickel structure is not revealed by leaching but rather is constructed during leaching by diffusion of Ni adatoms. (Although Ni adatoms do not dissolve there exists a case in the model whereby the Ni atoms may be removed. Should a Ni adatom be present at the surface where it is only neighboured by a single aluminium atom, which is itself selected randomly for dissolution, the Ni atom would become unattached. For simplicity, this Ni atom is removed from the simulation.) Equations 3a and 3b mean that atoms with fewer bonds are more likely to dissolve or diffuse. Clearly, adatoms diffusing to lower energy sites (i.e. sites where they have more bonds) will tend to become less likely to diffuse further. In this way Ni adatoms tend to cluster together at exposed surfaces and build up a nanoscopic structure.

Knowledge of the rates of all permitted transitions allows the time increment for each iteration of the model to be calculated. A residence-time algorithm is used by the model to model both of the above processes to model the evolution of the active catalyst.

2.2 Bortz–Kalos–Liebowitz (BKL) algorithm

The Bortz–Kalos–Liebowitz algorithm was used to determine the evolution of the spongy nickel on an event-by-event basis. It proceeds as follows.
1. The time is set to zero; $t = 0$.
2. Determine all possible transition states and calculate the cumulative function, $R_i = \sum_{j=1}^{i} k_j$ for $i = 1 \ldots N$, total number of transitions.
3. Get a uniform random number, $u \in (0,1]$.
4. Find the event to carry out, i, for which $R_{i-1} < uR < R_j$.
5. Find all transitions and associated rates, k_i that have changed due to the transition.
6. Get another random number, $u \in (0,1]$.
7. Update the time with $t = t + \Delta t$, where $\Delta t = -\log u / R$.
8. Return to step 2.

2.3 Crystal structures

The simulations are performed with the atoms arranged on a simple cubic orthogonal computational mesh. Since more than one phase maybe present, it cannot be assumed that all sites are suitable for an atom to be positioned in each phase, nor are neighbouring sites on the mesh necessarily suitable for diffusion within a particular phases. The phases considered with respect to crystal structure are $NiAl_3$, Ni_2Al_3 and $NiAl$. (Within the computational mesh, some vacancies are distributed randomly that permit solid-state diffusion as an event that can take place, although this is of negligible effect.)

2.3.1 NiAl₃
The $NiAl_3$ phase in an orthorhombic phase, although to permit the use of a cubic mesh the lengths of the lattice are distorted. This allows the approximation of the $NiAl_3$ phase to that of a cubic $AuCu_3$ structure.

2.3.2 Ni₂Al₃
The Ni_2Al_3 phase is less straightforward than that of $NiAl_3$. The structure of this phase can be derived from the CsCl structure by the removal of every third (111) plane of Ni atoms [7]. The simple cubic lattice is maintained by a slight contraction along the (111) direction.

2.3.3 NiAl
It important to include an NiAl phase in the model since this phase is resistant to the leaching of aluminium in the NaOH solution and exhibits different electrochemical behaviour [8]. The NiAl phase can be created and destroyed during the simulation via the diffusion of Ni and Al atoms although this phase is not included as a starting constituent of the simulations.

3 Results and discussion

Modelling the leaching of NiAl alloys to produce Raney-Ni catalysts using the kMC method leads to inherent advantages and disadvantages. kMC allows realistic timescales to be achieved that can be related to experimentation, whereas longer times are hard to achieve using molecular dynamics methods. There is no requirement by kMC for thermodynamic equilibrium; however mechanisms and associated energy barriers have to be known – in this case two have been identified. Using the BKL algorithm, the time increment alters as mechanisms become more or less prevalent, which can be beneficial. A drawback of using kMC for modelling the leaching of large NiAl powders is large CPU times. This requires efficient coding of the model, data structures etc. Finally, it is accepted that the Arrhenius law may not be fully adequate, such as when kinetic energy is accumulated when crossing a barrier and can be used to cross other barriers.

Below are very preliminary results from using the kMC model, examining the effect of composition and crystal structures of the phases present. In addition, the relative leaching behaviour of the two phases is assessed, including the orientation of the particular phases exposed to the aqueous NaOH solution.

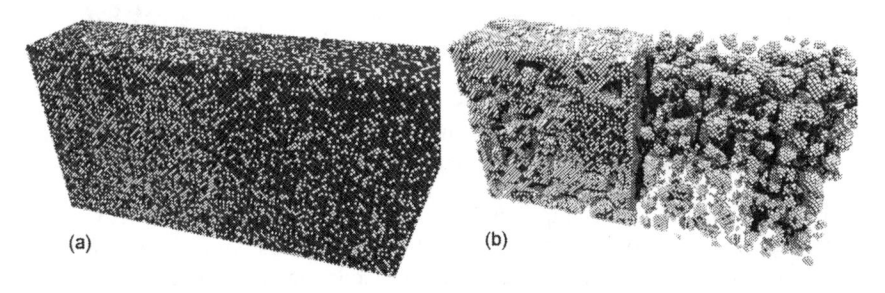

Figure 2: Solid solution of Ni-Al (Ni – light, Al – dark, FCC assumed). Mesh: left-hand side 60at.%Al, right hand side 75at.%Al; (a) initial mesh, and (b) after leaching.

3.1 Effect of composition alone

Shown in figure 2 is a hypothetical FCC solid solution of Ni and Al where the right-hand side of the mesh contains 15at.% more Al than the left. During the simulation, the 75at.% Al is subject to a higher degree of Al removal. The higher Al content in contact with the NaOH solution permits more rapid dissolution, in addition to a larger amount of permitted sites for the Ni adatoms to diffuse to, exposing further Al atoms.

The faster break-up of the higher Al-composition region in an advancing-front-type reaction and the greater cohesion of the lower Al-composition region can be seen in figure 2. This behaviour tends to agree with observation and lent some initial confidence in the use of the kMC method.

3.2 Effect of atomic arrangement

As stated previously, $NiAl_3$ and Ni_2Al_3 phases have a different crystallography to one another. Figure 3 incorporates the crystal structure of the $NiAl_3$ phase, whilst still approximating the Ni_2Al_3 as regions of solid solution. It can be seen from figure 3 that the crystal structure of the $NiAl_3$ promotes a structure to be formed that is much changed from the solid solution. The solid solution Ni_2Al_3 maintains the same form during leaching, although in this case there has been a build up of nickel at the entrance to adjacent developing nanopores. The effect of this would be that it would further reduce the Al removal rate of the Ni_2Al_3. The formation of these pores would inevitably effect the diffusion and thereby affect processes occurring [9].

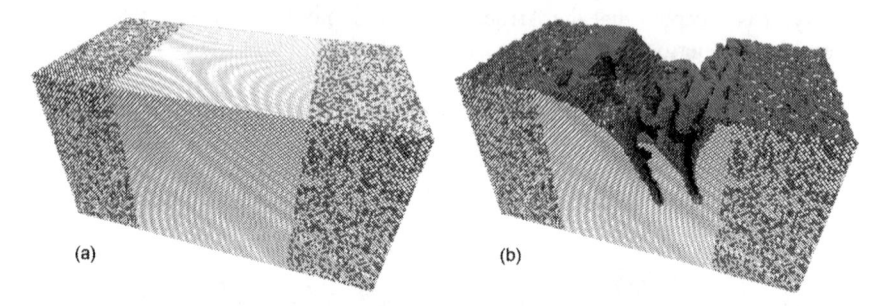

Figure 3: Approximated $NiAl_3$ crystallography in central region, hypothetical random solid solution assumed for Ni_2Al_3 at either edge. Ni – dark and Al – light; (a) initial mesh, and (b) after leaching.

Finally, using the assumptions outlined in section 2.3.1 and 2.3.2, and applying the same lattice parameter to both phases, it was possible to construct a simple cubic orthogonal computational mesh containing structured regions of both phases of interest. This is shown in figure 4(a).

Simply from observation, the $NiAl_3$ phase (right) in figure 4(a) should break up more quickly than the Ni_2Al_3 phase present (left), owing to the fact that there exists a layer of Al over the $NiAl_3$, whereas the Ni_2Al_3 phase is covered with Ni atoms. Initially, removal of aluminium from the Ni_2Al_3 would require diffusion of the Ni or removal of Al from the $NiAl_3$ phase at the interface of the two phases.

As expected, the $NiAl_3$ phase disintegrates much more quickly than Ni_2Al_3, which can be seen from above in figure 4(b). The leaching from the $NiAl_3$ phase reveals a structure that is comparable to that which characterises a Raney-Ni catalyst. Importantly, figure 4(b) also shows that FCC Ni forming over both phases.

3.3 Orientation of exposure

Until this point, leaching has only been simulated as occurring in [100]-type directions (i.e. from the top of the orthogonal mesh in figure 4 downwards). In an attempt to consider more orientations, a spherical particle is modelled to take account of multi-directional exposure to the NaOH solution. The starting spheres of the $NiAl_3$ and Ni_2Al_3 are shown in figure 5(a). The spherical regions of each phase shown in figure 5 represent particles that have radii of approximately 100 atoms. It is recognised that this is much smaller than the real particles, highlighting the need for improvements to both memory efficiency and code optimisation in order to model larger particles.

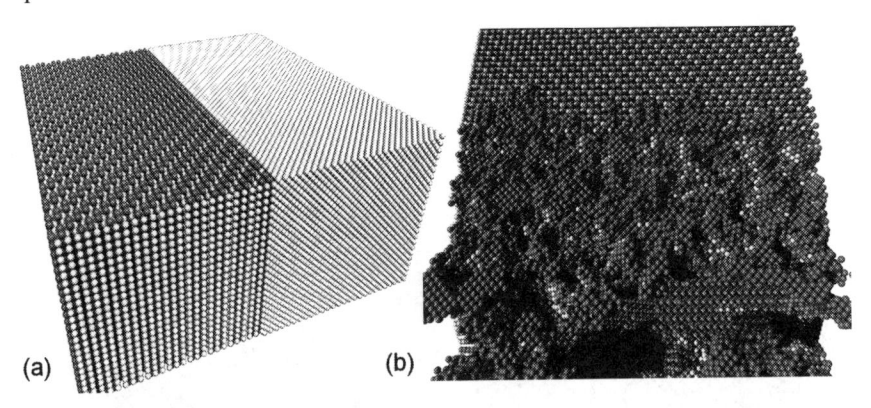

Figure 4: $NiAl_3$ and Ni_2Al_3 (left and top in (a) and (b) respectively) structures included. (a) initial mesh, and (b) after leaching: Ni – dark and Al – light.

The spheres represented in figure 5(a) have been leached from all exposed surfaces to produce the structures shown in figure 5(b). The resulting leached simulated structures that can be seen in figure 5(b) once again show a greater break-up for the $NiAl_3$ phase and a greater cohesion demonstrated by the Ni_2Al_3 phase. However, this difference is far less pronounced in this case than for the [100]-type dissolution in figure 4, indicating that the starting condition/position is critical at the start of a simulation. The surface of the leached spheres for both cases is predominantly Ni, although on-going experimentation suggests that a higher Al content would be expected.

Figure 5(b) shows that the Ni_2Al_3 particle contains a greater number of Al atoms within the evolving pores. This may be recently revealed Al atoms that have not been removed, but may also be the NiAl phase described in section 2.3.3, which would be identified in the simulation and prevent Al being removed from this phase in contact with the NaOH solution. In addition, the structural changes that take place in Ni_2Al_3 phase during caustic leaching is a topic of much debate [10]. Predictions made by a model with respect to these structural changes would be of much interest.

Figure 6 shows the removal rate (R) of Al and the total number of atoms exposed at the surface (N) for both cases in figure 5. For the Ni_2Al_3 particle, despite being haphazard, a fairly constant removal of Al was observed with a continually increasing surface area, shown in figure 6(a) (left). In contrast, the $NiAl_3$ particle in figure 5 showed a rapid rate of Al removal, which is completed earlier on reflecting the quicker break-up of this phase, figure 6(b) (left). The number of atoms at the surface of the NiAl$_3$ particle increases much more quickly than Ni$_2$Al$_3$, a constant level is reached after a relatively short time, figure 6(b) (right). It also appears that the NiAl$_3$ particle is beginning to coarsen (i.e. a slightly decreasing surface area) after further time by surface diffusion of Ni and coarsening leading to a reduction in the number of atoms present at the surface, shown in figure 6 (right).

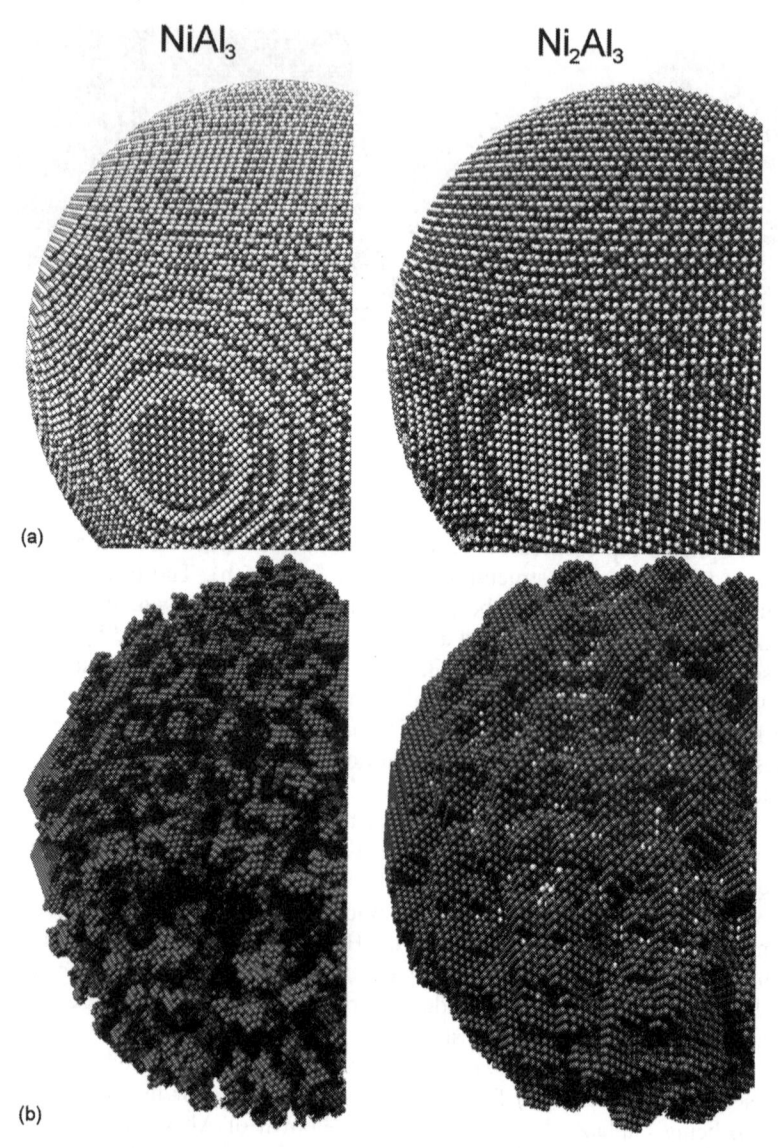

Figure 5: Simulated single-phase spheres of NiAl$_3$ and Ni$_2$Al$_3$ phase structure; (a) $t=0$, and (b) $t=200$s. (N.B. these are not representative of a multi-phase powder.)

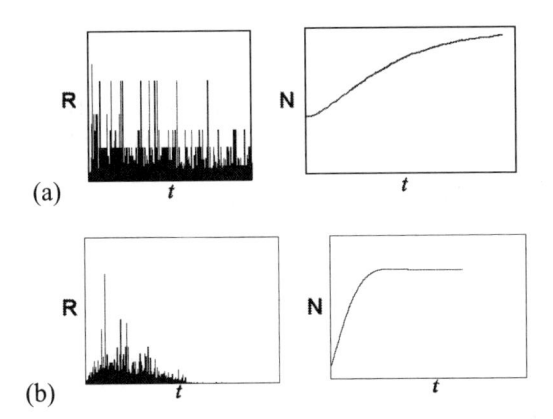

Figure 6: The rate of aluminium removal (atoms/s), R, and the number of atoms at the surfaces of the octants shown in figure 5 over time, t; (a) Ni_2Al_3, and (b) $NiAl_3$.

4 Conclusions

The prototype kMC model appears capable of simulating features of the leaching process assuming that all the relevant physiochemical processes taking place have been included and assigned appropriate rates. It has been shown that the leaching of hypothetical solid solutions appears logical. The $NiAl_3$ and Ni_2Al_3 phases that are present in cast-and-crush and spray-atomised NiAl powders can be approximately represented on a simple cubic computational grid using some simple assumptions and minor adjustments in lattice parameters. The simulated leaching of these phases can be seen to progress as would be expected and yield leached structures close to those observed for NiAl alloys leached with aqueous NaOH. Future work will concentrate on improving the quantitative nature of the model, attempting to link the predicted morphologies to catalytic behaviour and increasing the size of the computational domain to be able to deal with realistically sized particles.

Acknowledgements

The authors would like to acknowledge financial support from the EU from the IMPRESS Integrated Project (contract no. NM3-CT-2004-500635) and also several useful discussions with F. Devred and J.W. Bakker at Leiden University in Holland.

References

[1] Devred, F., Hoffer, B.W., Sloof, W.G., Kooymana, P.J., van Langeveld, A.D. & Zandbergen, H.W., "The genesis of the active phase in Raney-

[2]   Tanaka, S., Hirose, N., Tanaki, T. & Ogata, Y.H., "Effect of Ni-Al precursor alloy on the cathodic activity for a Raney-Ni Cathode." *Journal of the Electrochemical Society*, 2000. **147**(6): pp. 2242-2245.

[3]   Lee, G.D., Suh, C.S., Park, J.H., Park, S.S., & Hong, S.S., "Raney Ni catalysts derived from different alloy precursors (I) morphology and characterization." *Korean Journal of Chemical Engineering*, 2005. **22**(3): pp. 375-381.

[4]   Lee, G.D., Moon, M.J., Park, J.H., Park, S.S., & Hong, S.S., "Raney Ni catalysts derived from different alloy precursors (II) CO and CO_2 methanation activity." *Korean Journal of Chemical Engineering*, 2005. **22**(4): pp. 541-546.

[5]   Tanaka, S., Hirose, N., & Tanaki, T., "Evaluation of Raney-Nickel Cathodes Prepared with Aluminium Powder and Titanium Hydride Powder." *Journal of The Electrochemical Society*, 1999. **146**(7): pp. 2477-2480.

[6]   Erlebacher, J., Aziz, M.J., Karma, A., Nikolay, D. & Sieradski K., "Evolution of nanoporosity in dealloying." *Nature*, 2001. **410**: pp. 450-453.

[7]   Zacate, M.O., & Collins, G.S., "Site preferences of hyperfine impurities in Ni_2Al_3 phases." *Hyperfine Interactions*, 2002. **136/137**: pp. 647-652.

[8]   Lillard, R.S., & Scully, J.R., "Electrochemical passivation of ordered NiAl." *Journal of the Electrochemical Society*, 1998. **145**(6): pp. 2024-2032.

[9]   Saravanan, C., & Scott, S.M., "Theory and simulation of cohesive diffusion in nanopores: Transport in subcritical and supercritical regimes." *Journal of Chemical Physics*, 1999. **110**(22): pp. 11000-11011.

[10]   Wang, R., Lu, Z., & Ko, T., "The structural transitions during leaching of Ni_2Al_3 phase in a Raney Ni-Al alloy." *Journal of Materials Science*, 2001. **36**: pp. 5649-5657.

Dynamic simulation of a deposition process

J. P. Caire & A. Javidi
LEPMI-ENSEEG, Institut National Polytechnique de Grenoble, France

Abstract

A new solution for the computation of a current distribution with a moving boundary is presented. The numerical scheme is based on the capabilities of two commercial codes FEMLAB 2.3 and MATLAB 6.1. First a computation is carried out with FEMLAB which provides the electrical potential distribution at the initial time. The FEM problem and its initial solution are then saved as a MATLAB program. The MATLAB code containing the whole – draw, mesh, solve, plot – sequence is then modified to automate the iterative process. A movie is finally made from the successive solutions stored as JPEG pictures. A metal deposition process is presented as a simple test but difficult benchmark. This method is validated and can be easily extended to much more complex coupled electrochemical processes.
Keywords: FEM, primary current distribution, electrodeposition, moving boundary.

1 Introduction

The problems associated to evolutionary shape change are numerous and difficult to solve. They appear in various fields of electrochemical engineering such as electrodeposition [1,4,5,7,10-16], electro-machining [17], electropolishing [8], and corrosion processes [6]. Most popular commercial codes are now able to compute primary, secondary or tertiary current distributions for any geometry. If it is easy to compute the current distribution for a given geometry, it is very difficult and time consuming to deduce the time evolution of the moving interface related to the current distribution.

One of the first attempts to model the moving deposit profiles is probably due to Deconinck [1] in 1985. Since then, many authors tried to compute moving profiles by use of different numerical methods such as finite difference method (FDM) [6,10,12,14], boundary element method (BEM) [1,5,9], finite element method (FEM) [7,11,12], or specific methods [16]. All these authors used the same manual iterative process but Chauvy and Landolt [17] who developed a specific FORTRAN 77 code to compute the shape evolution of cavities produced by multistep electrochemical micromachining (the CREVICER code [18] developed for corrosion purposes does not have yet this capability).

The iterative process is simple. A first current distribution gives the thickness of the first layer deposited during the initial time step by solution of the Faraday's law. A new profile is then calculated and the geometry is modified consequently. Reiterating this manual process makes possible the determination of the shape evolution with time. However, this process is time consuming even for a very simple geometry since each time step requires a new − draw, mesh, solve, plot, compute the new profile − sequence. So there is a real need of automating the iterative process for industrial purposes or simply to check different initial configurations or to test corrosion scenarios for example.

This work intends to test a new solution based on the capabilities of the two popular commercial codes FEMLAB [2] and MATLAB [3]. The tests are here voluntarily restricted to electrochemical deposits corresponding to primary current density distributions. Calculations are simpler and such tests are more severe since the primary current distribution is known to give steep profiles and be very sensitive to the mesh quality [12].

2 The current distribution and the moving boundary

Figure 1 shows the displacement of the metal-electrolyte boundary due to an electrochemical deposit carried out during a time step.

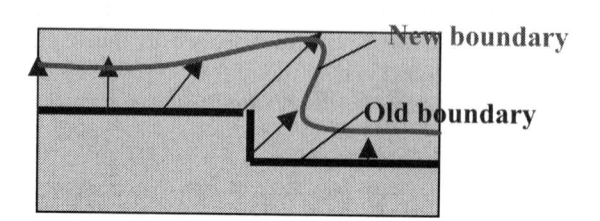

Figure 1: The moving boundary problem.

The local growth rate at any point i is governed by the Faraday's law:

$$\frac{d\vec{h}}{dt} = \sum_l \frac{M_l}{z_l F \rho_l} \vec{J}_l \qquad (1)$$

where J_l is the local mean current density (A/m^2), M_l the molar mass (Kg/mol), ρ_l the density (Kg/m^3) and z_l the charge of an ion l present in solution. In an industrial electrolyte each individual λ_l coefficient defined as:

$$\lambda_l = \frac{M_l}{z_l F \rho_l} \qquad (2)$$

is not necessarily known, so we assume that the global coefficient λ associated with the global current density J is known. Eqn. (1) is then simplified:

$$\frac{d\vec{h}}{dt} = \lambda \vec{J} \qquad (3)$$

Eqn. (3) can be discretized:

$$\Delta \vec{h}_i = \lambda \vec{J} \Delta t \qquad (4)$$

Let x_i and y_i the coordinates of a point i placed on the moving boundary. At the $(n+1)$th iteration corresponding to time $t + \Delta t$, the coordinates of point i are obtained from the coordinates computed at iteration n:

$$x_i^{n+1} = x_i^n + \Delta t \lambda J_x \qquad (5)$$

$$y_i^{n+1} = y_i^{n+1} + \Delta t \lambda J_y \qquad (6)$$

The time evolution of any point i is thus obtained from the local current density vector (J_x, J_y) deduced from the electrical potential V:

$$\vec{J} = -\sigma \vec{\nabla}.V \qquad (7)$$

Then the displacement of the points of the moving interface requires the preliminary solution of the Laplace's equation:

$$\vec{\nabla}(-\sigma \vec{\nabla}.V) = 0 \qquad (8)$$

When using a numerical code based on any numerical method, the first step consists in the solution of Laplace's equation at initial time. Then the current density J can be post processed from V and the new co-ordinates of each point of the border are computed using eqn. (5,6). A simple iteration scheme using eqn. (5,6) is necessary to compute the interface evolution.

Two main different methods can be considered to model the moving boundary evolution: (i) transform the electrolyte in metal by modifying the physical properties in the deposit area (ii) deform locally the mesh near the interface to take account of the move. Both methods were tested previously and did not give satisfaction. We noticed that FEMLAB preserves the topology of the related structure named "geom" in the code when the nodes situated on the moving boundary are manually shifted. Then it was possible to automate the moving of nodes and mesh the new geometry at each iteration. This method is described hereafter.

2.1 Electrochemical deposition in a trench

The method was checked using the very simple 2D scheme of Figure 2 which represents a simple electrochemical cell where a hole in the metal (grey) is covered by an electrolyte (white). This problem is frequently encountered in the semiconductor industry which uses electrochemical deposition processes in trenches [10,14,15]. In such conditions this very simple geometry provides a severe test for a numerical computation scheme since the primary current distribution exhibits two steep peaks at the borders of the trench.

2.2 The boundary conditions

The boundary conditions are presented in Figure 2. Dirichlet conditions are applied on anode and cathode, and symmetry conditions are applied on each side.

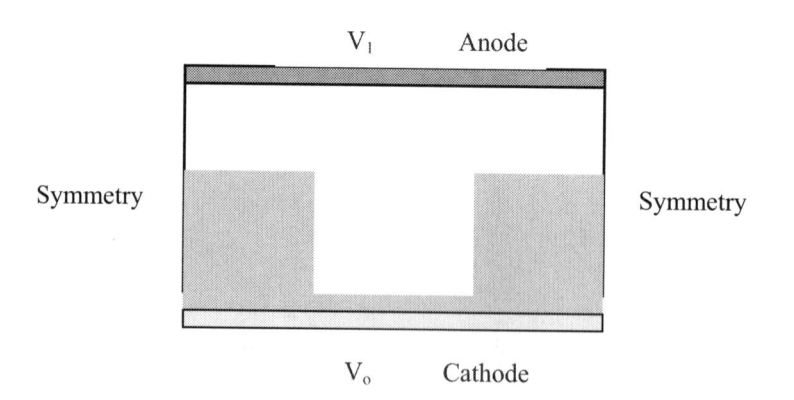

V_1 Anode

Symmetry

Symmetry

V_0 Cathode

Figure 2: The boundary conditions for a trench.

3 The iterative computational process with FEMLAB-MATLAB

FEMLAB is a code based on the finite element method (FEM), developed and marketed since 1986 by COMSOL [2]. The problem is solved in the following steps thanks to a user friendly graphical interface: draw the geometry, define the boundary conditions, give the physical properties for each subdomain, mesh, refine the mesh if necessary, solve, plot the solution.

FEMLAB was first used to draw the cell and to define the physical problem. For example, in this case, a first rectangle ABCD was drawn figuring the electrolyte (zone 1). The metal part (zone 2) was obtained by tracing a polygon starting from A and passing by the points E, F, G, H, I, J, D, A. Actually, segments were only used here to simplify the drawing, but one could just as easily use Bezier curves for a design including arcs.

In this scheme, it must be noticed that the end points of the segments are automatically selected as nodes by the mesh generator (the length of the segments constituting the moving boundary are not necessarily uniform).

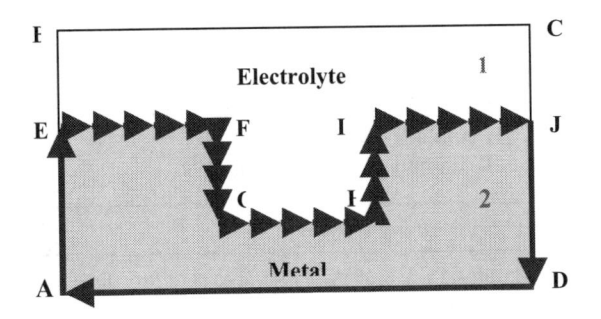

Figure 3: Drawing the trench with FEMLAB.

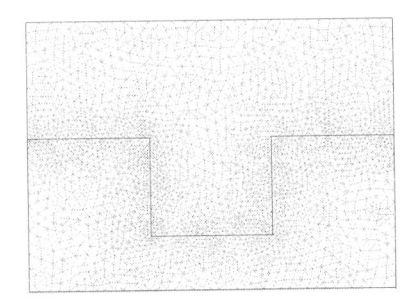

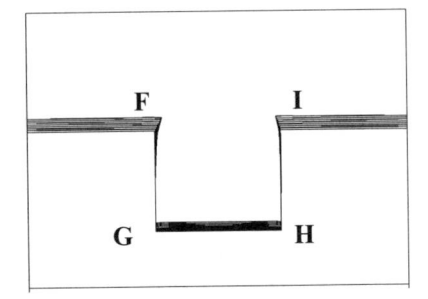

Figure 4: Initial Mesh.

Figure 5: First deposit steps.

Thereafter the algorithm moves only the nodes located at the ends of the segments appearing between the points E and J of Figure 3. This set of segments constitutes the border between electrolyte and metal, border which will be moved during the deposit process. Figure 4 shows the mesh given by the automatic mesh generator of FEMLAB after the grid initialization and one refining process. Figure 5 shows the five first successive steps that were hand made using the global algorithm described in Figure 6. It can be seen that the deposit is not perfectly symmetrical and the points F and I tend to approach, while the points G and H go up. These observations are discussed hereafter.

3.1 Automation of the iterative scheme

When the solution is obtained with FEMLAB it can be saved as a genuine MATLAB file. This file includes the whole structure describing the topology, the boundary conditions, the physical data and all the information necessary for the FEM solution. Since this ASCII file can be edited and modified, it is thus easy to add in the program the loop on the steps 4–7 of Figure 6 to obtain the iterative scheme and the periodic plot of the moving boundary solution stored as a JPEG file. Collecting the successive pictures it is then easy to make a movie showing deposit evolution with the time. Only stages 4 and 7 of the scheme must be programmed, all the other stages appearing without any change in the original MATLAB file.

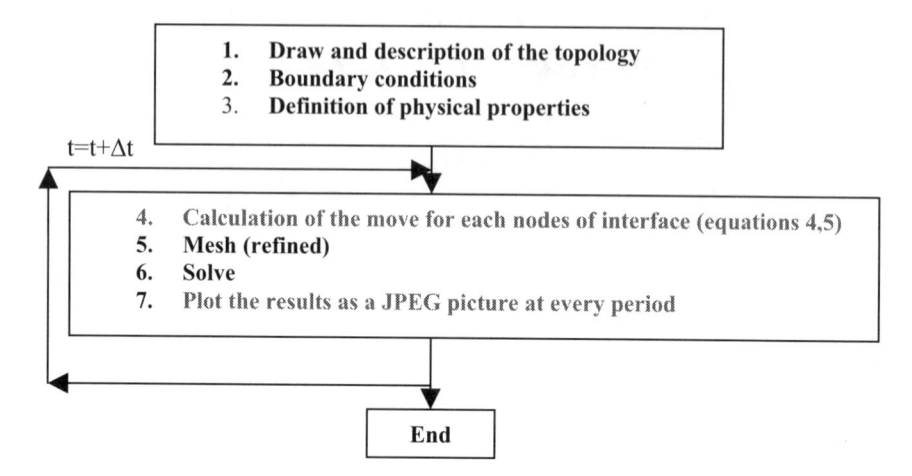

```
t=t+Δt
        1. Draw and description of the topology
        2. Boundary conditions
        3. Definition of physical properties

        4. Calculation of the move for each nodes of interface (equations 4,5)
        5. Mesh (refined)
        6. Solve
        7. Plot the results as a JPEG picture at every period

                          End
```

Figure 6: Iterative computation algorithm.

Due to the structure of the code, it is remarkable that any displacement of the points located at the moving interface does not change the boundary conditions, as long as the points remain in the same domain. Moreover, since the topological structure is not altered, the region where the deposit replaces the electrolyte receives automatically the physical properties of metal. This MATLAB equivalence of the FEMLAB solution has the advantage of giving direct access to the geometrical structure and to all the parameters and the physical data used in the computations in every point of each sub domain.

4 Numerical results

Figure 7 shows in black the deposit evolution for 200 iterations when the edges of the hole are tilted. It appears immediately that the side peaks grow more quickly than the deposit in the centre of the hole. It can be noticed that after 150 iterations a second small peak appears at the side of the larger one. This peak can be interpreted as a shadow effect of the principal peak. A corner effect is also seen at the edges of the central part of the hole. These effects are overvalued and would be less important in a real process, since this calculation of primary distribution of current neglects all the limiting effects, particularly the overpotentials.

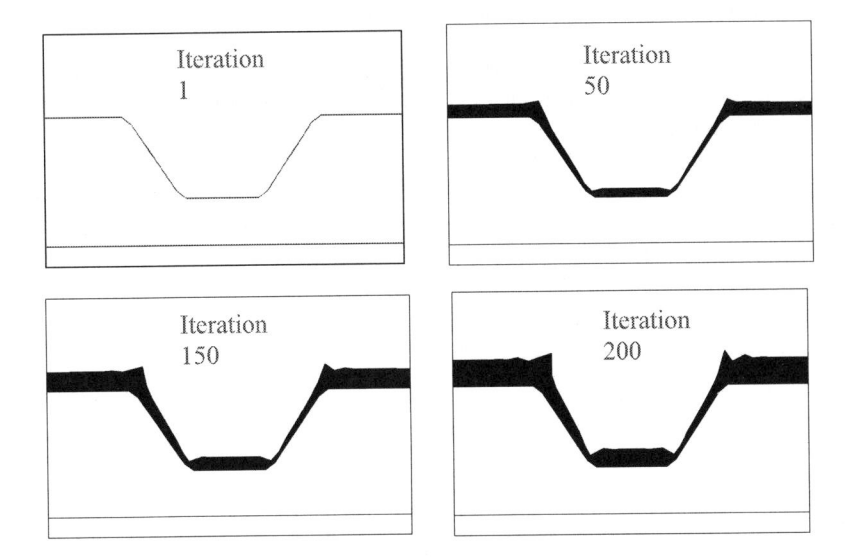

Figure 7: Evolution of deposit for 200 iterations.

Calculation and saving of the successive images on a PC - HP VECTRA 6800 last about only 20 minutes for the two examples presented above.

Thus when the methodology is developed, the user can easily carry out many tests very quickly. The same methodology can be used for research purposes to rapidly evaluate some assumptions.

5 Problems related to the algorithm

Actually, the algorithm presented in Figure 6 failed in some specific cases. Three problems were detected in the case of the trench with vertical edges.

5.1 Junction of two consecutive nodes

This problem appears when the length of a vector included in the moving boundary decreases during the iterative scheme. The nodes of the vector become closer and closer and can touch each other. This event is not improbable if we look at points I and F of Figure 5. When two nodes are confounded, the topology of the geometry is modified and this is not compatible with the present algorithm. Moreover, when two consecutive nodes come close, the automatic meshing generates locally smaller and smaller triangles, and then the number of degrees of freedom of the system blows up and the iterative process stops for lack of memory space. Then the user must draw a new geometry that integrates the fusion of the confounded nodes and restart the automatic process.

5.2 Unsymmetrical evolution of a symmetrical geometry

Our tests were systematically carried out on symmetrical cells in order to point out some possible numerical defects related to the algorithm. The Figure 8 exhibits a symmetry defect observed at the beginning of the computation when plotting the current density distribution along the path ab. This problem is obviously related to the discretization scheme which can produce small but significant numerical cut out errors on the large peaks appearing mainly in primary current distributions.

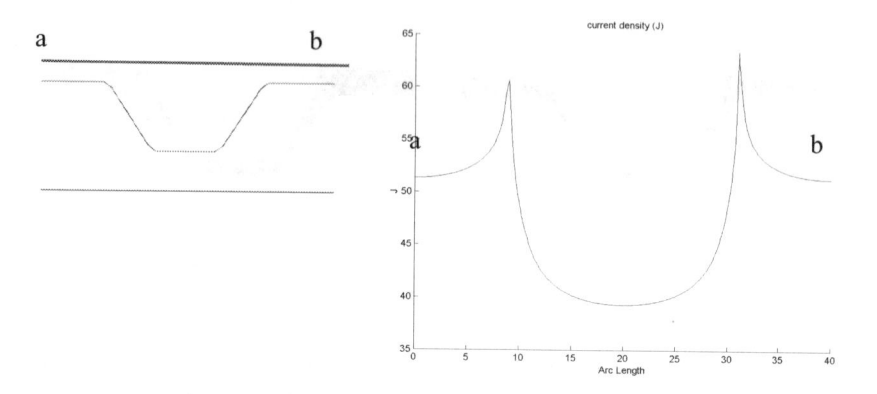

Figure 8:   Dissymmetry of current distribution along ab.

Such a cut out effect is unpredictable: it can be compensated from an iteration to the other or can be accumulated. It is inherent to the numerical method and cannot be fixed without an improvement of the precision of the numerical computation. It probably would be less apparent for the smoother peaks of the secondary or tertiary current distributions.

5.3 Wrong displacement towards the inner part of metal

This defect did not appear in the electro polishing tests. It was only noticed in the most severe tests of electrodeposition. This is due to a wrong displacement when the current flow lines come successively through metal, electrolyte and metal in the vicinity of points H and G of Figure 3. These irregular node displacements (1) as well as a lack of symmetry of Figure 8 were attributed to the finite precision of the computation. The problem was fixed by checking that every deposit displacement was well done in the proper direction, i.e. from metal to electrolyte.

6   Conclusions

This work has shown that it is possible to automate the tedious manual computation of a moving interface in primary distribution of current by using the

FEMLAB and MATLAB codes successively. It also revealed the limitations inherent to the numerical discretization. The numerical instabilities encountered with this simple algorithm could be fixed and the defects would not appear in secondary current distribution. So the iterative scheme is now well validated.

This work is obviously criticisable from a strict electrochemical point of view insofar it does not take into account the physical limiting phenomena. However, as the algorithm is now validated, it is possible to automate the treatment of coupled phenomena thank to the Multiphysics capabilities of FEMLAB. Calculations would be heavier, but in fact there is no additional issue. Taking account of nonlinear boundary conditions would either be straightforward since they do not interfere with the iterative algorithm described here. The first results allow considering more complex applications taking into account of current distribution coupled with heat, mass or momentum transfers. Now the problems related to the moving of nodes are fixed, the algorithm works properly and a movie describing the time evolution of a profile can be obtained in a couple of hours with a cheap usual personal computer.

References

[1] Deconinck J., PhD thesis, Vrije Universiteit, Brussel, 1985.
[2] FEMLAB, User's guide, http://www.comsol.com/
[3] MATLAB, User's guide, http://www.mathworks.com/
[4] Caire J.P., Chaînet E., Nguyen B., Valenti P., Study of a New Stainless Steel Electropolishing Process, SUR/FIN' 93, AESF Meeting, Anaheim, California, June 21-22, 1993.
[5] Qiu Z.H., Power H., An integral equation approach for the analysis of current density distribution controlled by diffusion, convection and migration, *J. of App. Electrochem.*, 27, pp 1333-1342, 1997.
[6] Zoric J., Rousar I., Thonstad J., Mathematical modelling of industrial aluminium cells with prebaked anodes. Part I: Current distribution and anode shape, *J. App. Electrochem.* 27, pp 916-927, 1997.
[7] Rabiot D., Caire J.P., Gerard F., Optimizing an electrochemical deposition process by use of design of computer experiments, Analusis, EDP Sciences, Wiley-VCH, 26, 6, pp 281-284, 1998.
[8] Shankar Subramanian R., Lu Zhang, Babu S.V., Transport Phenomena in Chemical Mechanical Polishing, *J. Electrochem. Soc.*, 146 (11) pp 4263-4272, 1999.
[9] Qiu Z.H., Power H., Prediction of electrode shape change involving convection, diffusion and migration by the boundary element method, *J. App. Electrochem.*, 30, pp 575-584, 2000.
[10] Georgiadou M., Veyret D., Sani R. L., Alkire R.C., Simulation of Shape Evolution during Electrodeposition of Copper in the Presence of Additive, *J. Electrochem. Soc.*, 148 (1) C54-C58, 2001.
[11] Gill W.N., Duquette D.J., Varadarajana D., Mass Transfer Models for the Electrodeposition of Copper with a Buffering Agent, *J. Electrochem. Soc.*, 148 (4) C289-C296, 2001.

[12] Caire J.P., Chifflet H., Meshing noise effect in design of experiments using computer experiments, *Environmetrics,* 13 :1, pp 1-8, 2002.
[13] Subramanian R., Venkat Z., White R.E., Simulating Shape Changes during Electrodeposition - Primary and Secondary Current Distribution, *J. App. Electrochem.*, 149 (10) C498-C505, 2002.
[14] Georgiadou M., Veyretz D., Modeling of Transient Electrochemical Systems Involving Moving Boundaries - Parametric Study of Pulse and Pulse-Reverse Plating of Copper in Trenches, *J. Electrochem. Soc.,* 149 (6) C324-C330, 2002.
[15] Moffat T. P., Baker B., Wheeler D., Bonevich J. E., Edelstein M., Kelly D. R., Gan L., Stafford G. R., Chen P. J., Egelhoff W. F., Josell D., Superconformal Electrodeposition of Silver in Submicrometer Features, *J. Electrochem. Soc.,* 149 (8) C423-C428, 2002.
[16] Lavelaine de Maubeuge H., Calculation of the Optimal Geometry of Electrochemical Cells- Application to the Plating on Curved Electrodes, *J. Electrochem. Soc.,* 149 (8) C413-C422, 2002.
[17] Chauvy P.F., Landolt D., Unusual cavity shapes resulting from multistep mass transport controlled dissolution: Numerical simulation and experimental investigation with titanium using oxide laser lithography, *J. App. Electrochem.*, 33: 135–142, 2003.
[18] http://www.virginia.edu/cese/research/crevicer/

Two-phase electrolysis process modelling: from the bubble to the electrochemical cell scale

P. Mandin[1], H. Roustan[2], R. Wüthrich[3], J. Hamburger[4] & G. Picard[1]
[1]LECA UMR 7575 CNRS-ENSCP-Paris 6, ENSCP, Paris Cedex, France
[2]Alcan, Centre de Recherche de Voreppe, Voreppe Cedex, France
[3]E.P.F.L. – L.S.R. – CH-1015 Lausanne, Suisse
[4]Transoft International, Fluidyn, Saint-Denis, France

Abstract

During two-phase electrolysis for hydrogen, aluminium or fluor production, there are bubbles which are created at electrodes which imply a great hydrodynamic acceleration but also quite important electrical properties and electrochemical processes disturbance. There are few works concerning the local modelling of electrochemical processes during a two-phase electrolysis process. Nevertheless, effects like the anode effect, particularly expensive on the point of the process efficiency, should need a better understanding. The goal of the present work is to present the modelling and the numerical simulation of the gas production, from the single bubble scale to the macroscopic electrochemical cell, during the two-phase electrolysis process. Bubbles are motion sources for the electrolysis cell flow, and then hydrodynamic properties are strongly coupled with species transport and electrical performances. The presence of bubbles modifies these global and local properties: the electrolysis cell and the current density distribution are modified. The present work shows theoretical modelling on both scales and also performance changes during the two-phase electrolysis processes.
Keywords: two-phase electrolysis modelling electrochemical bubbles.

1 Introduction

Gas release and induced fluid flow over electrodes exist in many electrochemical processes such as aluminium, fluorine production, water electrolysis and many other electrochemical processes. The hydrodynamic properties and the gas-flow motion in electrochemical cells are of great practical interest in electrochemical engineering science since the dispersed phase modifies the electrical properties of the electrolyte (as well as mass and heat transfers), and therefore modifies the macroscopic cell performances. In many cases, this phenomenon has to be avoided, but, in some other processes, the gas flow rate has to be controlled. This control is a great challenge, a multi scale one. The macroscopic performances are local gas concentration dependent, with a strong coupling with hydrodynamic and electric properties; but the gas volume shape and size are dependent of bubble scale properties, but also of the local hydrodynamic shear stress. The bubble formation during electrochemical processes and its electrochemical cell scale consequences is a few investigated fields [1–7], whereas the same topic has been intensively explored for heat transfer science or gas injection through porous walls for turbulence promotion and transfer increase [8–10].

Nevertheless the strong knowledge of the coupled phenomena is necessary: it is the case for electrochemical processes such as for gas production (CO_2, F_2, H_2, O_2, Cl_2, ...) [1–6] and other processes such as, for example, the chemical engraving (SPARK) [7].

In all these electrochemical processes, a coupling effect is particularly strong because bubble-dispersed phase acts like an electrical shield, the shielding effect depending on the density of the bubbles, which is namely the local gas volume fraction of the dispersion.

The present work shortly presents experimental set-up, bubble scale modelling and electrochemical cell scale modelling. The vertical gas evolving electrode configuration is more developed and primary current distribution evolution with electrolysis applied potential is presented and compared with experimental results.

2 Experimental set-up

A rectangular electrolysis cell (height =10 cm, length = 8 cm, width = 5 cm) has been chosen. The investigated working electrode was made with: a platinum pin, a tin oxide SnO_2 deposit upon a plane glass sample, a carbon one, and a tungsten one. Potassium hydroxide KOH 1 mM is added in the water as support salt and the final pH was 11. The alkaline water electrolysis is considered here as a reference process for the general two-phase electrolysis:

Anode: $OH^- = 0.25\ O_2 + 0.5\ H_2O + e^-$, $E° = 0.401$ V/NHE

The counter electrode is a tungsten wire put in a separated compartment (isolated from the electrolyte by means of sintered glass) to study only one electrode induced two-phase flow. The reference electrode used is the silver one put in a sheath of glass filled with a KCl 3 M solution, AgCl saturated. Electrochemical measurements (cyclic voltammetry, chrono amperometry and impedance spectroscopy) are performed with a Parstat 2273. The visualization, movies and photographs are performed with BIPOL Y11P camera completed with a modular endoscope H.S.W with illumination by halogen lamp.

3 The single bubble: from appearance at the electrode to departure from the electrode

In this part a single "in formation" bubble is considered alone in a motionless environment.

3.1 The bubble appearance

It is considered here one single electrochemical reaction, cathode or anode, leading first to a dissolved gas and second to bubbles nucleation. At the beginning the electrochemical product is first dissolved in the electrolyte if the electrolyte is not already saturated with dissolved gas. In fact beginning period is really difficult to model because of this uncertainty on the liquid electrolyte. For example it is known that there is always nitrogen and oxygen dissolved gas in laboratory electrolyte and then, because of the oxidizing capacity of oxygen, it is necessary to eliminate these atmospheric gas with, for example, the use of an argon flow before experiment. These attentions are usual in laboratory cells but not at industrial scale! Nevertheless, after this electrolyte preparation period, the criterion for the appearance of the first two-phase character is related to the thermodynamic Henry law: $X_k = P_k/H$. H is the Henry constant for the dissolved gas, electrolyte and considered temperature; P_k is the partial pressure of species k whereas X_k is the maximum molar fraction of dissolved gas k. For example, $H_{O2}(20°C) = 4.06 \ 10^9$ Pa, and then, before the use of argon flow, $P_{O2} = 0.21 \ 10^5$ Pa, and the electrolyte has an oxygen molar fraction $X_{O2}(20°C) = 5.24 \ 10^{-6}$. But to create pure oxygen bubbles, the partial pressure in equilibrium with electrolyte must be $P_{O2} = 10^5$ Pa and then X_{O2} must reach $2.46 \ 10^{-5}$ at electrode to satisfy this thermodynamic criterion.

The Henry constant increases with temperature; for example $H_{O2}(50°C) = 5.96 \ 10^9$ Pa. It is interesting here, in the case of the water electrolysis process study, to give the Henry constant values for hydrogen: $H_{H2}(50°C) = 6.92 \ 10^9$ Pa and $H_{H2}(50°C) = 7.75 \ 10^9$ Pa. These values are associated to pure water. But, often, electrochemical processes use support salt to increase electrical conductivity and reduce ohmic drop. In the condition of a large ionic strength $Is = 0.5 \ \Sigma_k(z_k^2 \cdot C_k)$, with z_k the charge of ion k and C_k its composition (mol l^{-1}), the Henry constant is modified [11]. In the present case Is=1 M and h=0.134 [11], so $H_{O2}/H_{O2,ref} = 0.73$. When the steady regime is reached, the dissolved species k composition at electrode is supposed to be constant and equal to the thermodynamic value.

The nucleation phenomenon, at the very beginning, and the transient description is difficult because of the initial conditions, the possibility of non-thermodynamic phenomena like over saturation, and the necessity to know the flow at the electrode vicinity. It is also difficult when a steady state regime is observed. If at the very beginning there is no flow, transport is purely diffusive and then $k_{OH} = D_{OH}/\delta(t)$, the OH$^-$ species transport velocity at electrode is fast: when $t \to 0$, $\delta(t) \to 0$. D_{OH} is the diffusion coefficient and $\delta(t)$ is the unsteady boundary layer thickness given by the Cottrell theory. Then the nucleation criterion is reached fast in absence of forced flow. If a forced flow is used (Couette flow, RDE or IRDE flows), the transport law is different and the nucleation phenomenon disappears. It can be only in small flow zones (for example nearthe electrode axis for a RDE configuration). In absence of electrochemically induced flow, the steady transport velocity is: $k_{OH}=D_{OH}/\delta$, with the constant boundary layer thickness δ (for example $\delta=1.61\ D^{1/3}.\omega^{-1/2}.v^{1/6}$ for the RDE configuration). At the steady regime, for a constant applied current density, it can be written for the applied current density j (A m^{-2}):

$$j/(nF) = k_{OH}.\rho_{electrolyte}/M_{electrolyte}.(X_{OH,bulk} - X_{OH,el})$$
$$= 4\ k_{O2}.\rho_{electrolyte}/M_{electrolyte}. (X_{O2,el} - X_{O2,bulk}) \quad (1)$$

The effective mass transport kinetic constant is called k_{OH} and k_{O2} respectively for the reactive and product; $\rho_{electrolyte}$ and $M_{electrolytet}$ are respectively the density and the molar weight of the electrolyte; n is the electron number exchanged in reaction (here n=1) and the Faraday constant is F = 96500 C mol^{-1}. A strong flow allows large limiting current density but has no incidence upon the dissolved oxygen molar fraction at electrode $X_{O2,el}$. For these forced flow study, the bubble nucleation is not evident if the oxygen bulk composition is zero $X_{O2,bulk} = 0$, in case for example of argon gas use. Even if a forced flow is not used, a flow is electrochemically induced due to the departure of bubbles and their motion exchange with electrolyte. It is really difficult after this bubble flow installation to know the effective transport velocity k, which is no longer the purely diffusive one or the forced flow one. This electrochemically induced flow is particularly strong for vertical or horizontal "floor" electrodes but almost negligible for horizontal "ceiling" electrodes. The electrode position and direction must be directly related to the gravity field direction.

It is also observed that the nucleation sites obey a regular mesh. It has also been observed that the nucleation process always take place at the same place. Though the high difficulty of this process study, there is some help from nature!

3.2 The bubble growth

A single bubble is here considered. The study of electrochemical bubble growth can use analogy with other scientific works: the gas injection through a porous surface or the phase change of a coolant flow for example in vapour generator or heat exchangers [1–2; 8–10]. The use of these other science works allows a better understanding but the analogy must be limited. Gas injection is often used to promote turbulence and heat or mass transfer. In this case, the gas flow is totally fixed parameter. In heat transfer, the analogy with the two-phase electrolysis is better but also different. If the primary flow is highest Reynolds number, the heat flux Q (W) is imposed in the same manner as the intensity I (A) is imposed, because the thermal resistance due to the metal separating primary and secondary flows is generally the smaller one. The gas at the electrode and in the electrode vicinity has a screening effect, interfacial (effective area for the electrical current reduced according with dimensionless screened area α, figure 1) and homogeneous (because of void fraction ε, thermodynamic and kinetic transport properties are modified, flow modified, etc) because the conducting electrolyte is replaced with gas, an insulator. It is also true in a reduced way for heat transport: the gas has an about 10 times smaller heat conduction coefficient and then it also screens the interface and modifies the flow properties. The analogy is strong for the research of primary current distribution. But for secondary or ternary current distribution the analogy is worse. Because the current density is increased due to the actual electro active area, the over potential are larger. Because the reactive electro active species arrive at electrode with about the same rate as the product species leave the place: the interface properties are then strongly coupled with homogeneous transport.

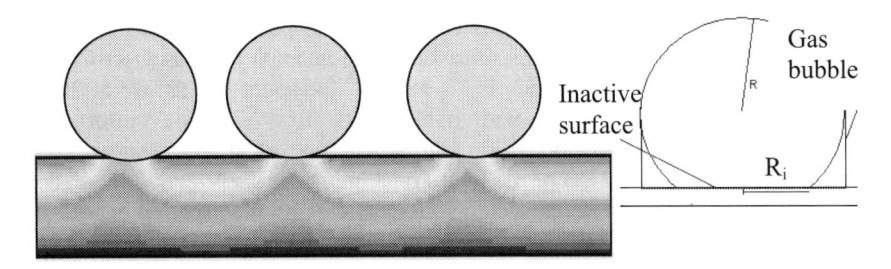

Figure 1: Current density contour in electrode due to bubble screening effect.

Until the dissolved gas concentration at electrode has reached the nucleation criterion, the first gas domains are observed at spatially periodical situated nucleation sites. The growth occurs in both normal to electrode and tangential to electrode directions. The growth kinetic is faster, like for crystals growth, in the tangential to electrode, the radial one. It is because, first; the dissolved species concentration is maximum at this electrode plane. This is also because the coalescence forces are confronted to the flow stress in the normal to electrode direction. So the gas domain should first be more developed in the radial direction from the nucleation site than in the normal direction. The volume is, at this time, defined as a plane disk shape. The electrode area is partially screened (figure 1), and the gas domain feeding can't be from the electrode contact area. The inactive area seems experimentally to be usually constant in location.

Then the growth of the gaseous volume continues but the mass exchange is performed with the unsteady contact area with electrolyte. Because the imposed current is constant, though the reduced electro active area, the effective applied current density j at the electrode-electrolyte contact area (called free area) increases and is $j_{eff}=j/(1-\alpha)$. The oxygen production continues at this free area. The dissolved gas molar fraction is generally supposed the saturation one (thermodynamic hypothesis), but the value can be larger because of the increasing current density and because over saturation processes can occur. Under steady regime assumption, this electrode dissolved oxygen production must be consumed by transport and gas domain growth at nucleation sites phenomena.

To know the rhythm of gas domain volume increase it is necessary to know both the intensive mass transport flux from liquid electrolyte to the gaseous bubble k (m s^{-1}) and the exchange surface area S. In fact these two factors are related because the unsteady bubble shape modifies the local flow and then the mass exchange.

The convecto-diffusive mass transport velocity is calculated from empirical correlation for the Sherwood number Sh= $k_{O2}.d/D_{O2}$. For a free spherical bubble (diameter d) in a uniform flow with relative velocity V_r, the correlation is: Sh = $2+0.66 \, Re^{0.5}.Sc^{0.3}$, for $10<Re<10^4$, with relative Reynolds number Re= $\rho.V_r.d/\mu$, ρ and μ are the electrolyte density (kg m^{-3}) and viscosity (Pa.s) respectively. Sc= $\mu /(\rho. D_{O2})$ is the Schmidt number. Other correlation Sh = $2 + 0.79 \, Re^{0.5} \, Sc^{1/3}$ is also used by Vogt et al [1–2;8], using an average velocity at the anchored at electrode bubble. In fact it is difficult to use rigorously these kinds of correlations: First because an internal circulation is possible inside the bubble volume. In this case correlations must take into account the gas viscosity and also the surface tension γ (which is 73 mN m^{-1} for water against air at 20°C). Second, because at electrode, with bubbles, the flow can't be uniform and simple, it seems to be better to introduce the local wall shear stress $\tau_p = \mu.\partial V_{tangential}/\partial normal$ (Pa) or the hydrodynamic stress $s = \partial V_{tangential}/\partial normal$ (s^{-1}). But the interaction between such a stressed flow and the growing bubble is a "to explore" situation under steady and unsteady assumptions. The local wall shear stress is also related to the entire cell flow, because of the elliptic nature of Navier-Stokes equations. The bubble shape is also an important parameter: it changes with the bubble feeding and characteristic length d increase. Nevertheless, the molar feeding rate is $<k_{O2}(t)>.S(t).(C_{O2,flow} - C_{O2,bubble})$. It is difficult to estimate this flux potential which implies both heterogeneous and homogeneous Henry saturation concentration.

The growing bubble shape is also difficult to describe. One important property is the contact angle θ which is determined with the triplet gas-electrolyte-electrode. For a given gas and a given electrolyte, this contact angle can be large or small. This contact angle property is generally related to the surface tension γ (N m^{-1}). These relatively well-known properties are nevertheless static properties in stagnant electrolyte. During the two-phase electrolysis, the flow, forced or purely induced, due to bubble, modify the bubble shape. Duhar and Colin have shown the modification of the anchoring angle in a Couette flow. Duluc et al have also shown the evolution with time of this angle. Then the actual unsteady anchored bubble shape is still an unknown only experimentally accessible with movie image processing.

Then, growing bubbles are supposed to obey an unsteady force equilibrium, between:
- the vertical Archimede force: $(\rho_{electrolyte}-\rho_{gas}).g.Vol$, with $Vol=\pi.h^2.(d/2-h/3)$ and $g= 9.81$ m s^{-2}.
- the flow direction friction force: $\iint_{wet} \tau_P.dS = <\tau_P>.Surf$, with Surf= $\pi.h.(2.d-h)$; $<\tau_P>$.is a difficult to access property directly related to the flow friction at the bubble wet surface.
- the capillarity friction force: $\int_{contact} \gamma.dl = <\gamma>.L$, with L= $2.\pi.d.[h/d.(1-h/d)]^{1/2}$; $<\gamma>$.is about γ the surface tension in the considered case.

Because the bubble shape is unsteady, the contact angle is unsteady [1–2;8–10] and these three forces are also unsteady.

3.3 The bubble "takeoff" and its replacement

At the growth beginning these three forces sum is dominated by the capillarity friction. But, when the bubble dimension increases, the Archimede and hydrodynamic forces increase faster than the capillarity one [1–2; 11–12]. So, for a given dimension d, these two forces begin larger than the contact one. The detachment can occur at the electrode contact but can also occur elsewhere at the bubble volume frontier. Duhàr and Duluc works [9–10] show that the detachment does not occur at the electrode but due to a critical elongation and finally the rupture phenomenon. Then, at the electrode, a gaseous volume is still present which also accumulate the neighbouring dissolved oxygen. The phenomenon takes place in the same place because of this begun nucleation.

The takeoff criterion is difficult to establish though it is important for the evolved bubble size prediction. In fact it is sure that detachment occurs for forces larger than the capillary one. But after this, a bubble surface elongation occurs and the rupture occurs only after.

3.4 The bubble displacement in the electrolyte

The single bubble now presented has leaved the electrode surface and its life now consists in rejoining the free surface of the electrolyte according with the gravity field $g = 9.81$ m s^{-2}.

The bubble is supposed to be instantaneously a sphere with the diameter d verifying:

$$\pi.d^3/6 = \text{last anchored bubble volume Vol}$$

This sphere is considered rigid, with a constant diameter if:

$$d < (\gamma/((\rho_{electrolyte} - \rho_{gas}).g))^{1/2}$$

Under this assumption, the Newton law for the single bubble in the electrolyte is:

$$M.dV/dt = \Sigma \text{ Forces} \qquad (2)$$

3.4.1 Vertical velocity

$$dV_{gas}/dt = \text{Archimede} + \text{friction}$$
$$= (\rho_{electrolyte}/\rho_{gas} - 1).g \pm 6/8.\rho_{electrolyte}/\rho_{gas}.C_x/d.(V_{electrolyte} - V_{gas})^2$$

If the bubble velocity V_{gas} is larger than the local electrolyte velocity $V_{electrolyte}$, then the friction force is negative (stagnant electrolyte for example). On the other hand, if electrolyte motion is large, then the friction force is a motion source for bubble and the term is positive.

The drag coefficient C_x is a function of the Reynolds number Re_p calculated with the relative velocity; $C_x = 24/Re_p$ for smaller velocities according with $Re_p < 0.1$. The evolution of the drag coefficient C_x has been given by Morsi and Alexander [13].

This equation has been solved with an explicit scheme and a time step 1μs or 0.5μS. When the sum of the forces applied to the bubble is constant, the bubble increases its velocity from zero to V_{lim} according with algebraic equation:

$$(1 - \rho_{gas}/\rho_{electrolyte}).g \pm 6/8.C_x/d.(V_{electrolyte} - V_{gas})^2 = 0 \qquad (3)$$

The limiting relative velocity is strongly diameter dependent whereas the gas chemical nature through ρ_{gas} is not important. One information very important is that the limiting velocity is reached very quickly, in time lower than 50μs, which leads to distance smaller than 5 µm. Then, the bubble vertical relative velocity will be generally supposed constant anywhere in the electrochemical cell. But the electrolyte flow has generally strong changes with space and then with bubble trajectory time. The knowledge of local electrolyte velocity $V_{electrolyte}$ is of great importance, but very difficult to model. The electrolyte flow is elliptic, strongly coupled with the electrochemical cell configuration and gas flow dependent. This flow description uses Navier-Stokes modelling and computational fluid dynamic software; it will be presented in next part. Bubbles are motion source for electrolyte.

3.4.2 Horizontal velocity

The single bubble obeys essentially vertical forces, Archimede and friction one. But, for large current density and bubble production, mostly for vertical or horizontal ceiling configurations, the bubble trajectory is experimentally not purely vertical. In the horizontal ceiling case, the bubble accumulation at the electrode implies coalescence and surface displacement. It is difficult to mathematically formalize the fact that for a gas evolving vertical electrode, the bubbles can't all stay at the electrode. The two-phase flow develops like a boundary layer, with an increasing thickness from bottom to top. The bubble is then submitted to a horizontal force, not explainable for the bubble alone, but due to interaction with electrode surface and others bubbles. Indeed, if the flow is purely vertical but with gradient (velocity zero at electrode), a slip force (also called Saffman force) is induced from small to large velocity values. This force is in fact a pressure force and the gradient is correctly described with the Bernoulli law. The result of this force is a distance of the bubble from the electrode. There is also interaction of the considered bubble with other bubbles: chocks or also pressure and wake effects. These interactions can imply coalescence phenomena. It is to notice here that the previous part has shown the small inertia of a single bubble. The limiting velocity given previously is relative with the local electrolyte velocity. Then, the more the bubble is far from the electrode (with electrolyte velocity zero) the faster and pressure effect is induced in this direction. There are few modelling attempts of these phenomena. Some models used the classical molecular interaction modelling in term of force $\mathbf{F}_h$ and associated potential field ψ: $\mathbf{F}_h = -\mathbf{grad}(\psi)$ with $\psi = -K\,[d/x\,]^n$. The potential law introduces two unknown constants K and n, and the parameter d/x, with d the bubble diameter and x the distance from electrode. Krepper [14] has also corrected and developed the Antal semi-empirical model, also with this molecular interaction analogy. This difficulty is the reason why the vertical gas evolving electrode is more difficult to describe than the horizontal floor one. The horizontal ceiling configuration, like in the classical Hall-Héroult configuration, is particular because phenomena are essentially horizontal on surface.

4 From the single bubble to the macroscopic two-phase electrolysis modelling

The previous part has shown that it is difficult to know the actual two-phase hydrodynamic flow property and particularly the boundary layer thickness in case of vertical electrode. However, this knowledge is important for the local properties (thermodynamic and transport) evaluation and then the macroscopic scale electrolysis modelling. This is the reason why a lot of authors are interested in horizontal floor gas evolving electrode configuration, like the IRDE (Inverted Rotating Disk Electrode). The following part presents electrolysis cell performance modelling, with first a reduced model and second the use of computational fluid dynamic software for electrolyte flow calculation.

4.1 Reduced model

Before heavy calculations with commercial computational software it is interesting to evaluate some easy to access properties of the two-phase electrolysis process. It is easy to calculate how many bubbles of a given diameter d are created per unit time for an imposed intensity I (A): I = j.S, with j the local current density (A m^{-2}) and S the electro active area; j/(n.F) is the local gas molar rate production; R.T/P is the molar volume of an ideal gas (such as O_2 or H_2) with pressure P=1.013 10^5 Pa, temperature T=300 K and gas constant R=8.314 J mol^{-1} K^{-1}; π.d^3/6 is the bubble volume. The created bubble rate is then:

$$dN/dt = 6.I.R.T/(n.F.P.\pi.d^3)(\text{evolved bubbles s}^{-1}) \quad (4)$$

The calculation of the electrolysis cell performances needs the knowledge of configuration and dimensions of cell and electrodes.

4.1.1 Horizontal floor electrode

The evolving electrode is supposed to be horizontal at the cell floor. Then the bubble trajectory is purely vertical if bubble interaction with other bubbles or cells boundaries is neglected. If the electrolyte height is H, the bubble, supposed to have instantaneously the limiting velocity V_{lim}, will have the residence time τ = /V_{lim} in the cell. After a fast transition, it can be supposed that the bubble number in the cell is steady and obeys:

$$F = dN/dt \cdot \tau = 6.I.R.T.H/(n.F.P.\pi.d^3.V_{lim}) \quad (5)$$

To simplify, the configuration is supposed one-dimensional (1D). The electrolyte height change H'-H due to the steady bubbles presence in the liquid domain:

$$H'/H = 1 + j.R.T/(n.F.P.V_{lim}) \quad (6)$$

This property is important because it is an easy to access experimental evidence but need very sensitive liquid level measurements: for a 1 mm diameter bubble, the relative elevation is lower than 2.5 % for a j=1 A cm^{-2} imposed current density. The volume change is generally neglected. One important parameter is also the void fraction ε:

$$\varepsilon = \text{bubbles steady volume/cell volume} = j.R.T/(n.F.P.V_{lim}) = (H'-H)/H$$

In the same way the void fraction is lower than 2.5 % for current density j= 1 A cm^{-2}.

The ohmic resistance of the electrolyte is then modified due to the presence of this insulating volume. This resistance is electrodes configuration dependent but also electrical conductivity σ dependent. The generally used relation is the Bruggeman one for a two-phase electrolyte, with an insulating second phase with small volume fraction ε:

$$\sigma = \sigma_0 \cdot (1-\varepsilon)^{3/2}$$

82 Simulation of Electrochemical Processes II

At the electrolysis beginning the electrolyte is stagnant; but when the steady regime is reached, the electrolyte has won motion from the evolving bubbles:

$$F_{electrolyte}.V_{electrolyte} = \Sigma_{bubbles} \, F_{friction} = N.\pi.d^2/8.\rho_{electrolyte}.V_{lim}^2.C_x(V_{lim})$$
$$= N.\pi.d^3/6.(\rho_{electrolyte}-\rho_{gas}).g$$
$$= I.R.T.H.\rho_{electrolyte}.g \, /(n.F.P.\, V_{lim}) \qquad (7)$$

This vertical motion is given to the electrolyte in contact with bubbles. But, this locally given motion is exchanged in all the electrolyte volume according with the phenomenological law of Fourier, for example for Newtonian electrolytes: $\tau = -\mu \, \mathbf{grad} \, V$, with τ the motion exchange density vector (Pa). In fact the electrolyte is confined and the flow mass conservation will force the electrolyte flow direction rotation: at the free surface, the bubbles exit the electrolyte but not the electrolyte! And then it has to rotate in the perpendicular direction. Then the electrolyte is stagnant only at the process very beginning but not at steady state; this is the reason why the two experiment families are interesting: unsteady and steady regime experimental measurements. Then, motion is both vertical and horizontal, and the perfect knowledge of the local electrolyte flow is possible to determine only with computational fluid dynamic programs. This will be presented in the last part.

4.1.2 Vertical electrode

The vertical electrodes are used for industrial production of gaseous fluorine F_2; H_2 bubbles are also created at the counter electrode. Like for the classical water electrolysis, both electrodes produce bubbles and it is really difficult to model the anode and cathode induced flows interaction.

Reference [7] shows experimental evolution of the electrochemical cell electrical performances during a two-phase electrolysis process. The imposed potential U is increased from the minimal value U_d for electrolysis. For the smaller potential values, the intensity I increase is almost linear (zone from A to B). With the bubbles production increase, the equivalent resistance increase and the linear evolution is lost and the intensity I becomes constant with value I^{crit} (zone from B to C). The potential increase leads to no intensity increase till the value U^{crit}. Between points C and D, the situation becomes unstable and with a small potential increase from U^{crit}, the intensity decreases rapidly. At the minimal intensity level, a new process occurs: the tool electrode is heated to a visibly large temperature. Bubbles must be, in this zone (between D and E), electrochemically produced, but also because of electrolyte phase change due to high temperature. Then it appears in this zone sparks. The numerous bubbles show an important electrolyte flow, according with a double vertices structure, at the electrode tip.

The case of the vertical gas evolving electrode is more difficult to model than the horizontal floor one because of the residence time. In the cell, the bubble residence time is no longer the same for bubbles emitted at electrode bottom and top: the residence time is also τ for the bottom emitted bubbles but it is 0 for top emitted bubbles. Because the limiting velocity is reached instantaneously, if the current density distribution at electrode was uniform, the bubbles steady number should be N/2. A lot of previously presented properties are usable here. But the difficulty is due to the determination of the actual bubbles trajectory, vertical but mostly horizontal. This is because the actual two-phase volume and the associated boundary layer thickness are difficult to know that the ohmic resistance and other macroscopic cells properties are difficult to model (see the previous part the difficulty to describe the horizontal force exerted upon the single bubble).

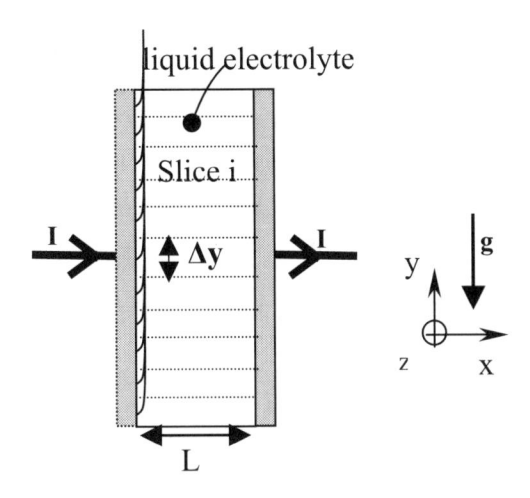

Figure 2: Vertical gas evolving electrode.

Figure 2 shows the vertical gas evolving electrode configuration. The Newman model for vertical electrodes uses a one dimensional vertical discretization and the local bubble number in each horizontal slice. This number is increasing from the bottom where N is about zero to the top where N is maximal.

The current density distribution is then clearly not uniform: the current density is maximal at the bottom and minimal at the top, around the average current density $j_{av}=I/S$. A simple recurrence algorithm, based upon bubble mass balance, has been programmed in C language to evaluate at slice i, the local vertical bubbles fluxes F_i and the steady bubbles number N_i:

$$F_i = F_{i-1} + 6.I_i.R.T /(n.F.P.\pi.d^3) \qquad (8)$$

Like in the previous part, the fluxes are immediately obtained. The local bubble number needs more information in term of local bubble velocity which is directly related to the local electrolyte velocity. If the bubbles size and bubbles relative velocity are supposed uniform, it is also possible to calculate the slice i bubble number, void fraction, resistance and then the primary current distribution under these assumptions. The numerical procedure is iterative.

$$N_i = F_i.\Delta y/V_i = F_i.\Delta y /V_{lim} \qquad (9)$$

The following results have been obtained supposing an electrolyte conductivity $\sigma_0 = 1$ S m^{-1}, an electro-active surface 25 cm^2 and a two-phase volume with L=10^{-2} m. The one phase process has the theoretical resistance 4Ω.

The height h=5 cm has been dicretized with 100 slices. The bubbles fluxes, bubbles number, local void fraction, local electrical conductivity and local current density profiles have been calculated.

Table 1:   Electrochemical cell macroscopic properties evolution with potential.

$U_{imposed}$ (V)	I (A)	$j_{average}$ (A m^{-2})	Resistance (Ω)	Elevation (m)
1.00E+02	2.45E+01	9.79E+03	4.09	7.10E-04
1.00E+03	1.99E+02	7.97E+04	5.02	7.10E-03
2.00E+03	3.09E+02	1.24E+05	6.48	1.42E-02
3.00E+03	3.49E+02	1.39E+05	8.61	2.13E-02
3.50E+03	3.51E+02	1.40E+05	9.97	2.49E-02

The table 1 gives the calculation results from small imposed potential $U_{imposed}$ to the critical potential which is here 3550 V with an associated intensity I=351 A. For imposed potential greater than this value, the void fraction ε reaches 100% at the electrode top and the calculation procedure is stopped.

Table 1 shows intensity non-linear increasing with imposed potential. The ohmic resistance increases due to the bubbles presence which also leads to an electrolyte level elevation associated with the steady gas volume in the electrolyte (initial height is h=5cm).

In the used algorithm, the critical point C is defined as the imposed potential which leads to a 100% gas fraction at the electrode top. But in fact, the current can always flows after this potential value, not at the top but at lower positions. This is not taken into account here and will be in future works.

The default of this model is the use of the geometrical length L between electrodes instead of the actual two-phase boundary layer thickness, and the simplification of the electrolyte and bubbles local velocities properties.

4.1.3 Horizontal ceiling electrode

This case is frequently met for example with the classical rotating disk electrode (RDE). It is also one of the major problems of the Hall-Heroult process for aluminium production; in this case In this case the working electrode is generally very screened due to the bubbles accumulation. It has been shown in the first part dedicated to the single bubble that the Archimede force was important for the bubble take-off from the electrode. Because of this horizontal configuration, this force can't lead to bubble motion, which can also be horizontal. Then friction force from the flow is alone to ensure this bubble motion. Other phenomena must be considered: the "on electrode surface" bubble motion; the meeting of other bubbles and the coalescence. The more the gaseous volume is, the more the height and then the flow friction. Experiments with this configuration are on progress.

4.2 Eulerian-Lagrangian model

As said in previous part, this kind of approach must be improved with a better description of the electrolyte and bubbles flows. But there are few two-phase flow modelling, mostly in the case of a second phase created at the boundary. A lot of works in fluid mechanics or combustion sciences are developed, but the second phase is generally created or injected in the calculation domain, not at the border, in interaction with boundary layers. The flow can also be turbulent: it is generally the case for large motion source and then for large bubbles creation. The Navier-Stokes electrolyte flow modelling is classically written:

Mass balance:
$$\partial \rho_e / \partial t + \partial(\rho_e . u_i)/\partial x_i = 0 \qquad (10)$$

Motion balance:
$$\partial (\rho_e . u_i) / \partial t + u_j . \partial (\rho_e . u_i)/\partial x_j = - \partial P/\partial x_i + \partial(\mu_e \, \partial u_i /\partial x_i)/\partial x_i + K_i \qquad (11)$$

with u_i the ith component of the local electrolyte velocity; ρ_e is the local equivalent density (kg m^{-3}); P is the local gauge pressure (Pa) and μ_e is the local equivalent dynamic viscosity (Pa.s). The local flow properties are calculated with computational fluid dynamic software using finite volumes method [15] such as Fluent® or Fluidyn®. Equivalent values are used to take into account the two-phase and evenly turbulent local flow character. The calculation of these equivalent thermodynamic and transport properties need to consider single bubble and group of bubbles: the equivalent density is directly related to the void fraction ε: $\rho_e = \rho_0 .(1 - \varepsilon)$ whereas the equivalent viscosity must be calculated for a given hydrodynamic stress and a given void fraction. K_i is the motion source term due to bubbles Archimede acceleration. The coupling between bubbles motion and this electrolyte motion source is calculated for each cells of the calculation domain mesh. The effort from the flowing bubbles to the electrolyte is mostly due to the sum of the friction terms over all the possible trajectories and the associated bubbles crossing the considered cell:

$$K_i = \Sigma_{trajectories} friction.q.\Delta t/Volume_{cell}$$

with friction the acceleration previously presented (m s^{-2}); q the bubble mass flow of one considered trajectory (kg s^{-1}) and Δt and Volume$_{cell}$ the residence time (s) and the volume (m^3) of the considered cell respectively. Each electrode discretization faces are associated with a bubbles emitted trajectory, or more precisely, its gravity centre. In the case of the vertical electrode configuration, if no horizontal force is introduced, these trajectories are exactly at the electrode which leads to quick accumulation problem. This is the reason why a horizontal dispersion force must be introduced to take into account the two-phase boundary layer thickness increase. Numerical results previously published [16–17] have shown that this horizontal force must be about half the Archimede one order of range to allow a good accordance with Tobias experimental results [5] obtained with a segmented vertical electrode.

To calculate primary current distribution, the usual CFD codes equations must be completed with an electric charge balance equation such as:

$$\partial(\sigma_e \, \partial \varphi /\partial x_i)/\partial x_i = 0 \qquad (12)$$

with $\sigma_e = \sigma_0.(1-\varepsilon)^{3/2}$ (S m^{-1}) according with Bruggeman law and φ the local potential (V).

This is generally easy to add diffusion equations in these commercial softwares.
This model is under the dispersed phase assumption: it must be used only for void fraction lower than 20%. This is the reason why the Eulerian-Eulerian model is generally preferred.

4.3 Eulerian-Eulerian model

The Eulerian-Eulerian two-phase flow model also uses both the mass and motion balances but for each phase:
Liquid mass balance:
$$\partial \rho_l / \partial t + \partial(\rho_l.u_{i,l})/\partial x_i = 0 \qquad (13)$$
Liquid motion balance:
$$\partial (\rho_l.u_{i,l}) / \partial t + u_{i,l}.\partial (\rho_l.u_{i,l})/\partial x_j = -\partial P/\partial x_i + \partial(\mu_l \partial u_{i,l}/\partial x_i)/\partial x_i + K_i \qquad (14)$$
Gas mass balance:
$$\partial \rho_g / \partial t + \partial(\rho_g.u_{i,g})/\partial x_i = 0 \qquad (15)$$
Gas motion balance:
$$\partial (\rho_g.u_{i,g}) / \partial t + u_{i,g}.\partial (\rho_g.u_{i,g})/\partial x_j = -\partial P/\partial x_i + \partial(\mu_g \partial u_{i,g}/\partial x_i)/\partial x_i - K_i \qquad (16)$$
K_i is the motion exchange between gas phase and liquid phase due to friction, lift, added mass or pressure variation. The local void fraction for a given cell in the mesh leads to the bubble number evaluation and then to the number of friction forces exchanged between the two phases. This model is rigorous but really heavier: four non-linear equations, strongly coupled and the convergence numerical conditions are generally difficult to found. Simplified two-phase flow models are generally proposed in computational fluid dynamic software: the "Mixture Model" (or "Algebraic Slip Model", ASM) and the "Volume Of Fluid" (VOF) model. The first one is mostly used for homogeneous gaseous phase with volume fraction from 10% to 50%, whereas the second is mostly dedicated for unsteady-free surface problems. This model can be used in the case of really large gaseous structures in the electrolyte, with clearly defined two-phase interface.

5 Conclusion

The two-phase electrolysis processes are of a great interest for industrial applications. They not yet have been rigorously modelled though it is now possible to compute the coupling between electrochemistry and two-phase flows. The present work has presented macroscopic and bubble scale modelling and has shown that both scales are coupled which define a multi-scale process. Bubble scales results have to be correlated to lead to a predictive model usable at the larger scales. The present work has focused its attention upon the vertical gas evolving electrode and has shown modification of primary current distribution in case of electrochemically induced two-phase flow. Future works will develop the strong coupling between phenomena and scales and between experimental evidences and calculations.

References

[1] Vogt H. and Balzer R.J., The bubble coverage of gas-evolving electrodes in stagnant electrolytes. *Electrochimica Acta*, 50, 10, pp. 2073-2079, 2005.

[2] Eigeldinger J. and Vogt H., The bubble coverage of gas-evolving electrodes in a flowing electrolyte. *Electrochimica Acta*, 45, 27, pp. 4449-4456, 2000.

[3] Javit A. Drake, Clayton J. Radke and Newman J., Transient linear stability of a Simons-process gas–liquid electrochemical flow reactor using numerical simulations. *Chemical Engineering Science*, 56, 20, pp. 5815-5834, 2001.

[4] Hine F., Murakami M. *J. Electrochem. Society.*, Vol 127, p. 292, 1980.

[5] Tobias C.W., The influence of attached bubbles on Potential Drop and Current Distribution at Gas-Evolving Electrodes. *J. El. Soc.*, 134, 2, 1959.

[6] Mandin Ph., Hamburger J., Bessou S. and Picard G., Modelling and calculation of the current density distribution evolution at vertical gas-evolving electrodes. *Electro. Acta*, 51, 6, pp. 1140-1156, 2005.

[7] Wüthrich R. and Bleuler H., A model for electrode effects using percolation theory. *Electro. Acta*, Volume 49, Issues 9-10, Pages 1547-1554, 2004.

[8] Vogt H., Aras Ö. and Balzer R.J., The limits of the analogy between boiling and gas evolution at electrodes. *I. J. Heat Mass Trans.*, 47, 4, p. 787, 2004.

[9] Duhar G., Colin C. *C. R. Mecanique* 331, 2003.

[10] Duluc M.-C., Stutz B. and Lallemand M., Transient nucleate boiling under stepwise heat generation for highly wetting fluids. *I. J. Heat Mass Trans.*, 47, 25, p. 5541, 2004.

[11] Janssen L.P.B.M., Warmoeskerken M.M.C.G., *Transport Phenomena Data Companion*, Arnold DUM, 1987.

[12] Devos O., Gabrielli C. and Tribollet B., Nucleation-growth process of scale electrodeposition—Influence of the mass transport. *Elec.a Acta*, 52, 1, p.285, 2006.

[13] Morsi S. A. and Alexander A. J., An Investigation of Particle Trajectories in Two-Phase Flow Systems. *J. Fluid Mech.*, 55(2), pp. 193-208, 1972.

[14] Krepper E., Lucas D. and Prasser H.-M., On the modelling of bubbly flow in vertical pipes. *Nuclear Engineering and Design*, 235, 5, p. 597, 2005.

[15] Patankar S. V., *Numerical Heat Transfer and Fluid Flow*, Taylor & Francis, 1980.

[16] Kiss L., Ponscak S., Effect Of Bubble Growth Mechanism On The Spectrum Of Voltage Fluctuations In The Reduction Cell, *Light Metals*, 2002.

[17] Janssen L.J.J., Behaviour Of Mass Transfer At Gas-Evolving Electrodes, *Electrochemica Acta*, Vol. 34, No 2, pp. 161-169, 1989.

Simulating electro-coating of automotive body parts using BEM

J. M. W. Baynham[1], R. Adey[1], V. Murugaian[2] & D. Williams[2]
[1]*C M BEASY Ltd, Ashurst Lodge, Southampton, UK*
[2]*IARC (The International Automotive Research Centre), Warwick Manufacturing Group, The University of Warwick, Coventry, UK*

Abstract

The simulation of electro painting has rapidly become a "must-have" tool for auto-manufacturers over the past three to four years. Various approaches have been used in such simulation, including both FE and BE.

In this paper details of the boundary element based "BEASY" Electro-coat Modelling software are presented which allow car-body parts to be simulated. Examples are presented showing application of the method. Drawbacks of the method are discussed.

1 Introduction

One electro-painting process in use today involves immersing a metallic (conducting) structure in a tank of water based paint, then passing a direct current from anodes (also immersed in the paint) to the metallic structure which forms the cathode.

Electrochemical processes at the cathode surface cause deposition of paint on the surface. Since the deposited paint has electrical resistance, the electrical fields in the tank are constantly changing as more and more paint is deposited.

Paint is deposited most rapidly where the current density is greatest. The metal surfaces inside any cavities initially receive no current, but as the paint film builds up on exterior surfaces, current begins to penetrate the cavities, and deposition occurs here too.

The difficulty of coating inside cavities is that the wet paint film is never a perfect insulator. Consequently current passing along the axis inside a hollow tubular cathode (for example) reduces steadily along the axis as current flows sideways into the walls.

The challenge for auto designers is to ensure that adequate paint film thickness is achieved inside such cavities, because film thickness directly affects corrosion resistance.

The incentive to use simulation is strong, since it allows design verification at an early stage of the overall design process. The costs of design changes early in the design process are significantly lower than later on. However the most significant advantage that simulation confers is compression of the Product Development process. The associated reductions in cost improve operating performance, and time savings enable manufacturers to shorten time to market and improve responsiveness to consumer requirements.

Simulation of electro-painting requires first a way to determine the currents flowing in the electrolyte (the paint). The boundary element method (BEM) has been applied to the very similar application of corrosion simulation since the late 1970s [1]. In both electro-painting and corrosion simulation the domain is the electrolyte surrounding the structure, and BEM is well suited to such situations because creating meshes on the surfaces is relatively straightforward.

Secondly, the simulation must provide a way to represent the paint film layer. From the electrical point of view the important property of the paint film is simply its resistance.

Finally, the simulation must somehow represent the build-up of the paint film thickness. In transient simulation this implies a time-stepping algorithm which uses current density to determine rate of deposition, and then incrementally increases the recorded film thickness.

2 Methodology

2.1 The BEM solver

The ability to solve the Laplacian using BEM is well documented, see for example [2]. The BEM is a mixed method, which means that both voltage on the surface (of the electrolyte) and current density in the normal direction through the surface are represented directly.

By contrast in the FEM domain method where only voltage is represented directly, the current density is obtained by differentiating the voltage field. For reasons of accuracy the differentiation is normally performed at positions inside the volume element (away from the sides) and the result is extrapolated to the surface.

The current densities produced by the mixed method BEM are therefore generally more accurate than those produced by the FEM. This means the BEM elements can be bigger than those used in a corresponding FEM model

In 3D the boundary elements are a mixture of triangular and quadrilateral shapes, each representing a part of the surface of the domain. In the BEM each degree of freedom is linked mathematically with every other degree of freedom.

For sheet metal cathodes, each side of the metal sheet is a surface of the electrolyte, and hence must have elements on it. This can cause problems when the metal sheet is thin, because the collocation, direct method BEM equations for a pair of nodes separated by the sheet thickness become more and more similar as the thickness reduces. For a "zero thickness" sheet the equations would be identical, so the matrices are singular and cannot be solved. For small but non-zero thickness the equations will be ill-conditioned, leading to problems when performing equation solution, and possible inaccuracy.

One approach to overcome this problem is to divide the electrolyte into multiple domains (or *zones*) so that opposite sides of the metal sheet are in different zones. This method works very well, but can lead to complication in model-building because artificial boundaries must be constructed inside the domain. An example of application of this method is shown in Figure 1 which shows a model of a "throw-box" (three metal sheets held some millimetres apart to form two cavities). The figure shows the boundaries of three zones, and how they fit together to form the complete model. It also shows the artificial boundary which covers the access from outside into cavity 1. In the model shown creating the artificial boundary was not difficult, but in a complex car part, the task is much harder.

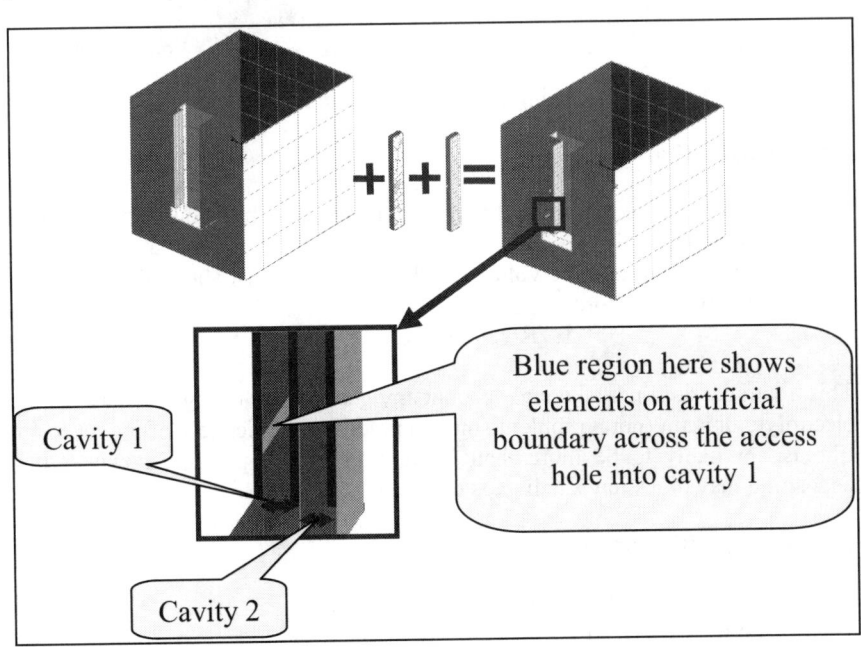

Figure 1: Multiple zone model of throw-box with two cavities.

If the metal sheet is approximated as having zero thickness, a different mathematical approach can be used when generating the equation for the node on the "other" side of the sheet. Consequently the equations for the pair of nodes are different, and the matrix is no longer singular. This technique, known in BEM as *dual elements* [3], is advantageous for the user since it is only necessary to create one element using a pre-processor. The *dual* can be generated automatically inside the BEM solver. An example of application of this method is shown in Figure 2, which again shows the throw-box model. Note the absence of elements across the entrance to the cavity.

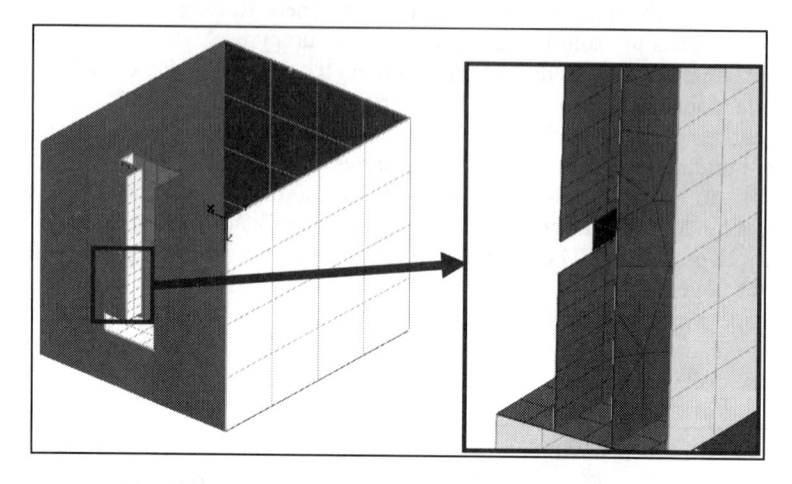

Figure 2: Single zone model using dual elements.

2.2 The paint film

The paint film can be represented very simply as a resistive boundary condition relating the current density flowing from the electrolyte into the metal, the voltage in the metal, and the voltage in the electrolyte just above the paint film by and equation of the form:

$$j_n \cdot R_{film} = V_{electrolyte} - V_{metal}$$

This treatment does not take into account any geometrical change of shape of the electrolyte domain (not a problem since in practise the thickness of the *dry* paint film is not likely to be more than 30 microns). It is possible however that deposition may block any small gaps between metal sheets.

2.3 The deposition process

The rate of deposition of paint may be expressed as:

$$dh/dt = C_{eff}(j - j_0)$$

where:

dh/dt is the rate of change of the (dry) film thickness
C_{eff} is the Coulombic efficiency of the paint, expressed as a volume of paint per Coulomb of charge
j is the current density flowing from the electrolyte into the metal
j_0 is the current density below which no deposition occurs.

2.4 The relation between thickness and resistance

The relationship between film thickness and film resistance can be expressed in tabular form. This allows non-linear variation, and makes it possible to account for the resistance of any pre-treatments.

3 The simulation process

For transient simulation the process is as follows:
- create influence matrices;
- apply cathode voltage to the sheet metal;
- solve for current density;
- integrate through time to obtain the change of film thickness;
- where film thickness is non-zero apply a resistive boundary condition;
- again solve for current density;
- repeat for the required duration of the simulation.

Clearly the magnitude of the time step must be controlled, since too big a step is likely to result in unrealistic distribution of film thickness. In practise it is found better to choose a delta(h) (i.e. thickness change) which is used to limit the time step.

For the direct method, the simulation process is similar, except that instead of integrating through time, the boundary condition applied to the cathode is switched between cathode voltage and target deposition current density.

Both methods provide the film thickness at the target time.

4 Application

4.1 Car door

The direct method of simulation has been applied to the car-door model shown in Figure 3.

The door was meshed at the IARC using Hypermesh, and exported as a NASTRAN bulk data file containing shell elements. This file was processed into a BEASY data file containing dual elements, and elements representing the paint bath sides and base, and the paint surface were added. Typically anode geometry would also be constructed, but in this case voltage boundary conditions were simply applied to some parts of the tank sides. Similarly the paint free surface is usually treated as a symmetry plane, so no elements are required on it. The total anode area was 21600 cm^2, and the anodes were assumed to be at 340 volts. The total cathode area was 43,540 cm^2.

The direct method was applied, using assumed paint properties, and wet film resistance 85MegaOhm.mm2 at film thickness corresponding to a 25 micron dry film thickness.

The paint film thickness at target time 180 seconds is shown in Figure 4 to Figure 8.

Figure 3: Door used in simulation.

4.2 Discussion of results

The film build on the outer surfaces on the "outer side" of the car door takes the expected form (Figure 4), that is: nearly constant thickness except near access holes into the interior of the door cavities.

Figure 5 shows paint thickness on the reverse of the metal sheets, but again viewed from the outside of the car. This figure shows a gradual reduction of thickness along the "axis" of the cavities, shown arrowed red in the figure. This gradual reduction is consistent with results of throw-tube experiments.

Figure 6 shows paint thickness as viewed from the inside of the car.

Figure 7 shows a detail of film build (using a different contour plot scale) on the reverse side of the metal sheets near the top left of the door when viewed from the outside of the car. The red arrows in the figure highlight regions where the film build is bigger than it is to either side along the axis of the cavity. The only way this can happen is if current is able to flow into the cavity nearby, and by looking at the same region from the inside of the car (Figure 8) the reason can be seen. These small holes (shown by red arrows on the right hand part of Figure 8) allow enough current to flow into the cavity to prevent the paint thickness from reducing too far.

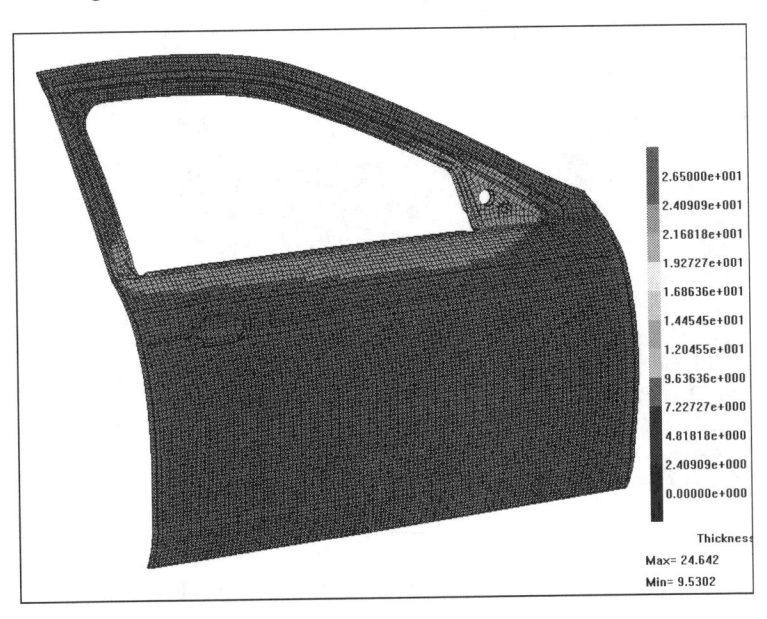

Figure 4: Distribution of film thickness, viewed from "outside" the car.

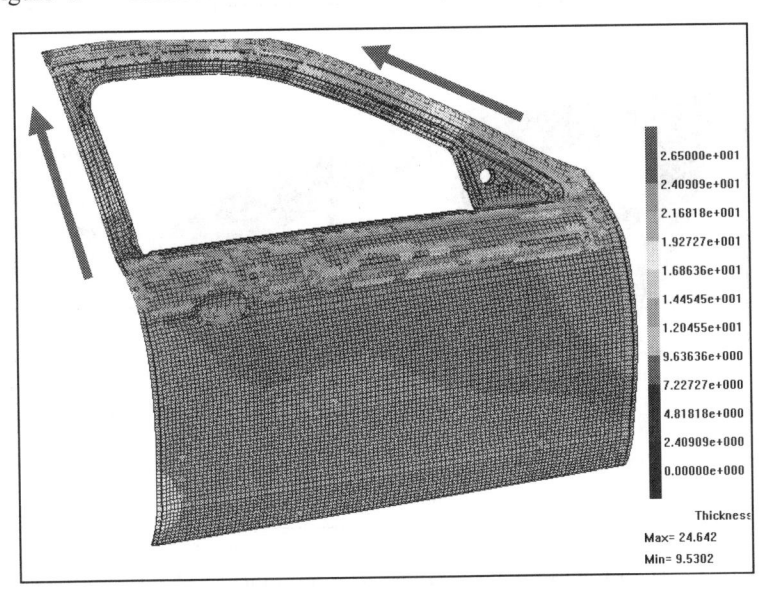

Figure 5: Distribution of film thickness on the "reverse side" of the metal sheets, viewed from "outside" the car.

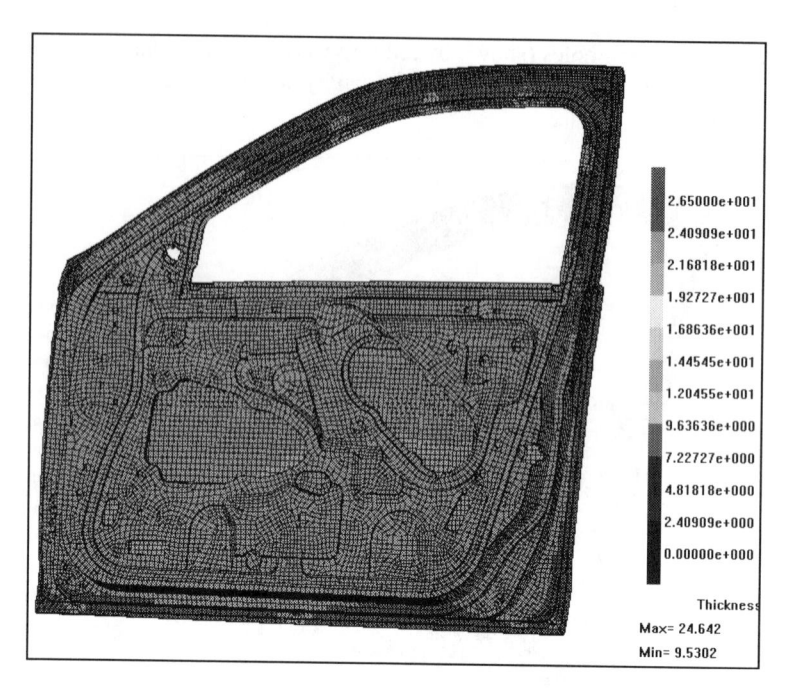

96 Simulation of Electrochemical Processes II

Figure 6: Film thickness, viewed from "inside the car".

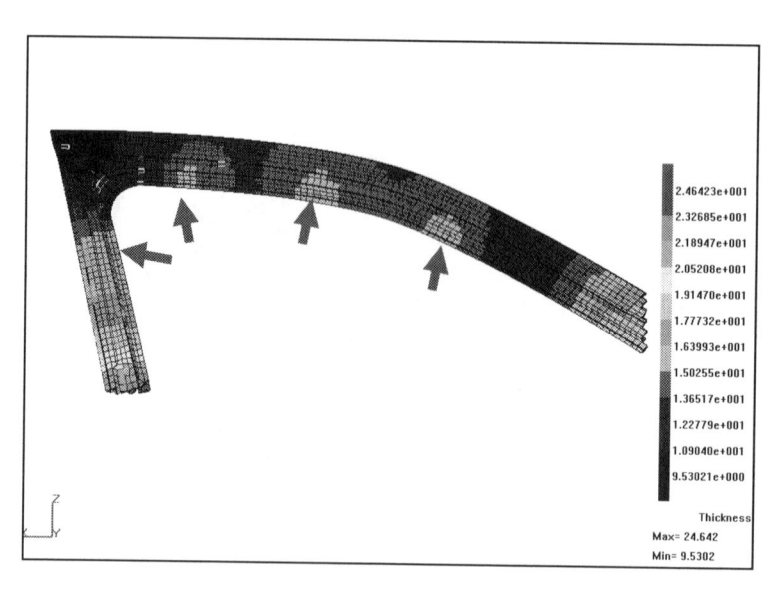

Figure 7: Distribution of film thickness on inner surfaces of the cavity, viewed from "outside" the car.

Simulation of Electrochemical Processes II 97

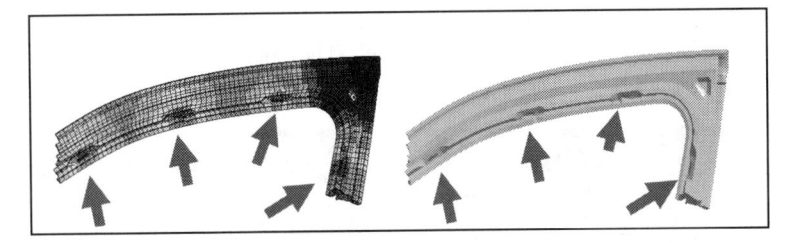

Figure 8: Distribution of film thickness on inner surfaces of the cavity, viewed from "inside" the car, and corresponding view of geometry.

4.3 Comparison with Experimental Measurements

Experimental measurements were made by the IARC and compared with simulation results from a model similar to, but less refined than that shown in the previous section. The comparison, shown in Figure 9, suggests good correlation between simulation and experiment.

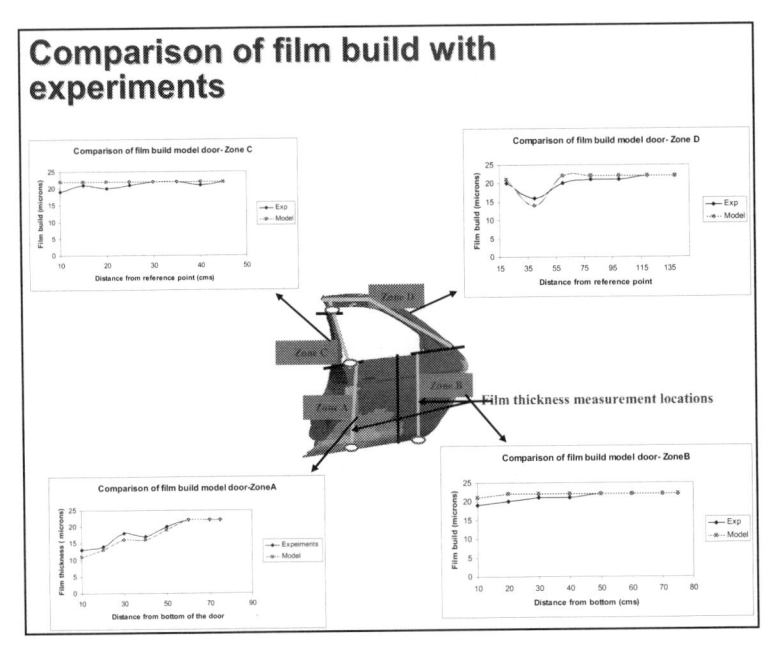

Figure 9: Showing comparison between simulation and experimental results.

5 Design changes

The method described using BEM is very suitable for simulation of design changes, because modification to add an extra paint access hole requires minimal change to the mesh. The change can be effected at a local element level, with no other changes required to the model. For example Figure 10 shows the mesh for a model before and after a "design change". The only change to the mesh is in the area shown circled.

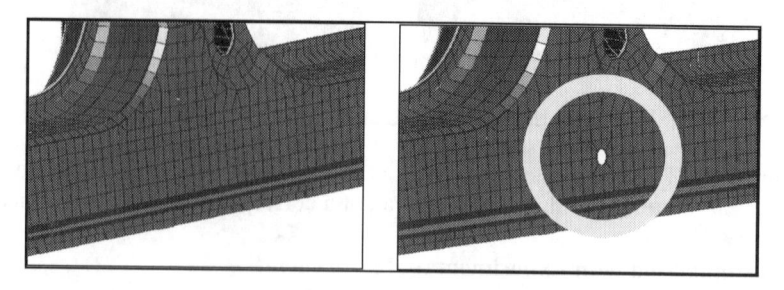

Figure 10: Mesh changes required to add a new paint access hole.

6 Drawbacks of the method

Although the idea of using a shell mesh of a car part as the mesh for BEM electro-coating simulation is attractive (in particular since such meshes *must* be created for structural simulation), there are some caveats as follows.

Firstly, the dual elements have zero thickness. This creates a problem if the elements are created on the neutral axis of the metal sheets, since there will be a gap between the dual elements where the metal sheets are joined together.

Secondly, at joints the metal sheets overlap slightly, and there is no electrolyte between the sheets. However the dual elements do not represent this situation since they will be separated by the thickness of the metal sheet.

These issues mean that in practise the geometry used to create the dual elements must be modified near joints, to remove overlaps and gaps. Figure 11 shows an example of how this geometry modification might be done.

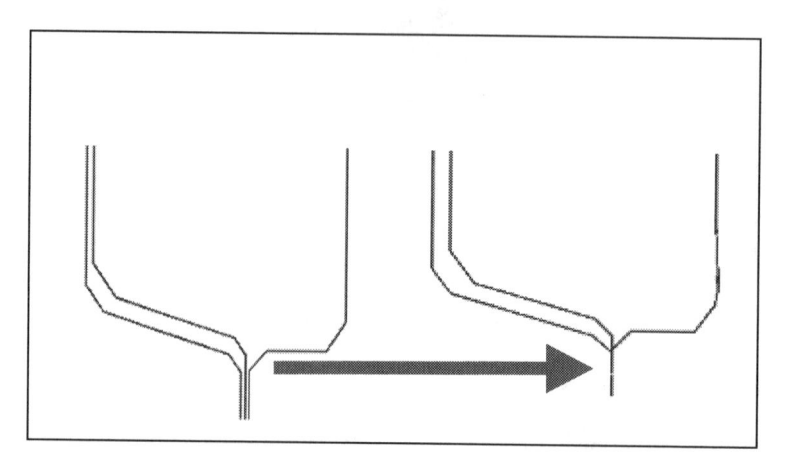

Figure 11: Geometry change to remove overlaps and gaps at joints.

7 Conclusions

The successful application of BEM technology to simulation of electro-coating of automotive parts has been demonstrated.

A notable feature of the methodology is the ease with which design changes can be incorporated for evaluation and the attendant virtual engineering performance benefits.

Acknowledgement

The authors are grateful to Jaguar and Land Rover for permission to show the door model.

References

[1] Danson, D.J. and Warne, M.A. Current Density/Voltage Calculations Using Boundary Element Techniques, Corrosion/83. 1983.
[2] Brebbia, C.A. The Boundary Element Method for Engineers, Pentech Press. London, 1978.
[3] Hong, H. and Chen, J. Derivations of integral equations in elasticity. Journal of Engineering Mechanics, 114(6), pp1028-1044, 1988.

Section 2
Cathodic protection systems

3D cathodic protection design of ship hulls

L. Bortels[1], B. Van den Bossche[1], M. Purcar[1], A. Dorochenko[1] & J. Deconinck[2]
[1]Elsyca N.V., Zellik, Belgium
[2]Vrije Universiteit Brussel, Department IR\ETEC, Brussels, Belgium

Abstract

This paper presents a 3D software tool for the design and optimization of cathodic protection systems for submerged structures. It provides the corrosion engineer with a powerful tool for managing operational costs, significantly reducing expensive commissioning surveys and costly repairs, adding major value to the cathodic protection business.

The software is entirely CAD integrated such that it can deal with 3D CP-configurations of arbitrary complexity with parameterisation of all geometrical dimensions. The CP model is based on the potential model describing the ohmic drop in the electrolyte (soil, water) with non-linear boundary conditions that model the electrochemical reactions at anodes and cathodes.

In this paper, it is explained why the Finite Element Method is used to solve the problem. As an example the protection level of a hypothetical marine vessel using impressed current cathodic protection (ICCP) systems will be investigated. In addition, the underwater electric potential (UEP) of the vessel will be calculated.

Keywords: cathodic protection simulations, ICCP, advanced software package, marine vessel.

1 Introduction

Cathodic Protection (CP) systems are widely applied to buried and offshore structures. Most often, these CP systems contain a series of impressed current and/or sacrificial anodes, sometimes placed at a remote distance from the structure. The entire configuration of the CP system and the structures necessitate and justify the use of numerical simulations.

First, the low accessibility of these structures makes installation, maintenance and repair very expensive. Also, the geometry of most steel structures that are subject to cathodic protection is too complex to allow analytical or even empirical estimations for the determination of the local protection level. Numerical modelling provides significant benefit by identifying insufficiently protected regions – possibly subject to corrosion, and overprotected regions – subject to excess gas evolution and hence coating disbonding. As a consequence, numerical modelling allows simplification and optimization of installation, maintenance and repair. Moreover, models provide reference values for measurements on operational sites, enabling to trace and solve any possible anomaly.

Most of the publications dealing with the computation of the CP of both buried and submerged structures are based on the well known Boundary Element Method (BEM) [1].

Orazem et al. [2,3] use a 3D BEM approach to compute the protection level of large coating defects on pipelines. Results are presented for a pipe segment of limited length (10 feet), in presence of a parallel anode system. Riemer and Orazem [4] produced results for a larger pipeline (> 6 km) with coating defects of varying size and investigated the ability of coupons in the vicinity of the defects to measure off-potentials. Adey [5] applied a full 3D approach to calculate the potential field in the neighbourhood of jacket joints under cathodic protection of sacrificial anodes. The present authors [6] used a 3D coupled multi-domain BEM approach to simulate the protection level of a buried pipe segment surrounded by a concrete vault.

Aoki et al. [7] applied the BEM to detect a coating defect on a ship hull. DeGiorgi [8] made a significant contribution to the modelling work of the CP of ships. Diaz and Adey [9,10] simulated the stray current corrosion of a vessel berthed to a steel dock protected by sacrificial anodes. The same authors presented a methodology based on boundary element techniques to determine the optimum anode configuration for shipboard ICCP systems.

The simulations presented in this paper are based on the Finite Element Method (FEM) [11], although it is well known that the Boundary Element Method (BEM) offers the advantage to reduce the dimensions of the problem and to obtain directly the current density distribution along electrodes.

2 Mathematical model

2.1 The equations

2.1.1 The conducting medium

CP protected structures (ships, tanks, pipes) are vast electrodes immersed in a semi-infinite electrolyte. Due to the nature of the conductive medium (soil, (sea)water) and its large scale, no concentration gradients occur in the electrochemical solution except for the well-known and very thin diffusion layer near the electrodes. Since the conducting medium is assumed to feature only charge transport with normal ohmic resistivity effects, the potential model holds, being described by the Laplace equation:

$$\overline{\nabla}(-\sigma \overline{\nabla} U) = 0 \qquad (1)$$

in which σ is the conductivity and U the potential field in the conducting medium. On insulating boundaries, the current density perpendicular to the surface is zero. The phenomena occurring in the diffusion layer and at the electrode interface are encompassed in the (non-linear) boundary conditions.

2.1.2 Coated metal surfaces

Protected structures are in general covered with a good quality coating. It is assumed that the latter can be described by a resistance that is placed in series with the pipe / soil polarisation resistance:

$$j_n = f(V - U - E_{corr} + \rho_{coat} L_{coat} * j_n). \qquad (2)$$

where j_n is the local current density, ρ_{coat} and L_{coat} are the resistivity and the thickness of the coating respectively. E_{corr} is the corrosion potential, V and U hold for the electrode and electrolyte potential respectively. Due to the high resistivity of the coating, curve (2) manifests a nearly linear behaviour.

Another approach to deal with coated surfaces is presented in equation (3). Here it is assumed that the polarisation of the surface is that of the metal. But a parameter θ is introduced that accounts for the fact that only a fraction of the total surface is active.

$$j_n = \theta f(V - U - E_{corr}). \qquad (3)$$

2.1.3 Anodes

The anodes are either sacrificial anodes or impressed current/potential electrodes. The electrode polarisation is often represented by an interpolation function of a measured electrode polarisation curve, relative to a mixed corrosion potential E_{corr}:

$$j_n = f(V - U - E_{corr}) = f(\eta). \qquad (4)$$

2.2 Numerical solution method

As the conductivity of the computational domain is constant, in principle the BEM is to be preferred over other discretisation methods as Finite Element (FEM), Finite Volume (FVM) and Finite Difference Methods (FDM), to solve the simplified charge conservation equation (1).

The initial version of the software was based on the BEM which was for performance reasons soon replaced with the more advanced Fast Multipole Method [12]. Both methods have the advantage that only the boundaries of the

3D model need to be meshed. The disadvantage however is that due to the intrinsic properties of both the BEM and FMM, the resulting system matrices remain quite populated, which severely limits the number of surface elements that can be used (about 40 000 on a standard PC). Therefore, a complete new software code has been developed which is based on the well-known Finite Element Method (FEM) [13]. Using the FEM, the complexity of the problems that can be solved (i.e. the total number of surface elements) has been increased with at least an order of magnitude (when compared to FMM). The non-linear system equations are solved using a Newton-Raphson iterative method [14], combined with an advanced iterative linear solver to solve the resulting system of equations at each iteration. It is important to mention that when the Potential Model is solved using the FEM, the technical problem is rather a volume mesh generation problem, while a solver based on the BEM struggles more with matrix assembly and system inversion.

2.3 Mesh generation

In order to make the use the FEM possible, a powerful and automated 3D volume mesh generator is required that produces the meshes needed for the computations [15]. The developed mesh generator is fully CAD integrated, generates an unstructured grid and works in a hierarchical way. First the meshes (line segments) on all edges are generated, based on the obtained edge grid, all surfaces are triangulated and finally a volume grid of tetrahedral is generated based on the generated surface grid. All operations are controlled using a set of general parameters such as the targeted size of an edge, surface or volume element and the preservation of curvature. The size of an element can be locally different (see Figure 3) e.g. on electrodes more elements are generated than on insulating boundaries.

The unstructured surface and volume grid generators control the transition from the local mesh size to the targeted one, based on an interpolation/extrapolation functions and a user defined growing factor (see Figure 3).

3 Description of the problem

As an example, the cathodic protection level of a hypothetical marine vessel (catamaran type) using 4 separate impressed current cathodic protection (ICCP) systems, each delivering 3 A, will be presented and investigated (see Figure 1).

The total length of the vessel is 120 m, with a width of 40 m (waterline). Two ICCP systems are located in the middle (inside) of the catamaran, one at the port side (MP), the other one at the starboard side (MS). The other two ICCP systems are located at both sterns (outside) and are denoted as anodes SP and SS.

The resistivity of the seawater is assumed to be 0.25 Ωm. The catamaran material used in the simulations is marine aluminium with polarization behaviour as measured by Kim et al [16]. The corresponding corrosion potential is -880 mV versus the Ag/AgCl reference electrode. The coating of the catamaran is considered to be non-ideal. The coating defects are assumed to account for 1.0 percent of the total surface area. To that purpose, the polarization curve for the bare aluminium, as obtained in [16], has been scaled with a factor $\theta = 0.01$ for the current density. The propellers are made from nickel, aluminium and bronze (NAB) and have a corrosion potential of -200 mV (versus Ag/AgCl). The polarization data are taken from the work by Hack [17]. It is assumed that the propeller and support are electrically continuous. The anodes are 7.5 cm in diameter and have a length of 1.5 m. The anode material is a typical metal oxide for which the data have been taken from a proprietary database. A detailed view of the propeller and support with nearby anode is presented in Figure 2.

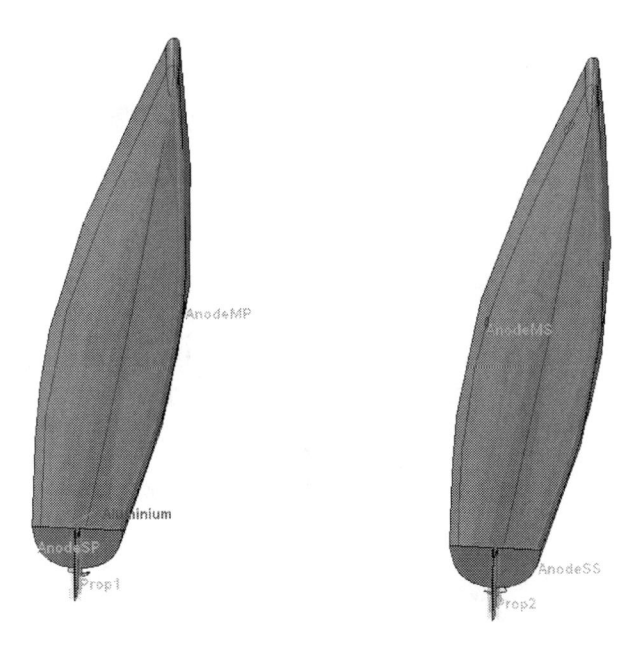

Figure 1: Layout of the catamaran with 4 ICCP systems: Anode MP, MS, SP, SS; (M = middle, S = stern; P = port, S = starboard).

4 Computational specifications

The total number of triangular surface elements used to solve the problem is about 200 000. The corresponding tetrahedral volume mesh consists out of nearly 2 500 000 elements. There are in total about 500 000 mesh points of which About 200 000 points are located on surfaces. Details of the triangular surface mesh near one of the propellers and at waterline level is presented in Figure 3. The total calculation time on a 2.2 GHz laptop is less than 6 minutes for 9 iterations in the Newton-Rapson procedure, after which a sufficient convergence level was reached.

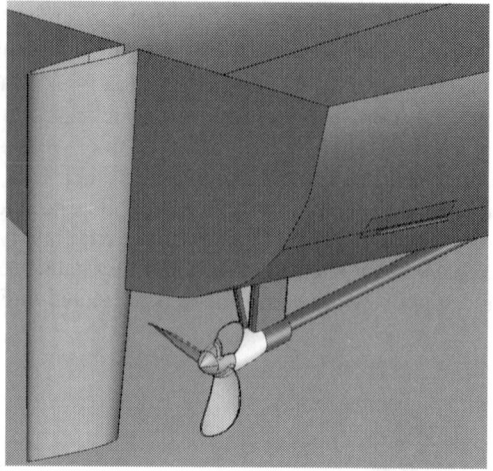

Figure 2: Detailed view of propeller and support with nearby anode (starboard site).

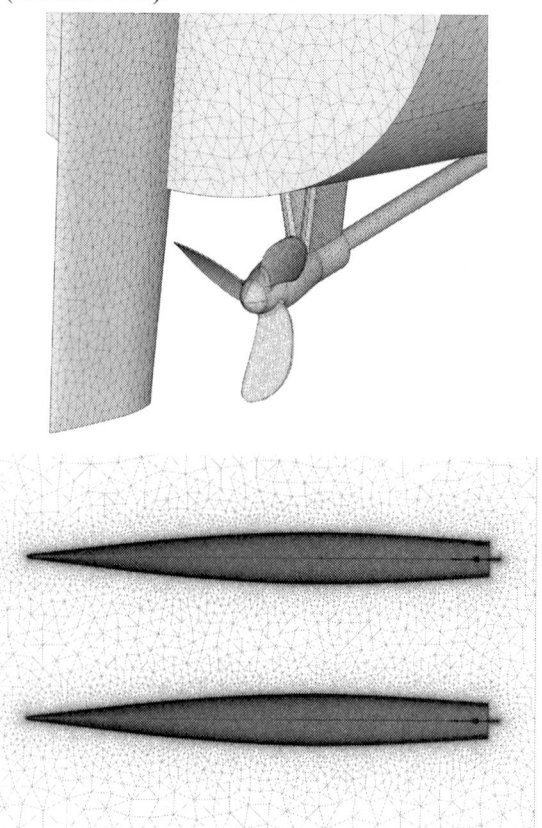

Figure 3: Zoom of surface mesh near one of the propellers (top) and at waterline level (bottom).

Figure 4:   "Off" potential distribution along hull - normal operation.

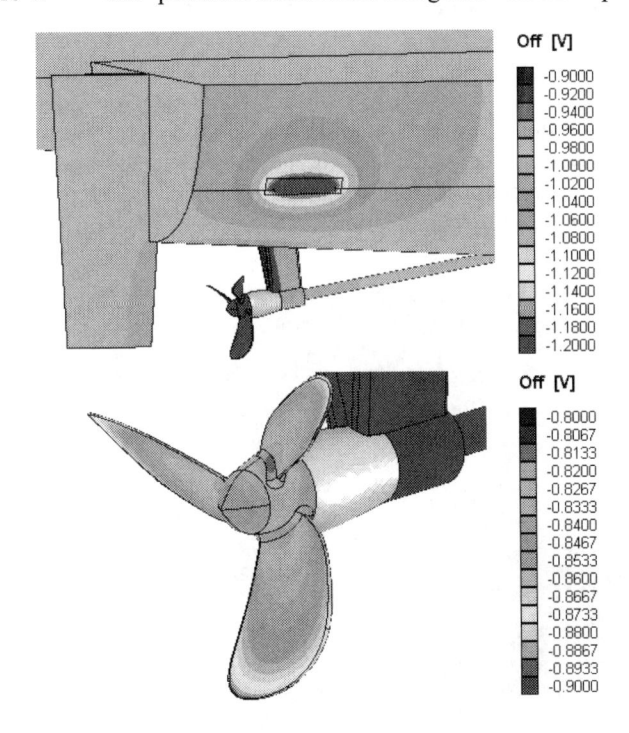

Figure 5:   "Off" potential distribution near anode (top) and propeller (bottom) - normal operation.

5 Results and discussion

5.1 Simulation of cathodic protection levels

In this simulation the normal situation is investigated. All four ICCP systems are active delivering a total protection current of 12 A. The calculated "off" potential distribution along the hull is presented in Figure 4. The average "off" potential is -977 mV which means that, taking into account the corrosion potential of -880 mV, an average polarization of the hull of -97 mV is obtained.

Figure 5 (top) shows the "off" potential along the hull near the stern anode at starboard side. From this picture it can be observed that there is a slight overprotection of the hull directly near the anodes with "off" potentials more negative than -1200 mV. Similar conclusions can be drawn from Figure 5 (bottom) presenting the calculated protection levels at the bare propeller. The obtained "off" potential at the hull near the propeller drops to -930 mV which means a polarization of only -50 mV which is below the minimum required limit of -100 mV.

5.2 Simulation of underwater electrical potential (UEP)

Besides for optimisation purposes, the software can also be used to calculate the underwater electrical potential (UEP) of the catamaran. The calculation of the electrical field in the seawater around the vessel is straightforward as it can be obtained directly from the gradient of the potential obtained in the nodal points of the FEM volume mesh. Figure 6 shows the calculated x-component of the electrical field on a cutting plane 2 m below the keel.

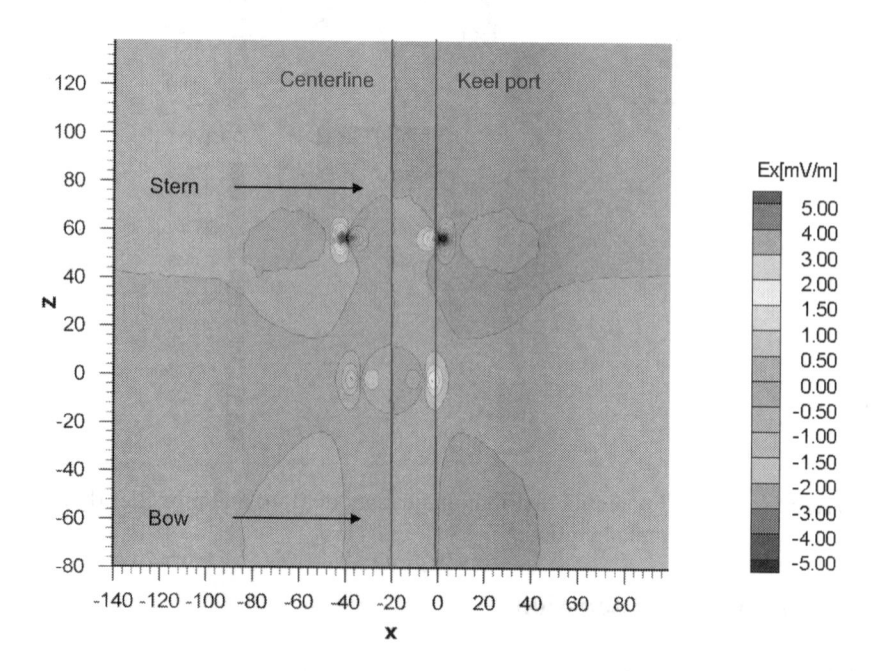

Figure 6: Underwater Electrical Potential UEP (Ex in mV/m) 2 m below keel.

6 Conclusions

In this paper a 3D software tool for the design and optimization of cathodic protection systems for submerged and buried structures has been presented. The software is entirely CAD integrated, and is based on the combination of a powerful volume mesh generator and the Finite Element Method.

Using this software, the protection level of a hypothetical marine vessel using impressed current cathodic protection (ICCP) systems has been investigated in normal operation. Optimisation is a logical step.

The software has also been used to calculate the underwater electric potential (UEP) of the catamaran.

References

[1] C.A. Brebbia, Boundary Element Techniques – Theory and Applications in Engineering, Springer-Verlag Berlin, Heidelberg, 1983.
[2] M.E. Orazem, J.M. Esteban, K.J. Kennelley, R.M. Degerstedt, Corrosion 53 (1997) 427.
[3] M.E. Orazem, J.M. Esteban, K.J. Kennelley, R.M. Degerstedt, Corrosion 53 (1997) 264.
[4] D.P. Riemer, and M.E. Orazem, Corrosion 56 (2000) 794.
[5] R. Adey, Topics in Boundary Element Research, vol. 7, Electrical Engineering Applications, chapter 3, Springer-Verlag Berlin, Heidelberg 1990.
[6] M. Purcar, B. Van den Bossche, L. Bortels, J. Deconinck, P. Wesselius, Corrosion 59 (2003) 1019.
[7] S. Aoki, K. Amaya, K. Gouka, Optimal cathodic protection of ship, Boundary Element Technology XI, pp. 345-356, 1996.
[8] V.G. DeGiorgi, Boundary Elements XIX, pp. 829-838, 1997.
[9] S. Diaz and R. Adey. A Computational Environment for the Optimisation of CP system Performance and Signatures, Warship CP2001, Shrivingham, UK.
[10] S. Diaz, R. Adey, Boundary Elements XIX, pp. 475-485, 2002.
[11] Elsyca CPMasterV2.0, User Manual, www.elsyca.com.
[12] F. Korsmeyer, D. Yue, K. Nabors, J. White, Multipole-Accelerated Preconditioned Iterative Methods for Three-Dimensional Potential Problems, Proceedings of BEM 15, pp. 517-527, 1993.
[13] O. Zienkiewicz, The finite element method in engineering science, McGraw-Hill, London, 1971.
[14] J. Deconinck, Current distributions and Electrode Shape Changes in Electrochemical Systems, Lecture Notes in Engineering 75, Springer Verlag Berlin, ISBN 3-540-55104-2, 1992.
[15] Athanasiadis A.N., Deconinck H., Int. J. Numer. Meth. Eng. 58 (2003) 301.

[16] Y.-G. Kim, Y.-C. Kim, Y.-T. Kho, BEM application for thin electrolyte corrosion problem.
[17] H. P. Hack, Atlas of Polarization Diagrams for Naval Materials in Seawater, 1995.

Validation plan for boundary element method modeling of impressed current cathodic protection system design and control response

E. A. Hogan[1], J. E. McElman[2], E. J. Lemieux[1], M. S. Krupa[1], V. G. DeGiorgi[1] & A. L. LeDoux[2]
[1]*Naval Research Laboratory, USA*
[2]*Naval Surface Warfare Center Carderock Division, USA*

Abstract

In the past 20 years, the US Navy has used the physical scale modeling (PSM) technique to design effective cathodic protection (CP) systems for the ships underwater hull, nickel-aluminum-bronze props and other hull components. In more recent years, a number of computational techniques have been devised in an attempt to fulfill this purpose. Physical models have proven highly adept at ICCP design, since modeled information provides a direct relationship to the actual hull and can be scaled up directly because of confidence in the physically measured data. Boundary element (BE) models have been correspondingly devised that mimic actual hull design and even the PSM layout, but because the BE method is a computational methodology, the calculated data requires systematic validation with a physical analog to insure confidence in the control response. BE literature has discussed design issues regarding mesh layout, intrinsic geometric complexities, accuracy of material response input, the predictive engineering design capability for zonal response, and assessment of electric field response. It does not significantly discuss the accuracy of the BE model calculated work predictive design capability, without the need for "tweaking," and ultimately a rigorous validation of both the mesh and resultant system design technique. This paper presents validation requirements, for any BE model, that is inherently robust enough to be used for CP design and control, and a proposed four-point methodology that will allow for the comprehensive validation of the BE model to predict the ICCP control responses and system performance behavior.

Keywords: boundary element, BEM, cathodic protection, ICCP, validation plan.

1 Introduction

The factors which affect the corrosion behavior and signature of an underwater ship hull are numerous, often highly complex and variable, and are in many cases directly related to the operational aspects of the ship. The seawater physio-chemical properties, biological growth and engineering aspects of the hull are all important components of the equation. The corrosion of the hull is most basically influenced by the nominal metallic components exposed to the seawater along with the chemical composition and properties of the bulk electrolyte (seawater), such as pH, dissolved oxygen, salinity and surface reactions. Directly controlling the kinetic behavior of these reactions are temperature, velocity and diffusion properties of the surfaces. Biologically, all surfaces foul with marine organisms, which further influence the surface properties of the metals involved.

Presently, the electrolytic PSM technique is required by the US Navy (USN) for the design of impressed current cathodic protection (ICCP) systems as well as for the verification of their operation [2,3]. The Naval Sea Systems Command (NAVSEA) has defined a design protocol [4] that describes the design criteria necessary for electrolytic PSMs. The PSM provides a means of testing the natural response of the system to changes in hull condition and coating damage. The limitations for the PSM technique lie more toward the mechanical and logistical elements, such as: (a) the size of components, and (b) the number of iterations required to test a vast number of hull conditions, control systems and environmental conditions.

The uses of boundary element (BE) computational methods are viewed as a potentially powerful tool for the design and analysis of corrosion prevention systems and resultant electric field generation for seawater structures. It follows, as these are calculated results, that the quality of the output is greatly affected by the detail of the model and the material input information [5]. In contrast with the electrolytic PSM, no equivalent protocol exists for building or validating BE models. In general, BE models are built using the dimensions of the full scale or physical scale items they are to represent. To provide the necessary engineering confidence, the validation process used for the computational model must also include steps to verify geometry, mesh, polarization curves, material response, zonal interactions, and system response. An accurate, validated BE model could greatly assist the ICCP design agent in evaluating and testing the control system architecture, system failure modes and system response under a wide variety of conductivities and temperatures. Without proper BE model validation, a high risk is assumed in control algorithm development, spatial ICCP relationships and assessment of overall life-cycle system operation and performance.

2 Four-step validation plan

A validation plan has been developed that breaks down the underlying components of the BE model and verifies each critical input aspect of the model for a defined ICCP configuration. After each validation step, the BE model is

then verified for use to accomplish specific analysis objectives. At the end of the complete validation process, the model will be established as being capable of providing critical data (i.e. current density distribution, potential profiles, etc.) on a given ICCP configuration. The validated model will also be a capable tool for the specified platform for the selection, sizing and locating of ICCP components and to determine the impact of future changes to the ICCP configuration or minor geometric features (major geometric changes would require revalidation of the model mesh). The validation plan consists of four steps as follows:
 (1) mesh and geometry validation,
 (2) polarization curve and material response validation,
 (3) design tool validation (optional), and
 (4) control algorithm validation.

2.1 Step 1: mesh and geometry validation

The first step requires validation of the model mesh. Geometry is critical to the function and response of ICCP systems. The mesh and geometry validation step ensures that the computational model is arranged properly and has the necessary geometric detail to accurately simulate shipboard conditions. Component sizes, locations, and current values for all anodes and cathodes from the PSM are provided to the computational modeler to be used as input to the BE model. In this step there is no need for characterization of the material responses. The anode and cathode currents are used to exercise the BE model to produce off-board electric signatures at the same distance electrolytic PSM measurements were taken. Comparison of off-board electric signature results that are within 20% root-mean-square (rms) error, result in a model validated for mesh and geometry. The resulting validated model can then be used reliably for simple failure mode studies without system response capability.

The 20% rms error comparison is based on several factors, some of which includes actual full scale measurement error from sensor accuracy and environmental noise. The derivation of this limit is not discussed in this work. This is not an arbitrarily chosen value. The acceptable error value will depend on the structure's end use and sensitivity of the sensors. It is possible that values for one type of vessel may be different than for other types of vessels or structures. The setting of the validation error criteria is part of the overall structure design/validation process.

2.1.1 Mesh details

The BE method is based on solving boundary Laplace integral equations and solves the equations in terms of potential flow. In the corrosion problem, only the vessel-electrolyte boundary needs to be included in the analysis. Electric field potentials and current densities are computed directly on the surface mesh [6]. The mesh can be defined by constant, linear, and quadratic line and surface elements. While quadratic elements provide the most precise results they currently should only be used in areas of high gradients, because the calculations can be time-consuming and memory-intensive. It is important, therefore, that BE models implement the proper combination of constant, linear, and quadratic elements, such that the model results are as accurate as possible within the time and memory constraints. Finding and defining this proper balance is not trivial. The effective BE mesh must be developed that represents all significant surfaces that can influence both cathodic demand and shielding of current. Areas where high surface potential field gradients are expected, need to have a more refined mesh to capture these effects (Figure 1).

Figure 1: Typical BE mesh for surface ship. Note the concentration of elements in the stern, at bilge keels and near bulbous bow.

2.1.2 Geometry details

In past modeling efforts, it was assumed that approximations of various components, such as a disk representing a propeller, were sufficient to define the geometry of a vessel. Upon further investigation, however, the shape and pitch of the propeller blades make a significant difference to both the on-board potentials, as well as, the off-board electric signature and computationally must be taken into account to yield accurate results. Details of all geometric features have been an issue in modeling on-board potentials. Early work identified the need for considerations of such features such as the bilge keels while more recent work has examined the geometric details and impact of small features, such as the angle and orientation of the bilge keel with respect to the hull [7]. Results showed there was no doubt that small geometric features are also of importance and necessitates the need for even more accurate and complex computational models. Another example related the complex shadowing effects where the computational modeling compares the performance of a static orientation with a stylized geometry representing spinning blades and compares the net on-board potentials [7]. Further, the representation of confined spaces, including ballast tanks and sea chests, and bow thrusters have proved important. Innovative means of defining confined spaces within the limitations of the computational modeling software must be employed. Dual elements were eventually designed by Computational Mechanics to address this issue of thin-walled areas that were exposed to electrolyte on both sides. The size, shape, and presence of grates also impacted the off-board electric signatures produced by the BE model. Figure 2, details some significant surface geometries that must be incorporated in the BE mesh to provide an accurate surface geometric representation of the structure.

As shown in Figure 3, the mesh differences between a stylized prop and the complexity of a detailed prop with multiple blades and contours is significant. The multiple blade structure offers more surface area for cathode demand. To provide the necessary surface area on the disk prop and maintain the correct geometric diameter the thickness of the prop had to be increased. The disk prop will not provide the geometric window effect or shadowing between blades nor the overall cathode geometric extent if the surface areas are equal. Off-board electric field data acquired from a PSM experiment is compared in Figure 4 with BE model electric field data using a disk propeller verses a bladed propeller. It is clearly shown in the data that the match to PSM data is much closer with the correct prop model geometry.

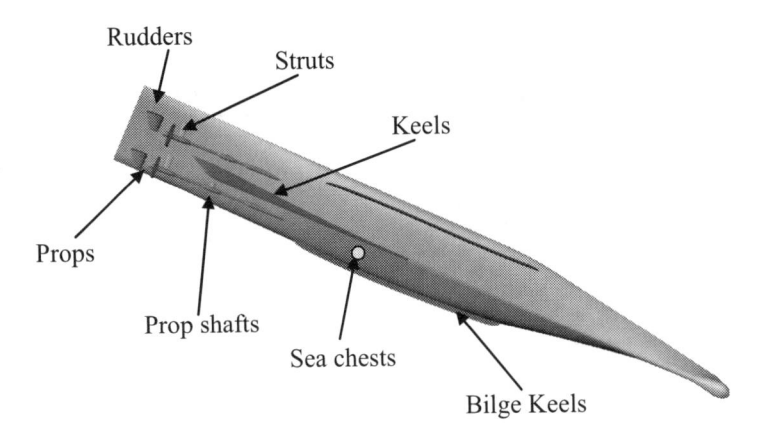

Figure 2: Surface model detail of known significant geometries.

Figure 3: Mesh of stylized disk prop (left) and mesh of actual prop (right) showing accurate number of blades and contours.

2.2 Step 2: polarization curve and material response validation

The polarization curve and material response validation step ensures that the computational model accurately reflects shipboard material characteristics and response, as well as relevant material interactions. During step two of the BE model validation process, the cathode and anode sizes, locations, anode currents, material polarization curves, and a complete hull potential profile from the PSM are provided to the computational modeler to be used as input to the BE model.

This information is used to exercise the BEM model to produce cathode currents, as well as off-board electric signatures at the same distance PSM measurements were taken. Comparison of the BE model and PSM off-board electric signature results within 20% rms error *and* the agreement of cathode currents within 10%, results in a model validated for polarization curve and material response. The resulting validated model can then be used reliably for complex failure mode studies with system response capability.

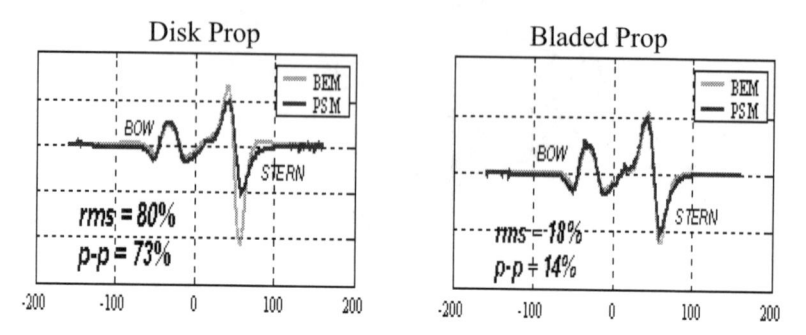

Figure 4: Example of differences in BE model electric field analysis as a function of prop geometric detail and compared to PSM measured data.

Typically, BE computational models utilize polarization curves, which are plots of potential vs. current density, to mathematically describe the dependence of potential on current density. The problem of course, then, is that the quality of the BE model and the validity of its results are directly dependent on the accuracy of the polarization curves utilized. However, the issue is further complicated, since the marine and ship materials not only have individual polarization characteristics, but these are significantly affected by galvanic interactions, surface kinetic effects, biological effects and electrolyte composition. The central electrolyte composition effect for Navy ships are salinity changes, but these effects are well understood and are not discussed in the current document.

2.2.1 Galvanic effects

Since ship hulls are comprised of a variety of materials, galvanic interactions are anticipated. The most common combinations include steel (hull plate and structure), nickel-aluminum-bronze (propellers), copper-nickel (sea chests) and zinc or aluminum (sacrificial anodes). Note, with respect to the latter, it is not uncommon to have sacrificial anodes in free flood spaces adjacent to the ICCP protected hull which can interact with the control response of the system. The net effect of these material combinations is to shift the polarization curves of the attached materials. These shifts can occur in either or both the current density and potential direction and is a strong function of the polarization resistance of the material, the relative surface area ratios and the proximity of the materials.

2.2.2 Surface kinetic effects

Surface kinetics, at least within the context of this document, is primarily a function of dissolved oxygen diffusion to the cathodic surfaces. Increased availability of dissolved oxygen to the cathode site results in higher cathodic reaction rates and increased current demand to support the oxygen reduction reaction, eqn. (1):

$$\frac{1}{2}O_2 + H_2O + 2e^- \rightarrow 2OH^-. \quad (1)$$

The effect on the polarization behavior is that the cathodic polarization curve becomes independent of current density at the limiting current density (i_L) and the point of oxygen depletion. The ships speed directly affects oxygen availability to the steel and results in a concentration polarization effect whereby i_L is shifted to right with increasing ships speed until oxygen diffusion and availability at the cathodic site is no longer the corrosion rate limitation. However, throughout the concentration polarization region increased i_L results in increased CP requirements. Note that this effect is mimicked in the PSM through bubbling of air beneath the model to simulate ship motion and increased oxygen availability.

Surface kinetics is also affected by CP deposits; cathodic polarization in seawater promotes the formation of calcareous deposits. Oxygen reduction and subsequent hydroxide formation at the cathode surface result in an increased pH at the cathode/seawater interface, such that calcium carbonate precipitates according to equations 2:

$$HCO_3^- + OH^- \rightarrow H_2O + CO_3^{-2} \quad \text{and} \quad Ca^{+2} + CO_3^{-2} \rightarrow CaCO_3 \quad (2)$$

At higher current densities, magnesium hydroxide can also form. Calcium carbonate and magnesium hydroxide are important, because they serve to limit diffusion of oxygen to the cathode. This results in enhanced oxygen concentration polarization and a lower net current density to maintain the same polarization level of CP [8–10]. The formation of these protective films significantly affects the material response and overall ICCP performance and control. This effect is included in the PSM through an initial ageing period in which calcareous films are deposited on the cathode surfaces until a specified maintenance current density is achieved. Any useful BE model must be able to accommodate into the polarization curves both the formation of calcareous films and the effects of ship speed on surface kinetics to make accurate predictions of ICCP performance and response.

2.2.3 Biological effects

Materials can change their polarization behavior just with exposure to seawater for extended periods of time, as shown in Figure 5. Super duplex stainless steel samples (AL6XN) were exposed to natural seawater and electrochemical tafel curves were conducted on the samples after 24 hours and 1 month exposure periods. The figure shows the significant change in the cathodic polarization behavior and electropositive shift in the corrosion potential. Notably, this effect is exhibited by a large range of alloys, predominantly passive alloys, and is termed "ennoblement". While the exact mechanism of ennoblement has not been identified, biological film growth is considered to be directly involved. This polarization curve is an example of how important it is to verify the material property curves being used in the BE model to account for the conditions of interest.

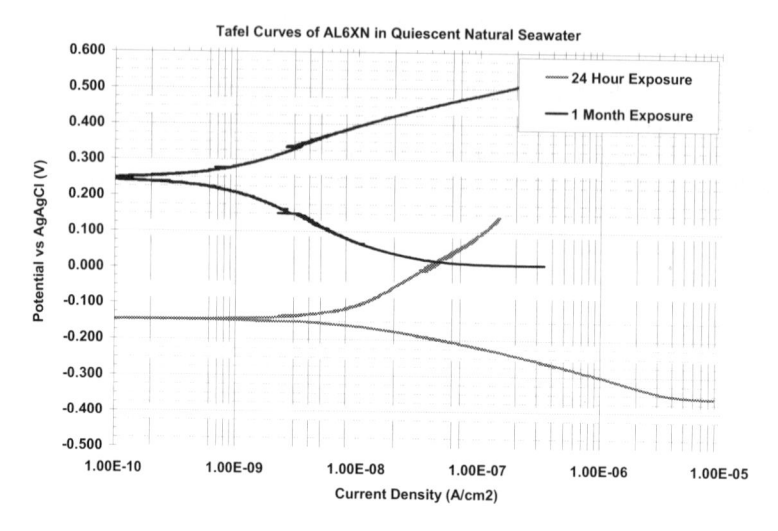

Figure 5: Electrochemical tafel curves on alloy AL6XN after exposure to seawater.

2.3 Step 3: design tool validation

The design tool validation step is considered optional as it allows for assessment of CP and signature requirements under various conditions outside of the initial ICCP system design. Cathode locations and overall percent damage; anode sizes, locations, and current values; and only the ICCP system controlling and sense reference cell potentials from the PSM are provided to the computational modeler to be used as input to the BE model. This information is used to exercise the BEM model, under different hull coating damage conditions, to produce an on-board hull potential profile, cathode currents, and off-board electric signatures at the same distance the PSM measurements were taken. Comparison of off-board electric signature results within 20% rms error, agreement of cathode currents within 10%, *and* agreement of hull potentials within ± 10 mV results in a model validated as a design tool. Completion of this validation step allows the user to make changes to the BEM model ICCP design, such as, anode or reference cell location changes, painted versus unpainted surface changes and paint damage condition changes. Additionally, the validated BE model can be used to evaluate minor changes in geometry and effects of adjacent sacrificial systems.

2.4 Step 4: control algorithm validation

A properly designed computational model should respond to system input appropriately based on the applied control algorithm, whether it is lead-lag compensation, proportional-integral-derivative (PID) compensation, etc. A basic control system is designed to provide an output that closely follows a desired value (see Figure 6). The system output should be stable, as well as, reach and maintain steady-state within the desired timeframe and fashion. Determining controller parameters is not always straightforward, as system transfer functions may not be available. In such cases, final parameter selection may be completed during initial system operation.

Accordingly, the final step to BE model validation requires the control system response to be evaluated. The control algorithm validation step ensures that the computational model can sufficiently predict system response with the desired control algorithm in place. As with step 3, anode sizes, locations, and current values in addition to controlling and sense reference cell potentials from the PSM are provided to the computational modeler to be used as input to the BE model. This information is used to exercise the BEM model under varying control parameters to produce a hull potential profile and off-board electric signatures at the same distance the PSM measurements were taken. Comparison of off-board electric signature results within 20% rms error and agreement of hull potentials within ± 10 mV result in a model validated for control. Completion of the control algorithm validation step allows for the BE model to be exercised under a wide range of scenarios and results in a platform specific computational model that allows for calibration with only ICCP system component inputs.

While correct control algorithm operation under ideal conditions is important to verify, it is equally important to determine how the controller and system respond under varying conditions, such as conductivity changes or component failures. Verifying correct control system response under such conditions is imperative to validating a BE model.

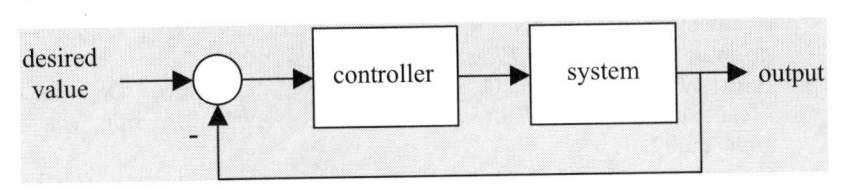

Figure 6: Basic controller scheme.

3 Summary

To develop confidence in a comprehensive computational model that addresses all aspects of the complex nature of the corrosion response for design, life-cycle analysis, and control system development, a rigorous validation process is necessary. The BE model must be shown to have the correct geometric detail, cathodic and anodic polarization inputs, correct response to changes in configuration and have the correct system response to test control algorithms. An error in any of these stages can lead to progressive error in the completed model. The validation process is progressive, such that the model can be validated at each step and success will provide a model that can be useful for specified tasks. A four step process has been provided that identifies each significant aspect of the computational model that needs to be validated.

References

[1] DeGiorgi, V.G. & Hogan, E.A., "Experimental vs. Computational System Analysis", *Simulation of Electrochemical Process*, WIT Press, pp. 37-46.

[2] Lucas, K.E., Thomas, E.D., Kaznoff, A.I. & Hogan, E.A., "Design of Impressed Current Cathodic Protection (ICCP) Systems for U.S. Navy Hulls", Designing Cathodic Protection Systems for Marine Structures and Vehicles, *ASTM STP 1370*, H. P Hack Ed., American Society for Testing and Materials, West Conshohocken, PA., 1999.

[3] Lucas, K.E., Thomas, E.D. & Hogan, E.A, "Physical Scale Modeling for the Design of Impressed Current Cathodic Protection (ICCP) Systems", *UMIST Advances in Corrosion and Protection Conference*, Paper # 199, Manchester, England, July 1992.

[4] US Navel Sea Systems Command, "Ships Cathodic Protection, Design Calculations, Design Requirements Manual", *NAVSEA TECHNICAL PUBLICATION: T-9633-AT-DSP-010/ALL USN*, Jan 2007. (In press)

[5] DeGiorgi, V.G., Lucas, K.E. & Thomas, E.D., "Scale Effects and Verification of Modeling of Ship Cathodic Protection Systems", *Engineering Analysis with Boundary Elements*. Vol. 22, 41-49.

[6] Moy, E. & Holmes, J., Mathematical models (Chapter 2). *Submarine Electromagnetics Tutorial (Unclassified)*, ed. Garduno, G., Rogalski, J., Izat P. & Kasper, R., NSWCCD-TR—2002/12, pp. 42-63, 2003.

[7] DeGiorgi, V.G. & Wimmer S.A., "Geometric Details and Modeling Accuracy Requirements for Shipboard ICCP Modeling", *Engineering Analysis with Boundary Elements*, Vol. 29, 15-28 2005.

[8] Hartt, W.H., Culberson, C. and Smith, S.W., "Calcareous Deposits on Metal Surfaces in Seawater – A Critical Review", *Corrosion*, vol. 40, 1994, p.609.

[9] Comeaux, R.V., "The Role of Oxygen in Corrosion and Cathodic Protection", *Corrosion*, vol. 8, no. 9, 1952, pp.305.

[10] Wolfson, S.L. and Hartt, W.H., "An Initial Investigation of Calcareous Deposits upon Cathodic Steel Surfaces in Seawater", *Corrosion*, vol. 37, no. 2, 1981, p.70.

Predicting the performance of cathodic protection systems with large scale interference

R. Adey[1], J. M. W. Baynham[1] & T. Curtin[2]
[1]*C M BEASY Ltd, Ashurst Lodge, Southampton, UK*
[2]*Computational Mechanics Inc, Billerica, Ma, USA*

Abstract

Understanding the interactions between cathodic protection systems is becoming more complex particularly in the marine and offshore industry. The search for oil and gas in deeper waters and remote environments has resulted in engineering designs with many components on the sea bed as well as at the sea surface. Traditional design methods are inadequate in these situations as interactions will occur between the Cathodic Protection (CP) systems of the components, and significant currents can develop in the connecting flow lines.

In this paper recent developments made in the BEASY Corrosion Modelling software are presented which enable these systems to be simulated. Examples are presented showing how computer modelling can play a major role in predicting the level of interaction and providing the key data for the design of the system and its integrity management.

1 Introduction

Boundary Element Methods (BEM) have been used to simulate the behaviour of cathodic protection systems since the late 1970s [1]. As the name implies, the method requires creation of elements, but only on the boundary (i.e. surfaces) of the problem geometry. The BEM is used to mathematically model the potential drop in the electrolyte represented by the Laplace equation. In many applications it is sufficient to couple these equations with the equations representing the electrochemical electrode kinetics on the metallic surfaces to form a system of equations capable of simulating the potential fields and current flowing in the electrolyte.

It was common in the early models to make the assumption that the internal resistance of the structure (through which the current returns) is very small compared with that of the electrolyte and the electrode kinetics. Therefore it was possible to ignore the metal resistance when formulating the equations and hence there was no IR drop in the return path. Pipelines present a typical situation where this assumption is not valid as over a long pipeline there is significant drop in the potential due to the internal resistance of the pipeline metal. In new large scale deep water Oil and Gas developments not only are long pipelines present but there are hundreds of individual flow lines and connections each introducing a new electrical connection path which must be considered.

Figure 1: Example of a modern deepwater Oil and Gas development.

Standard BEM modelling tools are inadequate in these situations as interactions will occur between the Cathodic Protection (CP) systems, causing significant currents (and therefore voltage drops) to occur in the flow lines connecting the components. In this paper recent developments made in the BEASY Corrosion Modelling software are presented which enable these systems to be modelled.

2 State of the art

Numerical methods have been widely used in the corrosion field since the early 1980s when IMI Marstons employed them to help model the performance of the impressed anodes to be used on the Conoco TLP platform for the North sea [2].

They have been successfully compared with physical scale modeling experimental results and data obtained from tests performed on full size ships [3]. DeGiorgi et al. studied a 3D model of a U S Navy CG class ship [4] using boundary elements. The potential profiles for reference cell readings of -0.85 Volts Ag/AgCl showed very good agreement between experimental and computational results. The difference between experimental and computational results was 6% for total current values, 5% for Amps to the propeller and 10% for Amps to the docking blocks. The general conclusions were that, *boundary element methods are a viable technique for determining marine corrosion parameters.*

Studies of interference between CP systems of a ship and nearby dock have been made [5], in which it was assumed the metal current paths had zero resistance.

3   Internal resistances

From a mathematical point of view the internal resistances can be viewed as a simple electrical circuit where the metal objects are connected by resistances or devices such as diodes. Figure 2

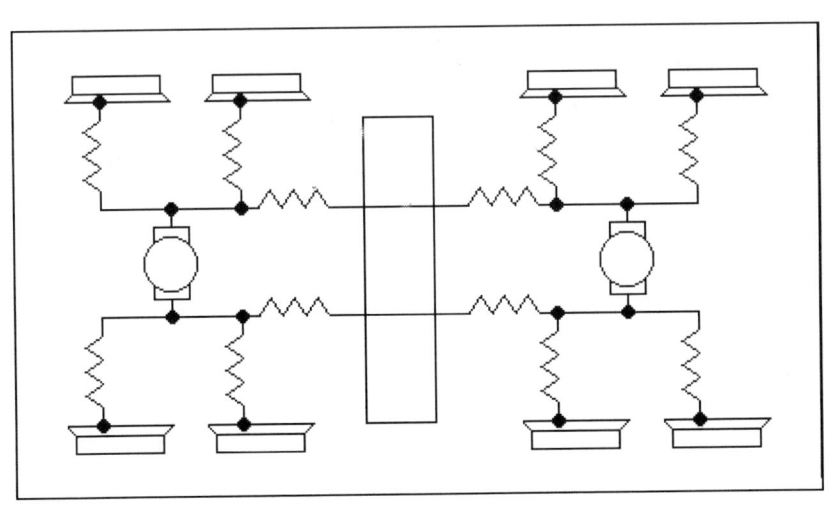

Figure 2:   Example internal resistance circuit.

Each of the objects in the circuit represents a metal object being protected by a CP system. The resistances are the resistances of the connecting pipelines or flow lines connecting the different components in the structure.

If we consider a simple case of a FPSO connected to a buoy by an offloading line. The FPSO may be protected by an ICCP system and the buoy by a sacrificial anode system. If they are not electrically connected then there is unlikely to be interference as the only electrical return path is though the sea water itself. If they are connected by the offloading lines an electrical circuit is created which allows for example current to flow from the ICCP anodes on the FPSO to the buoy and current to return along the offloading lines.

This can be expressed mathematically as:

$$I_{FlowlineConnectedToBuoy} = I_{BuoyAnodes} + I_{BuoyMetalSurfaces}$$

Similarly for the FPSO

$$I_{FlowlineConnectedToFPSO} = I_{FPSOAnodes} + I_{FPSOMetalSurfaces}$$

For equilibrium the current flowing along the Flowline to either the buoy or the FPSO must balance.

The currents flowing into or out of the metal surfaces on the buoy and the FPSO can be determined by integrating the normal current density calculated by the boundary elements over the metal surfaces of the structure.

Therefore the internal circuit resistance equations and the boundary element equations representing the sea water can be solved together to predict the potentials and currents flowing in the system.

3.1 User interface

In a complex situation there may be hundreds of individual circuits and connections to be modelled. Therefore a Graphical User Interface (GUI) has been developed to simplify the task of the user in creating the model and to provide a visual means of auditing the connections.

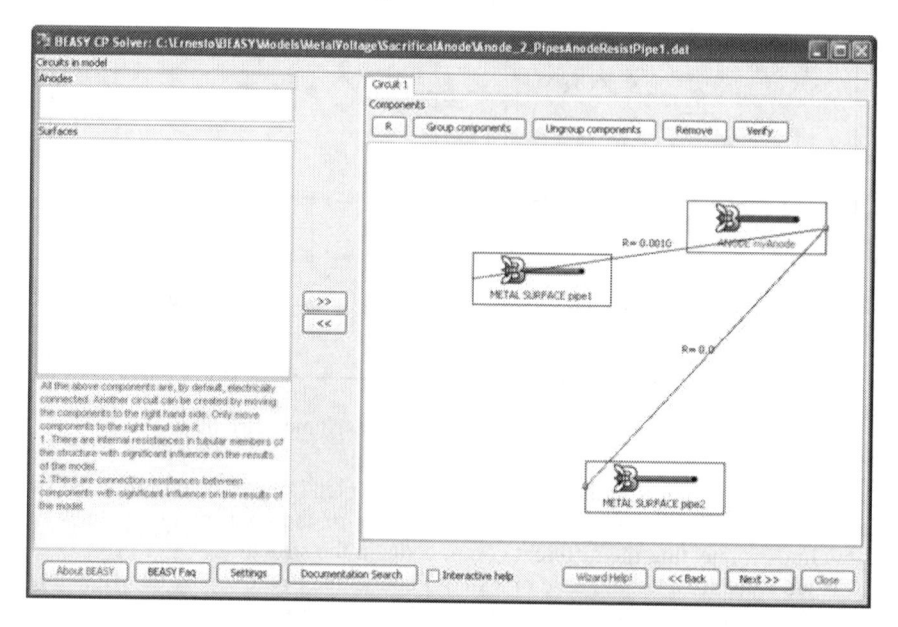

Figure 3: The Circuit editor enables electrical connections to be made between parts of the model and to specify their electrical resistance.

Figure 3 shows a view of the circuit editor. Metal objects can be selected and electrical connections made simply by picking them with the mouse and specifying the type of connection. Structures can also be grouped together if their electrical connections have negligible resistance.

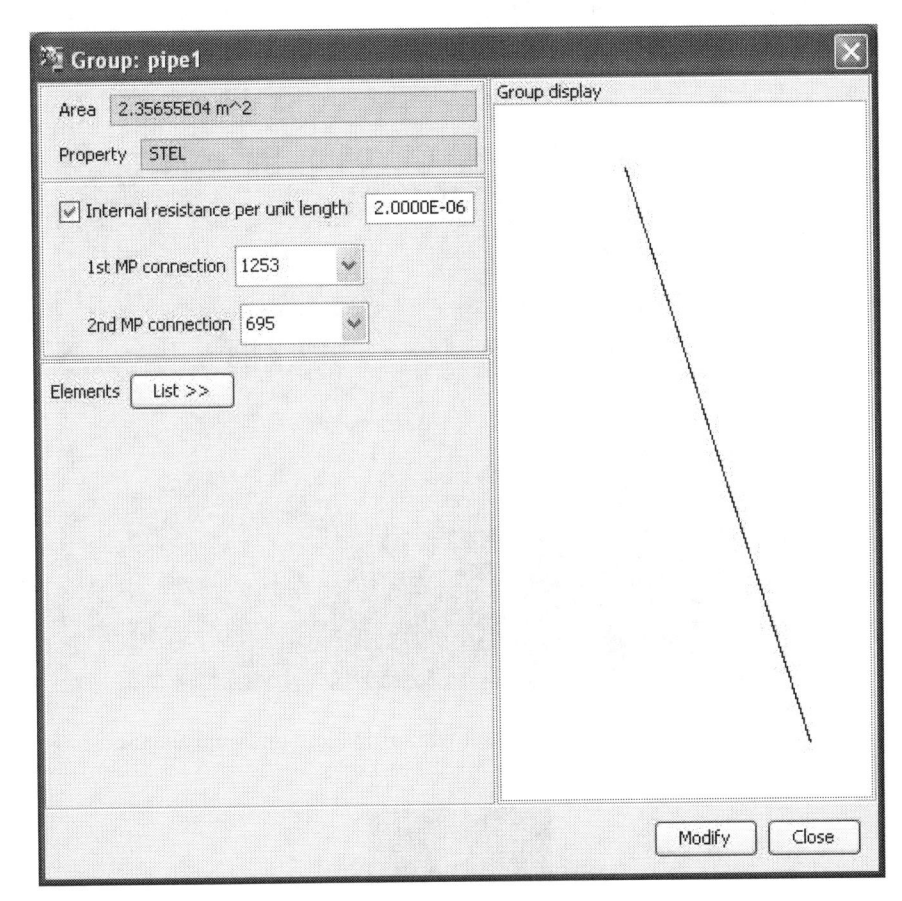

Figure 4: Pipelines properties can be viewed and specified by clicking on the object to display the properties window.

The internal resistance of pipelines and their connections can be specified in a similar way, as shown in Figure 4

4 Application

The technology has been used to simulate performance of a number of Oil & Gas facilities ranging from Jacket type structures with controlled anodes (i.e. diodes) to deepwater applications with subsea facilities and pipelines. In the application presented here the investigation focuses on an offloading buoy and its anchor chains.

The anchor chains present an interesting CP problem because there is significant attenuation of the current flowing along the chains due the very high internal resistance caused by the chain link to chain link contact. Therefore near the buoy the chain will receive significant current from the anodes but this will drop off significantly within 25–30m.

The model of the buoy and its anchor chains is shown in Figure 6.

Figure 5: Typical buoy layout with offloading lines and anchors.

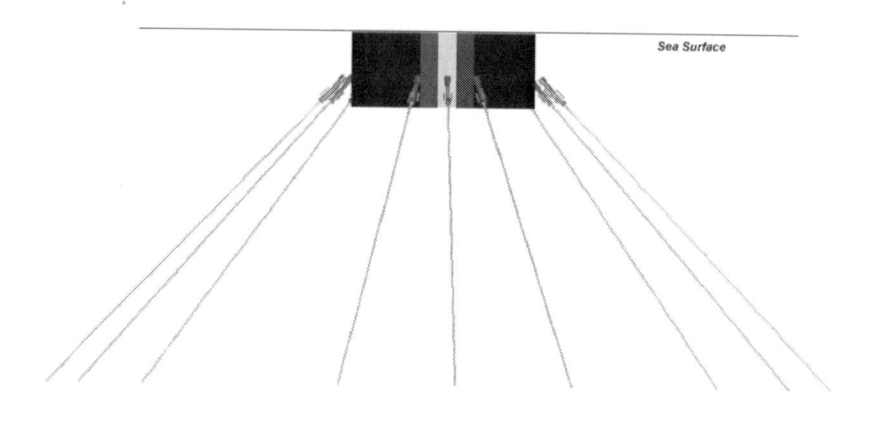

Figure 6: Model of the buoy and the anchor chains.

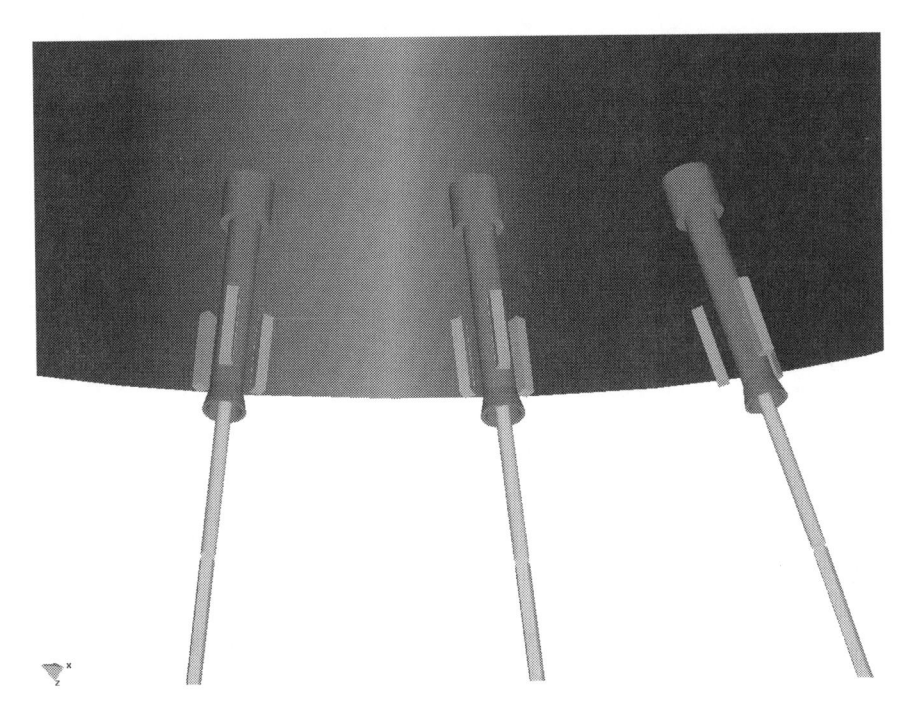

Figure 7: Detailed view of the connections between the buoy and the anchors.

Figure 7 shows the buoy and the anchor connections. The model includes a full 3D model of the connectors (Seen as cylinders in the model) including the internal geometry because the anodes not only need to protect the outer metallic surfaces but also the internal structure. The chains have been electrically connected to the buoy and have a high internal electrical resistance. In the model of the complete system the lines connecting to the FPSO would also be included as well as the FPSO and the subsea systems.

The predicted protection potentials provided by the CP system is shown in Figure 8.

The attenuation of the current on the anchors can be clearly seen in figure 9.

5 Conclusions

A method has been demonstrated which allows coupled calculation of the metal voltage drop in pipelines, the potential field in the electrolyte, and the electrode kinetics.

Since the metal voltage drops which occur in practise may be 100 mVolts or more, the determination of such coupled solutions with proper account taken of metal resistance is a very important part of CP design.

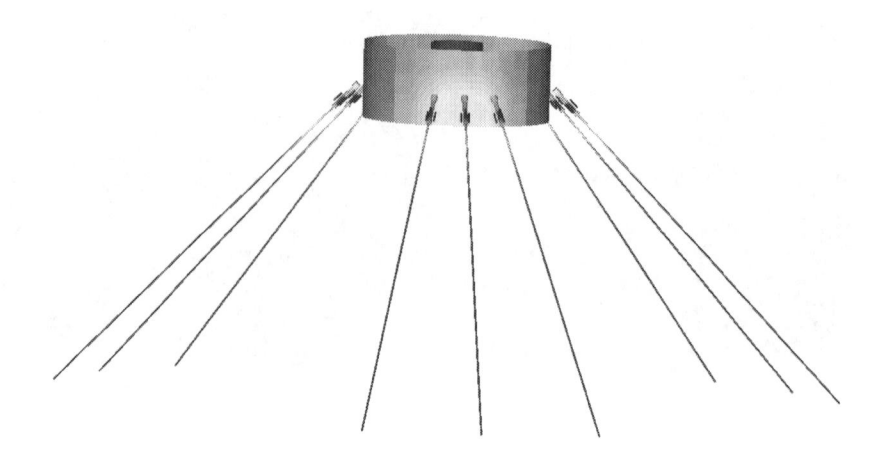

Figure 8: Protection potentials on the buoy and anchors.

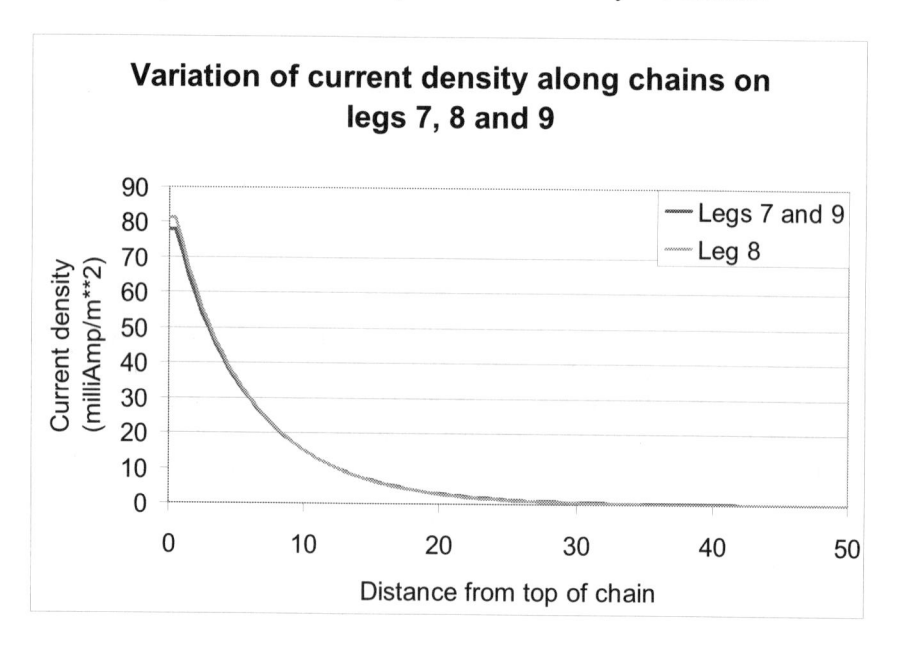

Figure 9: Attenuation of the current density on the anchor with distance from the buoy.

References

[1] Brebbia C.A, Boundary Element Method for Engineers, Pentech Press. London, 1978.

[2] Danson, D.J. and Warne, M.A. Current Density/Voltage Calculations Using Boundary Element Techniques, Corrosion/83. 1983.
[3] Thomas, E. D., Lucas, K. E., and Parks, A. R., Verification of Physical Scale Modeling with Shipboard Trails, Corrosion 90, Paper 370, National Association of Corrosion Engineers, Houston, TX, 1990.
[4] V.G. DeGiorgi, A. Kee, E.D. Thomas. Characterization accuracy in modelling of corrosion systems. Boundary Elements XV. Vol. 1: Fluid Flow and Computational Aspect. Editors: C.A. Brebbia, J.J. Rencis. CML Publications, Southampton, 1993.
[5] Adey R A, Santana Diaz E. Computer Simulation of the Interference Between a Ship and Docks Cathodic Protection Systems, Electrocor 2005, Cadiz, Spain. May 2005.

Cathodic protection application possibilities in activity building of overhead lines: protection of armature tower foundations

A. Muharemovic, I. Turkovic & S. Bisanovic
Faculty of Electrical Engineering, Sarajevo, Bosnia and Herzegovina

Abstract

This paper presents one practical method for the corrosion protection of armature tower foundations at overhead lines – the cathodic protection concept. The paper presents possible mechanisms of steel corrosion in concrete, as well as a discussion of the application of two dominant concepts: impressed current cathodic protection system and the system with galvanic anodes. The first system has the better performance and it is proposed as very efficient protection. In practice, for ensuring a significant saving in the supply of electrical energy from the distribution network, the application of solar energy system as support for impressed current cathodic protection system has been proposed. Also, real applications are been discussed with a presentation to satisfy need criterions and standards for each component of this system.
Keywords: cathodic protection, armature tower foundations, impressed current cathodic protection system, criterions and standards.

1 Introduction

With regards to the condition of usual building standards for making concrete (structure, additives), one can say that iron armature in concrete is corrosion inert. However, one can conclude that the corrosion of the armature in concrete is possible on certain conditions. The characteristic case is an iron armature in the footstep of tower foundations at overhead lines. Each footstep (four in total) has an iron anchor and an iron armature to make stronger concrete. On other hand, each footstep has its own ground (iron or copper). Contacts from different materials, as iron in concrete, iron in earth, copper in earth, results in the corrosion of the more negative metal.

The first case is when the ground of tower and the armature of the footstep are made of same material, for example steel. Then the steady-state steel potential armature is less negative than steel potential in the earth. This results in the corrosion of the ground as the steel in earth act as the anode.

The second case is in the phase of building of overhead lines when the armature in footstep of the tower foundations is made of iron and ground of tower is made of copper. In this case, the iron armature in the concrete takes the role of the anode.

The next case that is independent of the ground type is the phenomenon of corrosion caused by the presence chloride in concrete. Chlorides in concrete can be the result of technology in concrete production. Chlorides are often present near to the armature in the upper layers – near the surface of the earth.

It is very important is to standardize a practice in application of cathodic protection of the armature in the footstep of tower foundations. A real practical solution is to use an impressed current cathodic protection system when needed great protection currents. For uniform protection current distribution, it is appropriate to use ribbon anodes typically installed into slots cut in the concrete [5–7].

A shortage of quality criteria for the evaluation successful completion of cathodic protection of armature tower foundations and relevant standards, are main reason for initiation theoretical and practices analysis in application this type of protection on vital objects, as are overhead lines.

2 Possible mechanisms for steel corrosion in concrete

When steel armature in concrete is analysed, with the assumption that the presence of ground is neglected, one can conclude that is the most important reason steel corrosion in concrete is due to the presence of chlorides. With presence of chlorides, steel is depolarized even in an electrolyte with high pH values.

The potential of iron in phase destruction of passive layer is moved towards more negative values. Parts of the armature near to the surface earth with presence chlorides are in metal connection with armature from deep layers of construction (foundation) where chlorides do not reach. For this reason, those corrosion microelements under the influence of a chloride armature will be anodes, with potential -0.5 (V) to a copper sulfate electrode (CSE) [3]. This part will be subject to corrosion process and the armature from deep layers is a cathode with potential -0.1 (V) to CSE and it will not corrode. Surface zone and deep zone can change places.

Chlorides can be included in concrete in many different ways. Pebble aggregate can contain chlorides, as can water, and a large amount of chlorides come in concrete are due to the use of calcium chloride as catalyst. External sources of chlorides include the sea atmosphere, with contact with seawater, etc. Corrosion element can be formed very often although are some parts of steel have good ventilation. With great rate cathode and anode surface and with good cathode ventilation, on anode area can appear intensive corrosion process. In those analyses it should be calculated with concrete resistivity 100÷2000 (Ωm) [4].

When steel armature in concrete is analysed, with presence of ground made as strip steel coated with zinc, then, without chlorides, one can conclude that is value of steel potential in concrete about –0.2 (V) to –0.3 (V) to CSE, while value of strip coated with zinc potential is more negative for at least 0.2 (V) or more [2]. In this case anode role is assigned to ground. Corrosion current depends of total resistance between ground and armature in concrete.

When steel armature in concrete is analysed, with presence of ground made as copper rope, then corrosion galvanic element copper/steel appears. Steady-state potential of copper in electrolyte is about +0.3 (V) to CSE, while value of potential armature in concrete about –0.2 (V) to –0.3 (V) to CSE [3]. The potential difference (operational voltage) between those electrodes is at least 0.5 (V), resulting to corrosion of steel in concrete as anode (more negative element).

In all three cases can be concluded that is, on one side, armature in concrete is connected with ground (over earth surface), and other side of electrical circuit is closed through earth. This connection, over earth surface, can be realized by appropriate resistance R (concrete resistance if direct contact do not exist), or by direct contact armature and anchor footstep of tower [1].

3 Treatment of juncture copper/steel

Independent to installing ground closeness in regard on footstep of tower foundations one can conclude that transition resistance of ground toward earth is small (this is one main characteristic of good ground). It increases current of corrosion element copper/steel armature in concrete. In this element role of anode is taking over by steel armature in concrete [3].

Here is need to point on electrical contact of clamp anchor tower and armature footstep of tower foundations. If no mutual direct contacts all elements of armature or armature with anchor exists, in parts of armature additional anode places are made (place where corrosion current leave one part of armature).

With aspect of corrosion, a combination of making ground by copper ropes for armature footstep of tower foundations is not good solution. A good variant for armature in concrete is coat with zinc strip, but this is not good solution for corrosion of ground. Those facts shows that optimal solution is application of cathodic protection system both ground and armature in concrete.

4 Possible ways of corrosion protection

For application of cathodic protection both ground and armature in concrete (meaning that electrical continuity of armature and her connection with ground exists) two methods can be used [1]:
- impressed current cathodic protection system and inert anode;
- system with galvanic anodes.

A galvanic cathodic protection system consists of sacrificial anode(s) fixed to the underground metallic object and provides specified wiring for an inspection station installed near the surface of the ground. Galvanic systems have limited life spans during which the sacrificial anode will continue to degrade and protect the underground object. When the sacrificial anodes are no longer capable of protecting the object, they will lose their protection and begin to corrode. Galvanic anodes with very low operational voltage and concrete with very low electrical conductivity are unsuitable for this type of cathodic protection of steel in concrete and it has rare application.

The impressed current cathodic protection system usually provides electrodes of a much longer life span than a sacrificial anode. These systems include a rectifier that converts the alternating current power source to a direct current that is properly calibrated to provide the required protection. Since the power source is delivered to the electrode and is not generated by the degradation of the electrode, the power supply to the electrode may be recalibrated to provide additional power, when needed, as long as the electrodes are still functional. These systems are imperative in protection of armature tower foundations at overhead lines [2].

A big problem in application of cathodic protection on reinforced concrete constructions is high resistivity of concrete and small volume of electrolyte through protection current can flows to all parts of metal construction. Usually anodes are attached on concrete with layer cement mortar coating (Figure 1). Distance between anode and armature in concrete must be less then 1 (cm).

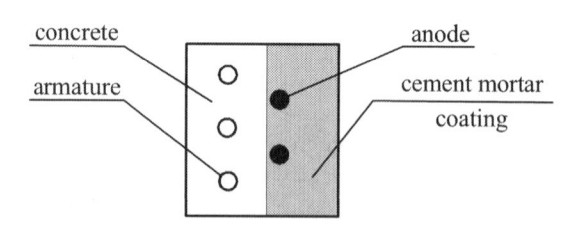

Figure 1: Concept anode and armature in concrete.

5 Real application

Often in practice anode system with titanium anodes (titan mesh) coated with mixed metal oxides (mixed metal coated titanium ribbon anodes) is applied. In Figure 2 principle scheme of anode system with mesh titanium anodes is illustrated (where do not take into consideration grounding system of tower).

Anodes are attached on surface concrete construction and covered with cement matter. If anode mesh is used, than uniform protection current distribution can be achieved, and mesh is easy to install on any surface. On this way installed titanium anodes can be calculated for life span more than 50 years.

In Figure 3 principle scheme of outputs from transformer rectifier for possible cathodic protection of armature tower foundations at overhead line and appropriate ground is illustrated. Positive and negative outputs must be enough for own supplying of each footstep of tower foundations. In electrical circuit must be installed regulation resistors in order to limitation of cathodic protection current. It is because that much less protection current of armature concrete then ground is needed. Ground of each tower foundation must have own secondary direct electrical circuit without regulation resistors [3].

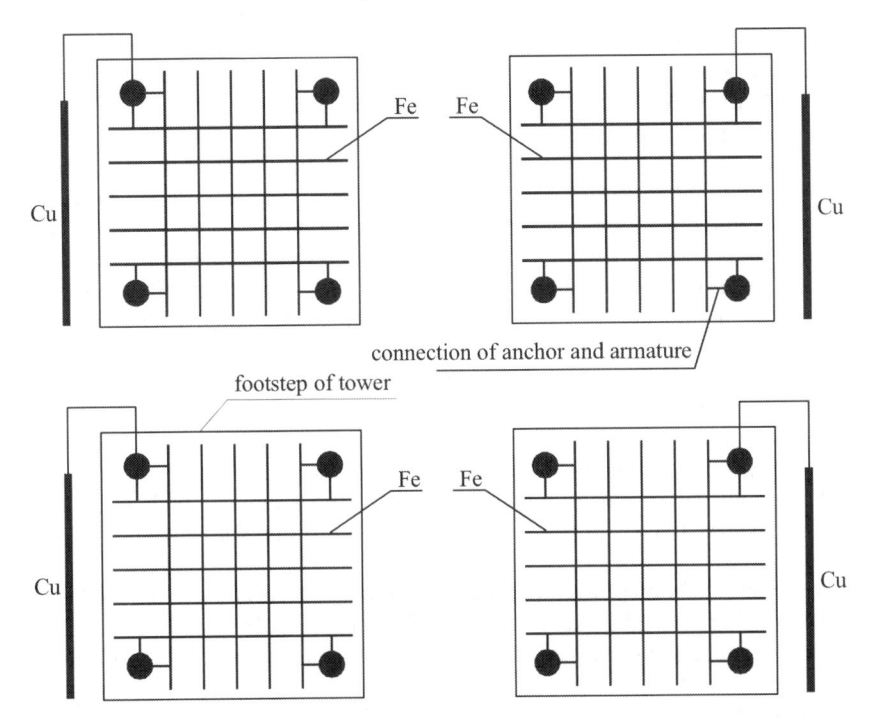

Figure 2: Principle scheme of anode system with mesh titanium anodes.

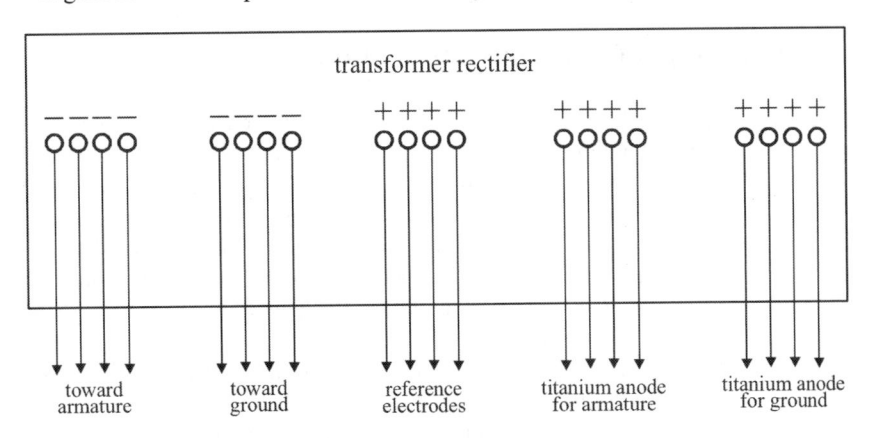

Figure 3: Principle scheme of outputs from transformer rectifier.

In regard on very expensive way for ensuring supply from distribution network 240 (V), 50 (Hz), one possibility is application of solar energy system. This system ensuring that battery has accumulated reserve energy minimum 48 (h).

Voltage characteristic of transformer rectifiers should be 48/60 (V) with minimal 8 outputs, total current 40 (A) and with operating current up to 15 (mA).

In box of transformer rectifier should be installed appropriate ammeters and voltmeters. Also, in this box should be brought outputs with permanent reference electrodes from each tower foundations separately [1].

In context of maintenance conditions, good solution is transformer rectifier with automatic regulation of output parameters – voltage toward maximum value of protection potential up to –2.0 (V) to CSE. In procedure of measuring protection potentials, for case that voltage droop component leave out is needed (in measured value), as well as that measuring of ON/OFF potentials can made, in transformer rectifier should be installed interrupter with appropriate time sets.

Anodes should be designed for life span up to 50 years. Ribbon titanium anodes should create anode system with appropriate plastic insulation at places of juncture. Length of anodes should be calculated according with dimensions of construction that is subject of protection. Protection current density of those anodes should be minimal 3 (A/m^2), where unit area by length minimum is 0.015 (m^2/m) [3]. Anodes must be robust for montage in concrete.

Each footstep of tower foundations should be equipped with own anode system. Alternative is installation of anodes in ring form with two sides supplying.

For continuous measuring permanent reference electrodes are used (for each footstep of tower foundations – single electrodes). Their position is on most distant locations (geometrically) from anodes. In application Ag/AgCl electrodes often are used.

Cable installation system with PVC/SVA/PVC cables and minimal cross section 10 (m^2) should be used.

6 Application of criterions and standards

In calculation procedure, minimal protection current density for armature in concrete of 5 (A/m^2) and for ground minimal 20 (A/m^2) should be used. Criterion for potential is as follows: after depolarization (switch off cathodic protection system) minimal shift on negative side should be 10 (mV) after 24 hours.

Cathodic protection system should be designed to satisfy following standards:

- NACE Standard RP 0177 Recommended Practice: *Mitigation of Alternating Current and Lightning Effects on Metallic Structures and Corrosion Control Systems*;
- NACE Standard RP 0187: *Design Considerations for Corrosion Control of Reinforcing Steel in Concrete*;
- NACE RP 0169-96: *Control of External Corrosion on Underground or Submerged. Metallic Piping Systems*;

- BS 7361 PART 1 Cathodic Protection Part 1: *Code of Practice for Land and Marine Applications*.

7 Conclusions

Of all the various anti-corrosion systems used, cathodic protection is one of the most efficient for armature tower foundations protection. In application of this type of protection on vital object, as overhead lines are, in this paper proposed impressed current cathodic protection system which is safe and economical method. This paper has as goal to offer motivation on wide application of this type protection since decisions are often made to replace the tower rather than install a more economic solution as cathodic protection. Utilities that overcome this problem by establishing a comprehensive tower building program will reap considerable economic benefits [8].

References

[1] A. Muharemovic, *Electric power system and environment*, Sarajevo, 1996.
[2] J. Morgan, *Cathodic protection*, Second edition, Published by National Association of Corrosion Engineers, Houston, Texas, 1993.
[3] W. Baeckmann, W. Schwenk & W. Prinz, *Handbook of Cathodic Corrosion Protection*, Gulf Publishing Company, Houston, Texas, 1997.
[4] A. Muharemovic, I. Turkovic, Aspects of cathodic protection solutions for city water wells Bacevo – Sarajevo, 9th *International expert meeting Power Engineering*, Maribor, Slovenia, 1999.
[5] A. Muharemovic, I. Turkovic & A. Kamenica, Analyse of the voltage component influence on the system measurement error of the stationary and protection potential, 5th *Libya Corrosion Conference*, Benghazi, Libya, 2005.
[6] A. Muharemovic, I. Turkovic & A. Kamenica, Distribution of the protection current and potential in system of cathodic protection with sacrificial anodes, 5th *Libya Corrosion Conference*, Benghazi, Libya, 2005.
[7] A. Muharemovic, I. Turkovic, Procedure for calculation and measurement of parameters ground system as recommendations, Study, Faculty of Electrical Engineering, Sarajevo, Bosnia and Herzegovina, 2002.
[8] http://tdworld.com/mag/power_sweden_confronts_groundline/.

Section 3
Experimental measurements and computer results

Dipole modelling and sensor design

S. A. Wimmer, E. A. Hogan & V. G. DeGiorgi
Naval Research Laboratory, Washington, DC, USA

Abstract

The truth values used for validation of computational models are measured values from either actual structures or experiments. The accuracy of computational models will depend on the accuracy in which key parameters can be measured. Therefore it is imperative that there is a clear understanding of what and how parameters are being measured. An incomplete understanding of the experimental process, including measurement sensors, will corrupt the computational model validation process.

In this work computational modeling has been used to further the understanding and assist in the design of a sensor used to measure off-board electrical fields. Previously data from a series of dipole models was generated using NRL's physical scale modeling experimental facility. Results were compared with both analytical and computational solutions. Variations were observed between results. Boundary element methods were used to extensively model tank geometry, water depth, sensor orientation and to some degree sensor geometry. It was determined that the sensor as designed was not adequate for the off-board electrical measurements required. In this work boundary element modeling is used to assist in the design of a new off-board electrical field sensor. Dipole models which consider the vertical and horizontal placement of half-cells on the sensor are used to quantify characteristics of the new sensor. Comparisons are provided between analytical, computational and measured results once the sensor design is finalized.

Keywords: electrical dipole, electrical fields, boundary element modeling, impressed current cathodic protection, physical scale modeling.

1 Introduction

The accepted process for validation of computational models is a detailed comparison of experimental and computational determined values. This approach is common to many disciplines. An underlying assumption is that experimental values are true. While the 'truth' of experimental values would appear to be the topic of a philosophical discussion, there are very real factors that may result in the measured value not actually being what the experimentalist intended to measure.

One estimate of the truth of experimental values is often the comparison of these values with analytical or theoretical solutions. This approach is often flawed because of geometric or environmental factors that are simplified in the theoretical or analytical solutions. So in addition to a comparison with analytical values, a detailed understanding of what is being measured and how it is being measured is important. These are things which must be understood before measured values can be used to validate computational modeling processes. If the baseline 'truth' is wrong, the validation process is not valid.

Measured values are a reflection of the real world phenomenon. There are, of course, issues with sensors and measurement systems. Slight chances in environment conditions can also result in larger variations in measured values. Differences in boundary conditions between computational, theoretical, analytical and experimental situations can result in large variations despite the minor nature of boundary condition differences. It is therefore important to evaluate and carefully determine the pedigree of any experimentally measured value. Pedigree means an accurate representation of the environmental and boundary conditions. The pedigree also involves understanding the physical methods used to determine the measured value. It is important to know what values the sensors used are actually measuring and what values reported (and may be commonly talked about as being 'measured') are actually calculated. If a value is calculated, it is important to understand details of the calculations and to know the underlying measured value is. Sensor type and sensitivities should also be known.

This work addresses some of the issues listed above in terms of understanding the physical aspects of the experiment. Computational models are used to establish reasons for variations between theoretical and experimental results. The authors began this investigation into sensor design because of unexplained differences between computational and experimental measured electric field values generated by dipoles [1]. Observed differences were greater than could be explained due to numerical errors or variations in boundary conditions between the two conditions. One aspect in the results that intrigued the authors was the good agreement between theoretical and computational results. Results from both theoretical and computational studies showed significant differences from experimental results.

In this work the computational studies of electric dipoles are compared with physical scale model experimental results measured using a newly design electrical field sensor. In previous work [1] the authors used computational modeling of a dipole to investigate the effect of testing tank geometry, sensor placement and tank wall material/boundary conditions on a simple dipole system. This work addresses issues related to comparison of analytical, computational and experimental electric field results from the previous work.

Computational and analytical results were nearly identical. The sources of variations were traced to sensor design. There are two tracks that could have been taken; to include detailed sensor geometry into the computational model thus resulting in agreement between experimental and computational or to evaluate sensor design. Sensor design was chosen to be evaluated in order to match experimental with theoretical. This is important since these types of sensors will be used in future evaluations of much more complex geometries with the dipole measurements being used as a calibration technology.

2 Geometry

The structure evaluated in this work is not a structure in the engineering sense. The geometry evaluated is a dipole. Electric dipoles are a concept that can be found in many textbooks such as [2]. Their geometric simplicity and the availability of analytical solutions make dipoles a good tool for validation of computational and experimental techniques. Electric dipoles are also important because the dipole concept is used for far field modeling of ships and other structures with discrete electrical sources, such as the features of an impressed current cathodic protection system for a ship. Multi-pole (dipoles connected in series like a multi-span beam) models are typically used to determine electrical fields at a distance such that the influence of geometric details is negligible. Details of this modeling approach can be found in [3].

3 Dipole model

Computational, analytical and experimental results were obtained for the far-field voltage values for a simple dipole. Model orientations and axis directions are shown in Figure 1. The dipole moment strength was 11.725×10^{-3} A-m. The source and sink were spaced 250 cm apart at a depth of 82.398 cm below the waterline in a cylindrical tank. The tank diameter is 10 m and is made of galvanized steel. The tank has a 30 mil neoprene liner. The depth of water in the tank is 264.16 cm. The tank water had a scaled conductivity of 1/40 full strength natural seawater. The tank water measured resistivity was 740 ohm-cm. Scaling water conductivity is a standard process in physical scale modeling (see Section 5, Experimental Method). Fresh water is added to a volume of natural seawater until the desired conductivity is reached. An electric field sensor is passed under the dipole centerline at a depth of 47.5 cm. The sensor provided two curves of different potential values; the vertical (z-direction) and longitudinal (x-direction) potentials. This configuration is duplicated in the computational and analytical models.

4 Computational and analytical methods

The computational models used in this work were created in MSC PATRAN [4]. The geometry was then translated using an NRL written program to create input files to for the commercial boundary element code BEASY-CP [5]. BEASY-CP provides a solution process for LaPlace's equation. Details of the application of BEASY-CP, and boundary elements in general, to electrical field problems can be found in [6]. The results are post-processed using an NRL customized program that extracts and translates pertinent data for plotting using TECPLOT [7].

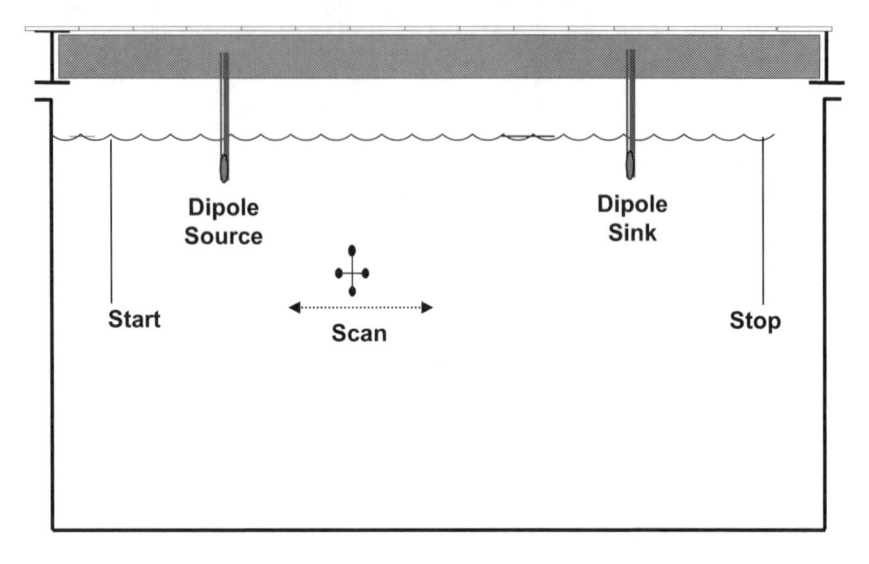

Figure 1: Dipole setup.

The boundary element model of the tank and dipole is shown in Figure 2. The dipole source and sink are modeled as truncated poles in the shape of cylinders (2 mm diameter, 3 mm height). The poles and the tank are meshed using 9-noded quadratic quadrilateral elements. A normal current flux density equal to the dipole strength divided by the surface area of the pole (±24.88 mA/cm^2) is applied as the driving force at the source and sink seen in Figure 2. The neoprene liner is represented by zero current flux boundary condition along the tank wall and floor. Internal node points were located along the sensor path.

Analytical solution results are calculated using the FN Remus Characterization Suite [3]. This is a commercial code which calculates fields resulting from dipole or multi-pole models. This model consists of discrete and sea conductivity. The dipole geometry was input directly into the code. A grid of internal points is defined at the sensor path location for calculation of results.

A comparison of calculated, computational from the boundary element model and analytical results are discussed in Section 6, Sensor Design Issues.

5 Experimental method

Physical Scale Modeling (PSM) was used to generate measured dipole electrical field values. PSM is an established process based on the physics of electrochemical response which uses scale models to produce information on structures. Structures which have been tested at the NRL Key West facility range in geometric complexity from dipoles to real ship geometries with detailed appendages such as rudders, bilge keels and propellers with moving blades (as seen in Figure 3).

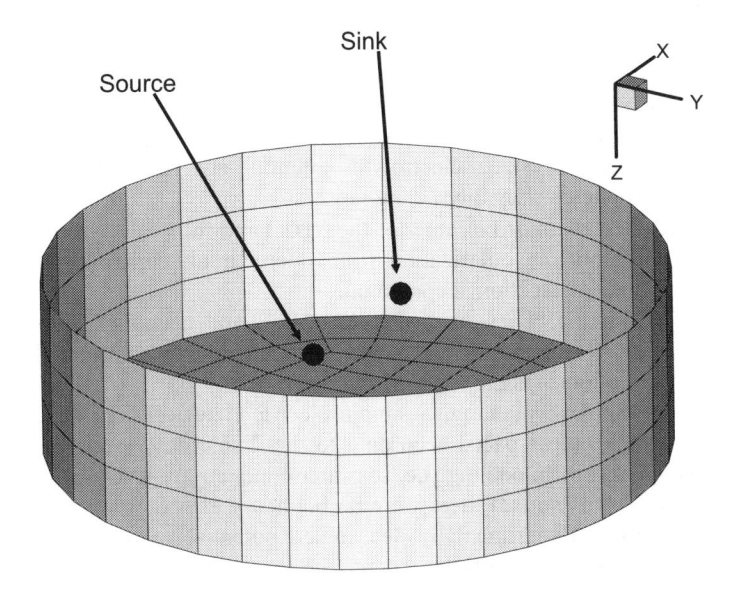

Figure 2: Boundary element model.

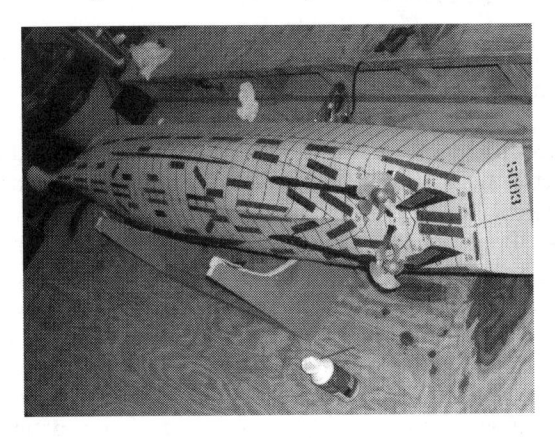

Figure 3: Example of a near-exact scale model used in PSM.

PSM has been extensively used in the study and design of shipboard Impressed Current Cathodic Protection (ICCP) systems. For both PSM and the computational techniques, cathodic protection is modeled in the steady state condition and follows Ohm's Law:

$$E = I(R_P + R_{OHMIC}) \quad (1)$$

where, E = potential (Volts), I = current (Amps), R_P = polarization resistance and R_{OHMIC} = electrolyte Ohmic resistance. R_P can be highly nonlinear and is influenced by environmental conditions. The three basic assumptions in PSM technique are:
- The wetted surface areas and geometry are exact and scaled such that $A_{STRUCTURE} = A_{MODEL} (k)^2$ where A is surface area and k is the scaling factor.
- The current density relationship, $i_{STRUCTURE} = i_{MODEL}$ is true. This means that model size, electrolyte dilution and polarization resistance components obey the scaling law.
- $R_P = \Delta E / i_C$ must be same for the model and full scale system, where ΔE represents the polarization from open circuit corrosion value to the cathodic protection set potential.

PSM technique relies on accurate scaling and accurate reproduction of geometric features at the smaller scale. Model potential, current density and scale factor relationships are described in detail in [8,9].

For the dipole evaluation, a simple electric field test was designed using source and sink poles spaced according to the dipole description given earlier. The scale chosen for modeling (i.e. seawater conductivity level) was 1/40. This scaling was chosen so that there were no boundary effects from the presence of the tank walls. Experimentally determined potential values are presented in Section 6, Sensor Design Issues.

6 Sensor design issues

A schematic of the original sensor used in previous work is shown in Figure 4. The sensor consisted of 4 Ag/AgCl half-cells for measurement of potential and was fabricated from a fiber composite. The two curves of differential potential values are the differences of measured values from the half-cells. Dipole generated electric field measurements obtained with this sensor are shown in Figure 5. There is significant variation between calculated and experimental measurements. Prior work investigated variations in sensor orientation as the cause of these variations [1]. Variations in pitch, yaw and roll which were considered to be possible by the experimentalist were evaluated. Even though changes in the calculated peak electrical potential occurred with variations in orientation, the changes did not eliminate the variations observed in Figure 5.

The next step was to focus on sensor design. Once the sensor was examined as a structure rather than seen as 'just' a sensor, design deficiencies became obvious. It has been observed, both experimentally and in computational modeling [10], that minor geometric features can have a significant impact on electrical field values. Shadowing, or blocking of current flow, has been noted for several ship features. The bilge keel was found to be necessary to model both in PSM and computational models, despite its relatively small size compared with the hull structure. It is known to shield areas of the hull from current flowing from anodes. The differences in a model with a bladed propeller and a solid propeller model also deal with the shadowing of regions by the solid structure [11]. Taking these concepts into consideration the experimentalists clearly saw the possibility that portions of the sensor structure were shielding the half-cells.

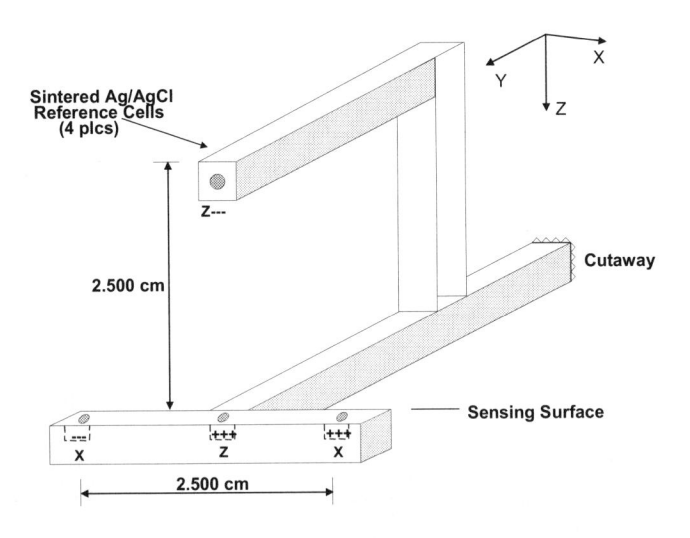

Figure 4: Original electric field sensor.

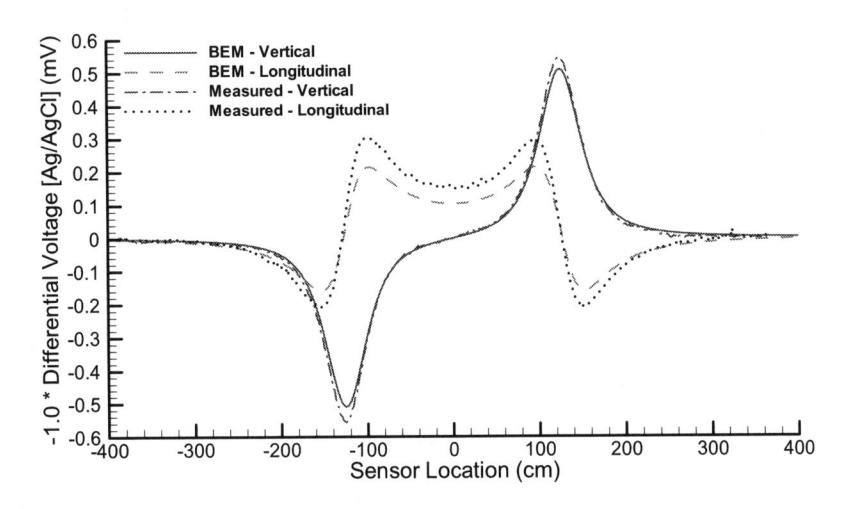

Figure 5: Boundary element (BEM) results versus measured PSM results from the original electric field sensor.

A new sensor was designed that incorporated concepts of shielding and shadowing of regions by the structure. The new sensor consists of 4 A/AgCl half-cells mounted so that they are co-planer. There is no massive structure avoiding any shielding issues. The new sensor is shown in Figure 6.

In order to test the new sensor, a repeat of the dipole experiment was conducted. There were slight modifications to the dipole set up. The dipole moment strength was 11.725×10^{-3} A-m. The source and sink were placed 250 cm apart at a depth of 75.0 cm below the water line. The tank water level was 262.5 cm. Tank water was again 1/40 scaled natural seawater. The measured resistivity of the tank water was 741 ohm-cm. The same tank was used previously with the same neoprene lining material.

Computational and analytical solutions were repeated with the new dimensions, dipole strength and material properties. Measured and calculated longitudinal and vertical electrical potential profiles are shown in Figure 7. The change in sensor design made a significant difference in observed variations. Maximum differences are reduced to 6.8% for vertical and 4.7% for longitudinal potentials.

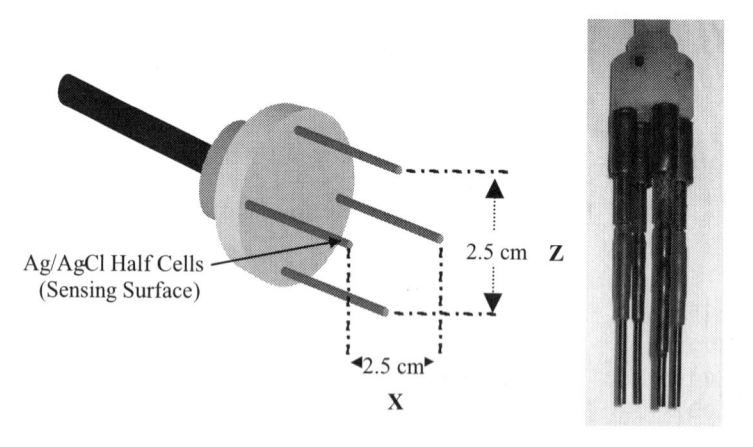

Figure 6: New electric field sensor.

One computational parameter variation run was completed. The nominal distance between half-cells in the new sensor is 2.5 cm. To determine the effect this dimension has on sensor measured values, the horizontal distance was defined as 2.6cm while maintaining the vertical distance between half-cells at 2.5 cm. The maximum difference was reduced to 2.1% for longitudinal potential (reduced from 4.7%). There is no change in horizontal potential differences since no change was made in horizontal dimensions. Therefore accurate sensor fabrication is essential.

There are other issues which may have contributed to the variations between measured and calculated data. The calibration of the electrical field sensor at the tank wall has an influence on measured values. Water stratification, while not typically thought of as occurring in experimental facilities, is a phenomenon that occurs in even shallow waters. Adjustments in raw measured data are also always an issue with any sensors. Even though the agreement is not perfect between experimental and calculated values, the new sensor design improves the fidelity of measured values. Lessons learned in ship analyses were successfully applied in the sensor design process.

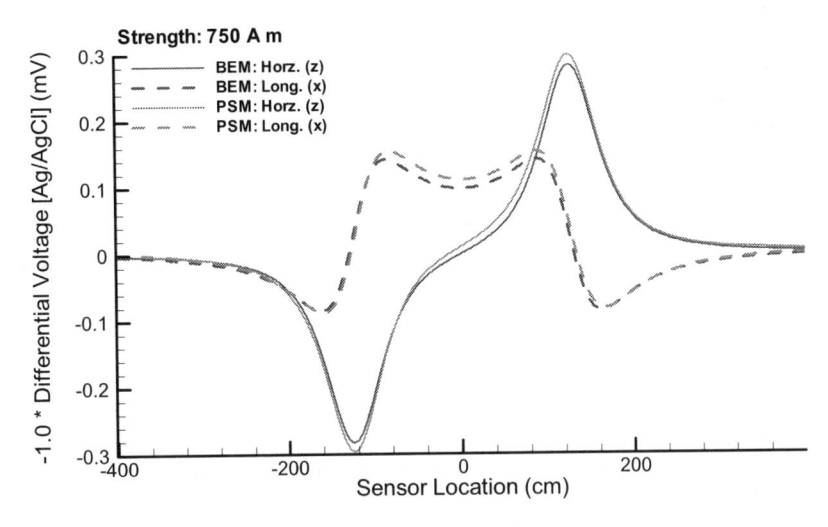

Figure 7: Boundary element (BEM) results versus measured PSM results from the new electric field sensor.

7 Summary

Variations between computational, analytical and measured electrical fields for a dipole submerged in scaled seawater were observed in previous work [1]. The source of variation was traced to the design of the sensor used for the experimental measurements. A new sensor was designed with the specific goal of eliminating the geometric issues related to the older sensor design. A comparison of measured and calculated dipole generated electrical fields indicates the new design has met these goals.

The current work in which computational methods are used to verify the accuracy of measured data may seem in conflict with other work by the authors, specifically processes for computational model validation [12]. Rather than think in linear terms of computational, analytical or experimental tracts of study, one should think in terms of understanding a physical problem. In striving to be able to predict physical phenomenon for complex structures there is a triad of knowledge which must be obtained and used. Physical experiments must be understood in terms of simplifications, boundary conditions, sensor capabilities and limitations and environmental factors. Computational methods must be understood in terms of underlying mathematical theories, computational implementation, and implications of modeling decisions (boundary conditions, loads, simplifications, etc). Analytical or theoretical solutions must be understood in terms of assumptions versus real complex structural conditions and the regions of applicability. Experiments, computational methods and analytical methods are tools and must be judicially applied to further our understanding.

References

[1] S. A. Wimmer and V. G. DeGiorgi, "The Intricacies of Modeling a Simple Dipole, *Proceedings BETECH 2003*, WIT Press, 213-222 2003.
[2] O. D. Jefimenko, Electricity and Magnetism, Electret Scientific Company, Star City, WV, 1989.
[3] FN Remus Characterization Suite User Guide, Frazer-Nash Consultancy, Dorking, Surrey, UK, Jan. 2001.
[4] MSC PATRAN 2001 User's Manual, MSC Software Corp., Los Angeles, CA, 2001.
[5] BEASY User's Manual, Computational Mechanics Intl., Billerica, MA 2000.
[6] V. G. DeGiorgi, "Corrosion basics and computer modeling," Chapter 3, *Industrial Applications of the Boundary Element Method*, Computational Mechanics Publications, 81-93 1993.
[7] Tecplot, v. 9 User's Manual, Amtec Engineering, Inc., Bellevue, WA, 2000.
[8] Lucas, K.E., Thomas, E. D., Kaznoff, A. I. & Hogan, E. A., *Designing CP Systems for Marine Structures and Vehicles*, STP-1370, ASTM, 17-33, 1999.
[9] V. G. DeGiorgi, E. Hogan, K. E. Lucas and S. A. Wimmer, "Shipboard Impressed Current Cathodic Protection System," Chapter 2, *Modeling of Cathodic Protection Systems*, WIT Press, 13-44 2006.
[10] V. G. DeGiorgi, E. D. Thomas and K. E. Lucas, "Scale Effects and Verification of Modeling Of Ship Cathodic Protection Systems," *Engineering Analysis with Boundary Elements*. Vol. 22, 41-49 1998.
[11] V. G. DeGiorgi and S. A. Wimmer, "Geometric Details and Modeling Accuracy Requirements for Shipboard ICCP Modeling," Engineering Analysis with Boundary Elements, Vol. 29, 15-28 2005.
[12] E. Hogan, J. McElman, V. DeGiorgi, E. Lemieux, A. LeDoux, M. Krupa, "Validation Plan For Boundary Element Method Modeling Of Impressed Current Cathodic Protection System Design And Control Response," ELECTROCOR 2007, in press.

Transport phenomena in an electrochemical rotating cylinder reactor

F. Tomasoni[1], J. F. Thomas[1], D. Yildiz[1], J. van Beeck[1]
& J. Deconinck[2]
[1] *Von Karman Institute for Fluid Dynamics, Belgium*
[2] *Department of Electrical Engineering,
Vrije Universiteit Brussel, Belgium*

Abstract

Electrochemical processes are at the heart of a wide variety of both basic and advanced industrial activities. In many industrial electrochemical applications high flow rates are applied such that the electrochemical process is *convective-transport*-dependent. In order to obtain the mass and charge transfer from the solution to the surface of the electrodes the flow field characterization is therefore essential.

In view of understanding in detail such processes, this paper presents the preliminary characterization of the complex flow field that takes place in an electrochemical rotating cylinder reactor (short aspect ratio $\Gamma = 1.7$ and wide gap $\eta = 0.16$). For this purpose a combined experimental, numerical and theoretical approach is followed. The experimental characterization is performed using time-resolved Particle Image Velocimetry (PIV), a non-intrusive and laser-based experimental technique. Special attention is paid to the image processing and possible optical problems. The investigation is performed for different flow conditions, from moderate (1700) to high (13600) Reynolds number.

The experimental results have been compared to the velocity profiles obtained using RANS and DNS simulations.

The theoretical part of this study is to provide new analytical models based on the solution of the momentum boundary layer equation and of the mass and charge transport equations. The analytical velocity profile agrees well with the experimental data.

Keywords: rotating cylinder electrode, PIV, RANS, DNS.

1 Introduction and motivation

Electrochemical processes are governed by multi-component mass, heat and charge transport in laminar and turbulent flow, that are often multi-phase due to gas evolution at the electrode. Furthermore the mass transfer boundary layer is about one order of magnitude smaller than the hydrodynamic boundary layer and this, with other side effects (such as side reactions, surface contamination effects, gas evolution), yields to an increase of the difficulty of the problem.

Besides that, in many industrial electrochemical techniques the electrode moves with respect to the solution. These systems are called *hydrodynamic electrochemical processes*. In these processes generally hydrodynamical and electrochemical coupling is implied, because the electrochemical phenomenon is *convective transport* dependent. The flow field characterization is therefore essential in order to obtain the mass and charge transfer from the solution to the surface of the electrodes.

The objective of the present work is the preliminary characterization of the complex flow field that takes place in an electrochemical rotating cylinder reactor.

The flow characterization (bulk velocity, vorticity, statistics) is obtained using time-resolved Particle Image Velocimetry (PIV §2). A continuous laser is used as a light source and the acquisition of the images is realized via a high speed camera. The investigation is performed for different flow conditions, from moderate to high Reynolds number. Numerical simulations (RANS §3.1) were run before deciding the PIV parameters. The results provided by DNS simulation §3.2 have then been compared with the experiments, with the aim of validating the used DNS code, for which the capability of capturing an unsteady and complex solution represents a very important task. The aim of the theoretical section §4 is to provide new analytical models based on the solution of the momentum boundary layer equation, necessary for the future development of the equations for mass and charge transport. In section §5 the results of the flow field characterization are presented and discussed, while the profiles provided by the different approaches are compared. The paper is closed in section §6 by the conclusions and the future development of this work.

2 Experiments

The experiments have been performed on a cylinder reactor, which geometry can be described in terms of non dimensional quantity: the radii ratio is $\eta = r_i/r_o \approx 0.16$ and the aspect ratio is $\Gamma = h/d \approx 1.7$, where $d = r_o - r_i$ is the gap size. The top, bottom and outer cylinder walls are in Plexiglas, in order to allow optical access.

The velocity measurements have been performed from two measurement planes: the *horizontal plane* (Figure 1(a)), located in the middle of the cylinder height, and the *meridional plane* (Figure 1(b)) which includes the axis of the cylinder.

The cylinder was filled with a solution of demineralized water at room temperature and seeding particles (Polyamide particles) previously mixed.

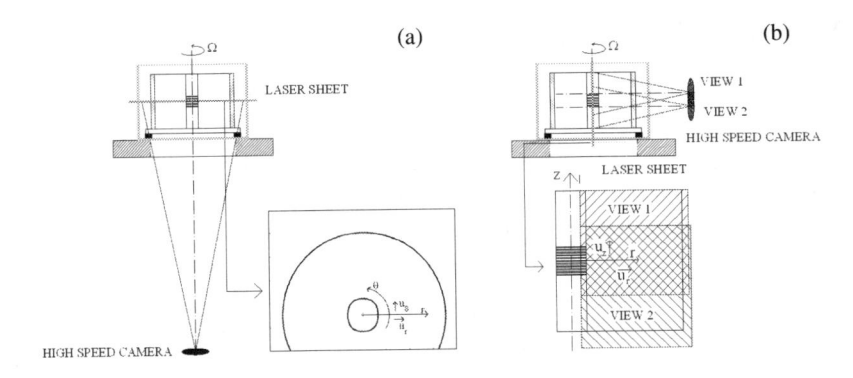

Figure 1: PIV: experimental setup. (a) horizontal plane measurement, and (b) meridional plane measurement.

The light source used in order to make visible the particles was a continuous laser (Coherent INNOVA 70C). By means of a spherical and a cylindrical lenses a narrow ($\approx$ 1 mm) and intense (2 MW single wavelength) light sheet has been obtained.

To acquire, display and record digital images, the Phantom v7.1 high speed camera has been used. The system is composed of: CMOS video sensor (800x600 pixels, 256 grey color) able to acquire images at a rate from 100 up to 160000 pps and a PC equipped with a dedicated high speed camera software (Phantom 606). The maximum number of images acquirable consecutively was 1455.

The inner cylinder was driven by a stepper motor (Maxon motor). The angular velocity of the inner cylinder has been varied from $\Omega_i = 100$ rpm to $\Omega_i = 800$ rpm, in order to characterize the flow field that takes place in the cylinder reactor. The Reynolds number based on the gap $r_o - r_i$, $Re = \Omega_i r_i (r_o - r_i)/\nu$, varies from 1700 to 13600.

2.1 Measurement method: PIV

Time-resolved Particle Image Velocimetry (PIV), a non-intrusive and laser-based experimental technique, has been used to experimentally characterized the flow field. Special attention is paid to the image processing and possible optical problems.

To compute and predict the separation time between the images, according to the basic PIV constraints, RANS simulations have been performed and from the resulting velocity field, the adequate separation time has been estimated.

Each acquisition contains 1454 velocity fields. A minimum non dimensional observation time $\overline{T}_o$ of 10 cylinder revolutions, where $\overline{T}_o = T_o \Omega_i/(2\pi)$, is achieved in all the measurements.

3 Numerical simulation

3.1 RANS simulation

The CFD software used to perform Reynolds Averaged Numerical Simulation is FLUENT, a flow modeling solver. The main objective is to have a reliable estimation of the velocity field, in order to use it as "predictor" for the experimental measurements. A second goal is represented by an a-posteriori validation of the Fluent results.

For this purpose steady-state, 2D, axisymmetric with swirl simulations have been performed. The grid used has 75×100 points in the radial and axial direction respectively. The rotational speed has been varied in the range of the studied velocities. In all the simulations the flow is considered turbulent and the $k - \varepsilon$ model is used.

3.2 DNS simulation

The numerical tool used in order to perform Direct Numerical Simulation is SFELES, a spectral/finite element code [1]. The simulation is performed on a 3D computational domain. A symmetric and a slightly asymmetric grids were used for the computation. The grids used have $100 \times 120 \times 64$ points in the radial, axial and azimuthal direction respectively.

The case analyzed is the one at $Re = 1700$. The unsteady simulation is realized with a time step $\Delta t = 5 \cdot 10^{-3}$ s, for a physical time of $T_{\mathrm{DNS}} = 384$ s. The simulation required 55 hours running on 16 processors. The convergence is achieved approximately after 2000 iterations.

4 Theoretical analysis

4.1 Flow modeling: boundary layer modeling

The knowledge of the hydrodynamic boundary layer allows the study of the mass boundary layer and then, mass transfer at the electrode. Therefore the characterization of the studied boundary layer is essential for future developments. Thus, the aim of this section is to derive an expression for the mean tangential velocity profile in this layer.

Starting from the inner cylinder walls and moving in the radial direction till the center of the gap, the velocity profile is considered divided in three main regions: the wall region, the inertial sublayer and the core region. The subdivision is symmetric respect to the center of the gap.

The boundary layer equation is obtained starting from the continuity and the momentum equations in the azimuthal direction in cylindrical coordinates. Under the assumptions of axial symmetry and that the gradient of the axial stresses is small compared with the radial gradient of the azimuthal stresses, one obtains:

$$r^2(\bar{\tau}_{r\theta} + \tau_{r\theta}^t) = const = r_i^2 \tau_w, \tag{1}$$

where τ is the viscous shear stress tensor, τ^t is the turbulent momentum flux tensor (the so-called *Reynolds stresses*) and τ_w is the wall shear stress.

4.1.1 Near wall behavior
In the vicinity of the wall the asymptotic behavior of the velocity profile has been often assumed linear with respect to the distance from the wall [2, 3].

In order to be coherent with what is observed in (1) and assuming that in the vicinity of the wall only the viscous terms play a role, it has to be stated:

$$r^2 \bar{\tau}_{r\theta} = r_i^2 \tau_w. \tag{2}$$

Integrating the latter equation, with the adequate boundary condition, one gets:

$$u_\theta = \Omega r - \frac{u_\tau^2}{2\nu} \frac{r^2 - r_i^2}{r}. \tag{3}$$

Observing equation (3), one can see that the tangential velocity close to the wall does not follow a purely linear trend. The linear behavior is therefore an approximation that is applicable only very close to the wall, that is when r is very small, so that $(r + r_i)/r_i \approx 2$.

4.1.2 Inertial sublayer
In the inertial sublayer, often called logarithmic layer, the viscosity plays at most a minor role. Therefore, in this region, equation (1) reduces to:

$$r^2 \tau_{r\theta}^t = r_i^2 \tau_w. \tag{4}$$

In order to obtain the time-smoothed velocity profile it is necessary to introduce a model for the turbulent momentum flux. In analogy with the Prandtl mixing length, a *new* model is here proposed:

$$\tau_{r\theta}^t = -\rho k^2 (r - r_i)^2 |r \frac{d}{dr}(\bar{u}_\theta/r)| r \frac{d}{dr}(\bar{u}_\theta/r). \tag{5}$$

To close the equation, the continuity of the solution between two adjacent layer can be imposed. For example the expression of the average tangential velocity obtained imposing the continuity at $y^+ = y_{si}^+$, with the velocity obtained in (3), holds:

$$\bar{u}_\theta(r) = \Omega r -$$
$$- \frac{u_\tau}{k} \left[\frac{r}{r_i} \log \left[\left(\frac{r - r_i}{r} \right) \left(\frac{r_{si}}{y_{si}} \right) \right] + 1 - \frac{r}{r_{si}} + \frac{u_\tau k}{2\nu} \frac{r_{si}^2 - r_i^2}{r_{si}^2} r \right], \tag{6}$$

where $r_{si} = y_{si}^+ \nu / u_\tau + r_i$ and $y_{si} = r_{si} - r_i = y_{si}^+ \nu / u_\tau$.

5 Results and discussion

The tangential velocity profile represents the primary component of the motion in the RCE. Nevertheless, the topology of the flow is more complex. As it can be observed from Figure 2, the motion is not purely circular (the streamlines are not close in circles) as a radial velocity component exists.

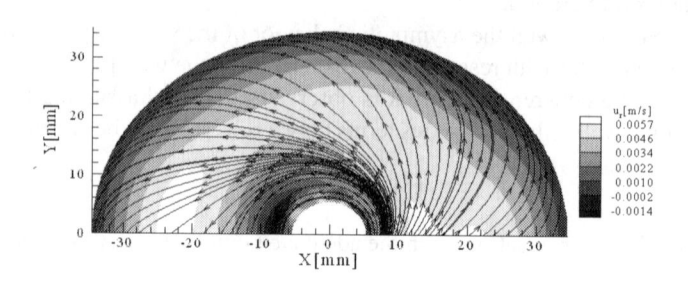

Figure 2: PIV (horizontal plane): Re = 1700. Average radial velocity contour.

It is possible to relate this behavior to the existence of an outgoing jet from the inner cylinder boundary layer, as it can be seen in Figure 7(a). For all the Reynolds numbers analyzed, in the average vector field, two big counter rotating cells are recognizable in the lateral plane.

A deeper understanding of the flow behavior can be obtained by analyzing the instantaneous velocity fields. In Figure 3(a), where an instantaneous field is reported for $Re = 1700$, one can observe two rollers. Observing the time evolution of the two cells, it appears that three different configurations are possible: a symmetric position of the rollers with respect to the middle height of the cylinder and an asymmetric one (one bigger than the other in both the direction). The latter configuration appears more frequently in the observed time history, so that the average velocity field keeps memory of the periodicity of the flow.

The situation is different for $Re = 13600$, Figure 3(b), where small turbulent structures are distributed in all the field and they coexist with bigger cells. The two coherent structures that appear in the average field find therefore their origin into an average-operation.

The average specific angular momentum $\mathcal{L} = u_\theta r / \Omega_i r_i^2$ (Figure 4) is constant in the core of the gap. As explained by [4], this means that, far from the wall, the flow is well mixed. As the Reynolds number increases the $\mathcal{L}$ value increases, till the value $\mathcal{L} \approx 0.45$, which has been found by [3] and [5], for fully turbulent flow conditions.

The normalized RMS (Figure 5) exhibits a peak close to the inner cylinder wall, where it exhibits always the maximum value near 10%. A maximum normalized RMS of 7% characterizes the central region of the gap. In all the cases the velocity fluctuation profiles are maximum at the middle height of the reactor that is the interface region between the two rollers.

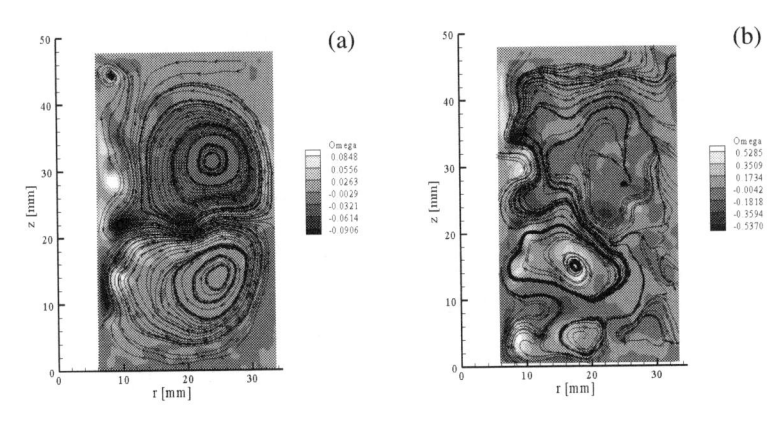

Figure 3: PIV: vorticity, instantaneous field: (a) Re = 1700, and (b) Re = 13600.

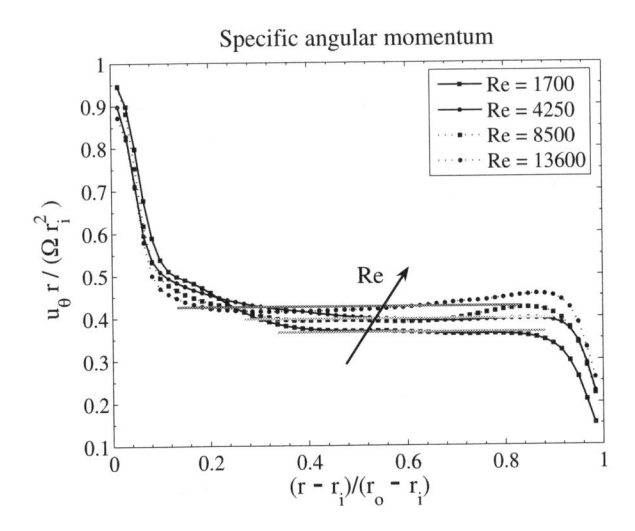

Figure 4: Average specific angular momentum $\mathcal{L} = u_\theta r / \Omega_i r_i^2$.

5.1 Comparison: PIV and analytical models

An iterative method is employed in order to compare the analytical velocity profiles with the experimental data. The validation of the viscous sublayer velocity profile is particularly difficult, because the latter region extends in a very small region close to the wall (less than 1 mm) where the PIV measurements risk to be compromised.

As it is possible to observe in Figure 6, the analytical velocity profile agrees well with the experimental data.

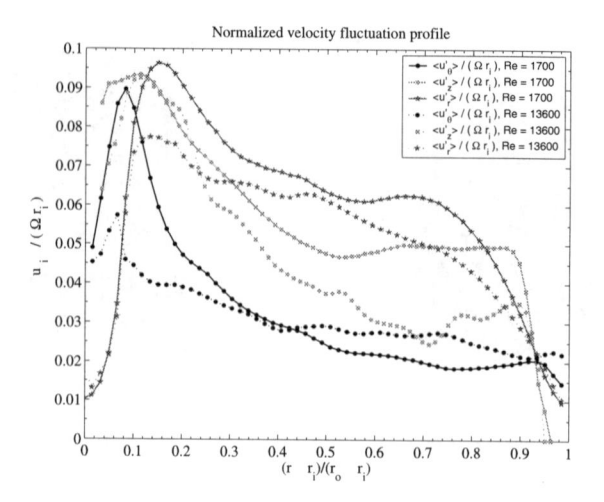

Figure 5: Normalized RMS.

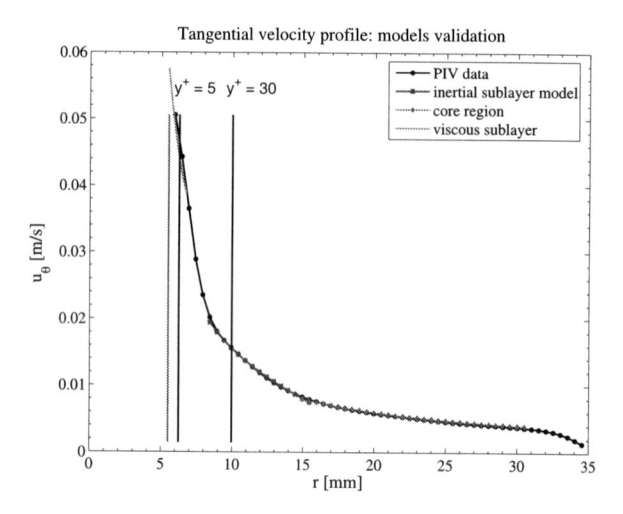

Figure 6: Average tangential velocity profile: models preliminary validation with PIV data.

5.2 Comparison: PIV and numerical simulations

The numerical simulations are compared both in a qualitative (topology) and quantitative way to the experimental results, for $Re = 1700$. From Figure 7, it is possible to see that in the average field, all the data show two counter rotating cells. While in the experiments and in the DNS simulation with a slightly asymmetric grid, the two rollers are not symmetric with respect to the middle of the cylinder height, in the case of the RANS simulation and the DNS simulation with a symmetric grid, the coherent structures are symmetric. The asymmetry of the two rollers can be explained again by looking to the instantaneous behavior of the flow. For $Re = 1700$, the flow is not fully turbulent and an unstable regime seems to be established in the cell. A small perturbation is sufficient to trigger the asymmetry of the two rollers. In the DNS simulation, the asymmetry of the grid induces such a "perturbation". From the quantitative comparison of the velocity profiles, it appears that the maximum difference between the experimental and the numerical data is of the order of 20%.

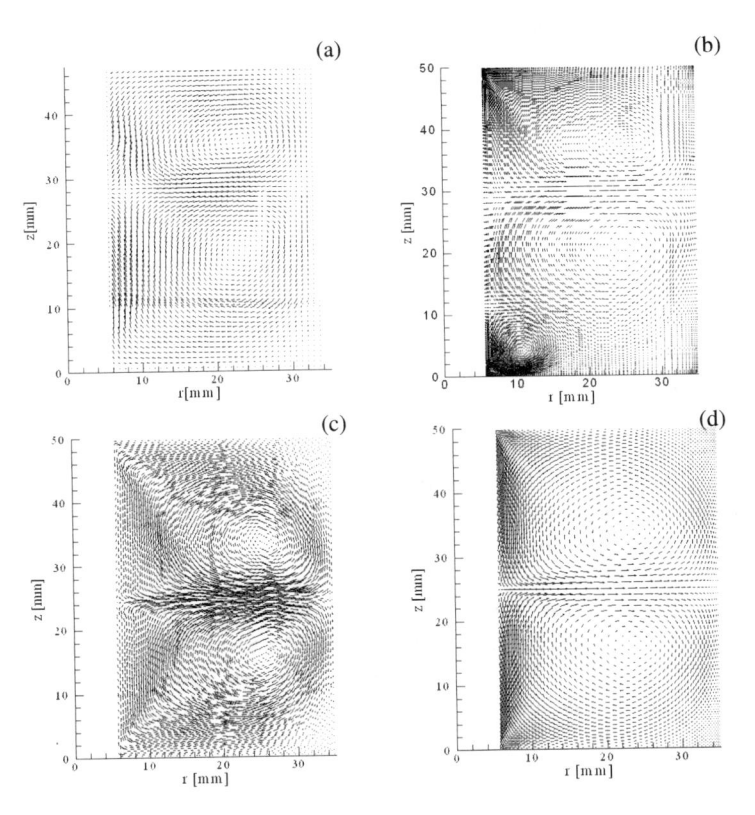

Figure 7: Average vector field, (a) $Re = 1700$. PIV, (b) DNS not symmetric grid, (c) DNS symmetric grid, (d) RANS. One vector out of two is plotted.

6 Conclusions and future work

The average flow field shows, in the range of the Reynolds studied, two big counter rotating structures in the meridional plane. Those cells are characterized by radial inflow at the cylinder endwalls, while an outgoing jet constitutes the interface line between the rollers themselves. Furthermore, a core region of constant angular momentum (the specific angular momentum found is 0.37-0.45) is observed. This indicates that a region of well mixed flow exists in the center of the gap.

In both the flow condition nevertheless, the normalized RMS never exceeds 11% as peak value and 7% in the core region.

The results of the numerical simulations are in qualitative agreement with the measurement results, but an influence on the grid can be observed. For $Re = 1700$, a symmetric grid provides two symmetric rotating structures, while a slightly asymmetric grid gives two non symmetric rollers. The grid asymmetry acts therefore as a perturbation that triggers the asymmetry of the two structures.

The preliminary validation of the analytical models for the average tangential profile in the boundary layer gives satisfactory results. Having completed the validation of the models of the velocity, it will be possible to develop the model for the concentration profile in the mass boundary layer.

Acknowledgement

We acknowledge the support from the Instituut voor de aanmoediging van innovatie door Wetenschap & Technologie in Vlaanderen (IWT, contract nr. SBO 040092).

References

[1] Detant, Y. et al, Sfeles manual. Technical report, VKI, VUB, BYU, 2006.
[2] Bilson, M. & K., B., Ćomparison of turbulent scalar transport in a pipe and a rotating cylinder. *Third International Conference on CFD in the Minerals and Process Industries*, pp. 493–498, 2003.
[3] Lathrop, D., Fineberg, J. & Swinney, H., Transition to shear-driven turbulence in couette-taylor flow. *Physical Review A*, **46**, pp. 6390–6408, 1992.
[4] King, G., Li, Y., Lee, W., Swinney, H. & Marcus, P., Wave speeds in wavy taylor-vortex flow. *J Fluid Mech*, **141**, pp. 365–390, 1984.
[5] Lewis, G. & Swinney, H., Velocity structure functions, scaling, and transitions in high-reynolds-number couette-taylor flow. *Physical Review E*, **59**, pp. 5457–5467, 1999.

Simulation of the electrical transient during the porous anodizing of pure aluminium substrate

F. Le Coz, L. Arurault & R. S. Bes
CIRIMAT-LCMIE, Université Paul Sabatier, Toulouse, France

Abstract

Electrical transients were recorded during the anodizing of highly pure aluminium in phosphoric electrolyte, carried out in potentiostatic mode (25-150V) or in galvanostatic conditions (20-1000A/m^2). The experimental reproducibility is satisfactory, according to the low standard deviations.

For the galvanostatic mode, the voltage experimental transients show a "bell shape", characterized by some significant parameters (S_0, t_m, V_m, V_{ss}). Two mathematical relations were then proposed to simulate the voltage transient curves considering two parts, i.e. before and after the maximum (V_m, t_m). The validity of these computational simulations was next checked by comparison with the corresponding experimental curves. All the corresponding fittings of voltage transients are in good agreement, especially for the first part of the experimental curves, within the current densities range.

Then, these computational simulations were correlated with the corresponding experimental FEG-SEM plan-views. By analogy with the nucleation phenomena during the metal electrodeposition, the "bell shaped" curves could be interpreted by the initial formation of a highly resistive oxide layer, followed by the subsequent appearance of the nanopores. The pores formation was in part explained showing that the V_m experimental values obtained with the galvanostatic mode are closed to the previous critical voltage value U_c initializing the anodic dissolution phenomenon.

Keywords: electrical transients, anodic film, simulation, nanopores, anodizing.

1 Introduction

Anodizing of aluminium and its alloys is an established electrochemical process, discovered in 1855 by H. Buff and developed in 1911 by De Saint Martin.

Industrially, it rapidly acquired great importance due to the various applications (against corrosion, decoration…) in different fields like aeronautic, architecture, etc. About twenty years ago, there was a renewal of interest for this process due to the possibility, in special operational conditions, to obtain highly nano-ordered templates based on the Anodic Aluminium Oxide (AAO) [1,2]. Moreover, recent research works [3-5] showed that a later impregnation of metal or oxide, followed by the removal of the AAO matrix, allowed to prepare nano-objects, like plots or wires, of which the sizes depend directly on the previous geometrical characteristics of the AAO nanopores. The porous anodic film growth on aluminium substrates in general, and the self-nanostructuring of the AAO templates in particular, is still today considered to be a complex phenomenon, until now not well understood in spite of the wide range of academic experimental studies. But the previous research works experimentally show for example the great importance of the surface pretreatments and of the initial electrical transients of anodizing to prepare convenient AAO templates.

From this point of view, despite previous preliminary works [6-10], further efforts are necessary to explain the initial electrical transients obtained either under galvanostatic anodizing mode or in potentiostatic anodizing conditions, as well as to correlate simultaneously these electrical transients with the surface nanoporosity of the aluminium substrates.

The aim of this study is to develop mathematical relations simulating the electrical transients during the anodizing of highly pure aluminium in acidic electrolyte, thereby extending our previous works [11,12] about the control of the initial surface state and of the final nanoporosity characteristics of the anodic film. In particular, these computational simulations will be compared to the corresponding experimental measurements, and correlated with the nanoporosity using FEG-SEM views of the aluminium substrates.

2 Experimental

The substrate material is highly pure aluminium (99.99%Al). The aluminium sheets (0.1x40x40mm) were first degreased for one minute in an aqueous alkaline bath containing NaOH (5g/L), $Na_2CO_3,6H_2O$ (5g/L), $Na_3PO_4,12H_2O$ (10g/L), $Na_2SiO_3,5H_2O$ (1g/L) and $NaC_6H_{11}O_7$ (10g/L), then etched in aqueous NaOH (25g/L) for one minute and neutralised in aqueous HNO_3 (20%v/v) for 2 minutes. Each step was conducted at ambient temperature, and the samples were rinsed in distilled water.

The aluminium sheet was then used as anode and a lead plate (3x40x40mm) as counter-electrode in a cell regulated at $-1.5°C$. The electrolyte was made up of an aqueous motionless H_3PO_4 (8%wt) solution. All chemical were analytical grade products (PROLABO) and the aqueous solutions were prepared using deionized water.

The input direct current or voltage was imposed by a NEMIC LAMBA generator (GEN 300-5), while a potentiostat/galvanostat (PGP 201 TACUSSEL-RADIOMETER) controlled by a microcomputer was used to record the curves as a function of the time (U = f(t) or J = f(t) respectively).

The micro and nanostructures, especially of the porosity of the samples, were observed by Field Emission Gun Scanning Electron Microscope (FEG-SEM JEOL JSM 6700F).

3 Results and discussion

3.1 Potentiostatic anodizing mode

Figure 1 shows the electrical transients obtained under potentiostatic mode, i.e. at a constant potential drop (U) between the working and the counter-electrodes. This figure reveals two main types of curve shapes. The first one, obtained when U < 120V, is a decreasing curve whereas for higher voltages, the current density increases drastically with the time. In this second case, the anodizing process becomes rapidly out of control due to the high value of current density (>500 A/m^2), while the aluminium sample is rapidly damaged.

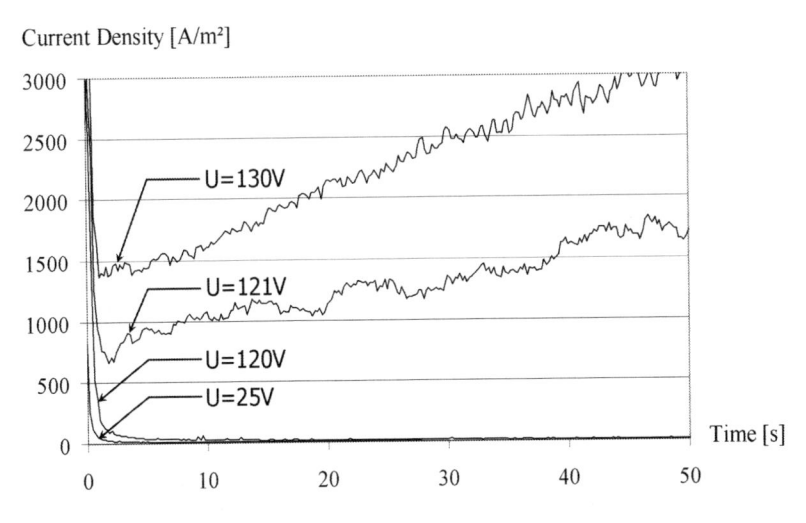

Figure 1:   Electrical transients obtained in potentiostatic mode (U=25V, 120V, 121V and 130V).

The critical voltage value (here U_c = 120V) distinguishes between two domains: the anodizing process (for U < U_c) and the anodic dissolution (for U > U_c). The anodizing in these operational conditions can be assimilated to a "hard anodizing" due to the use of a strong acid solution (pKa_1 = 2.05 at 0°C [13]) at a low temperature (−1.5°C), the resulting surface porosity being usually lower than 10% [14]. On the contrary, at higher voltages, the anodic dissolution induces the direct oxidation from the aluminium metal to the aluminic ion (Al^{3+}), preventing the formation of an anodic film and causing the sample damage.

3.2 Galvanostatic anodizing mode

3.2.1 Preliminary definitions

For the galvanostatic mode, the voltage evolution as a function of time shows qualitatively a "bell shaped" curve (Figure 2), characterised by some quantitative experimental data:
- the initial slope of the voltage transient: $(dV/dt)_{t\to 0} = S_0$
- the coordinates of the maximum: t_m and V_m
- the steady-state voltage V_{ss} at $t = +\infty$

For the simulation, the transient voltage-time curves are considered as composed typically of two parts, i.e. before and after the maximum (V_m, t_m).

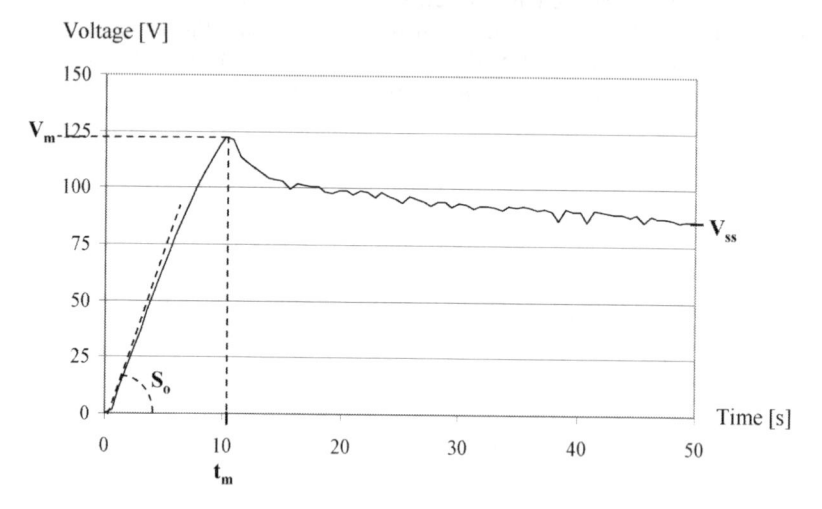

Figure 2: Typical voltage transient obtained in galvanostatic mode ($J=300A/m^2$).

3.2.2 Reproductibility

Three voltage-time transients, obtained for strictly identical anodizing conditions (300A/m^2), show that the initial slope S_0 (S_0 = 13.6±0.7V/s), the maximum voltage V_m (V_m = 122.3±0.3V) and the maximum time t_m (t_m = 10.5±0.3s) remain constants, their relative deviations being less than 5%, while the steady state voltage V_{ss} varies by 11% (V_{ss} = 85.8±9.3V). According to the low standard deviations obtained, the experimental reproductibility is considered to be good.

3.2.3 Response before the maximum ($0 \leq t \leq t_m$)

At first, the initial slope S_0 was studied as a function of the current density J. The results clearly demonstrate that the initial slope is directly proportional to the current density (Figure 3).

By similarity with the previous work of Patermarakis et al [7], the following relation is now proposed to simulate the first part of the electrical transient, i.e. to the maximum voltage:

$$V(t) = P_0 \cdot t - P_1 \cdot t \cdot \exp(\lambda_1 \cdot t) \qquad (0 \leq t \leq t_m) \qquad (1)$$

where P_0, P_1 and λ_1 are constants for a fixed value of the current density, under the considered anodizing conditions.

The derivative of this relation is:

$$(dV/dt) = P_0 - P_1 \cdot (\lambda_1 \cdot t + 1) \cdot \exp(\lambda_1 \cdot t) \qquad (2)$$

At $t \approx 0$, the initial slope is then:

$$(dV/dt)_{t \to 0} = S_0 = P_0 - P_1 \qquad (3)$$

The corresponding fits are in very good agreement with the experimental curves for current densities from 20 to 1000 A/m². As an example, Figure 4 shows the excellent fitting of the experimental transient obtained at 300 A/m², the values of P_0, P_1 and λ_1 being respectively 14.0V/s, 0.0001V/s and 0.98s^{-1}.

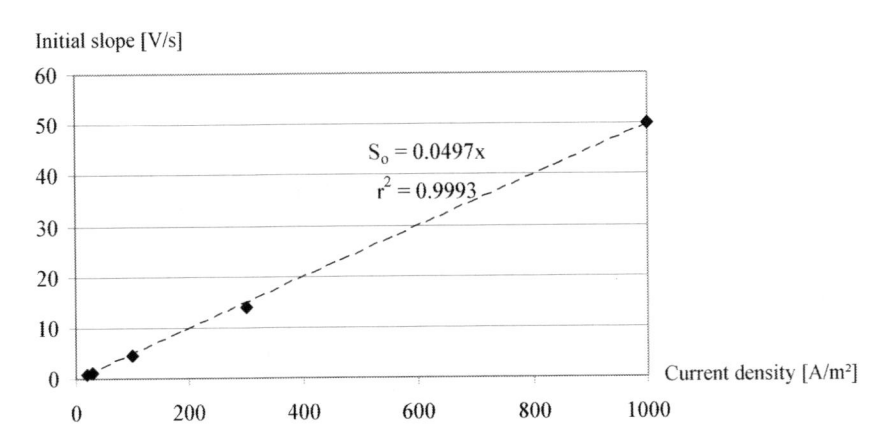

Figure 3: Evolution of the initial slope S_0 as a function of the current density J.

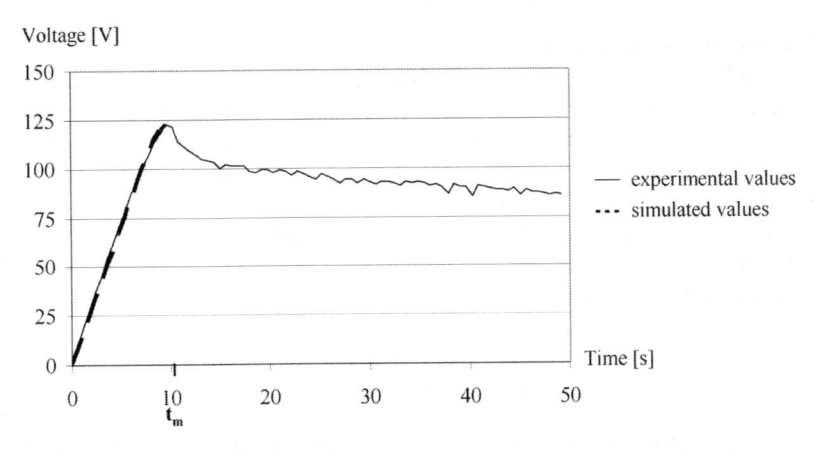

Figure 4: Fitting (for $t \leq t_m$) of the experimental voltage transient (J = 300A/m²).

3.2.4 Response after the maximum ($t_m \leq t < \infty$)

The relation used to fit the second part of the voltage transients in the 20 to 1000 A/m^2 range, is based on the previous works of Goad and Dignam [6] and Wu and Hebert [9,10]:

$$V(t) = V_{ss} + (V_0 - V_{ss}).\exp(-\lambda_2.t) \qquad (t_m \leq t < \infty) \qquad (4)$$

with $V_0 = S_0.t_m$ and λ_2 is a constant for a fixed value of the current density, under the considered anodizing conditions.

Figure 5 provides an example of the experimental transient obtained at 300 A/m^2 compared with the corresponding fitting curve ($\lambda_2 = 0.35s^{-1}$, $V_{ss} = 85.8V$ and $V_0 = 1284.1V$). In this second part, the fitting curve is closed to the experimental one apart from the first seconds after the maximum voltage V_m.

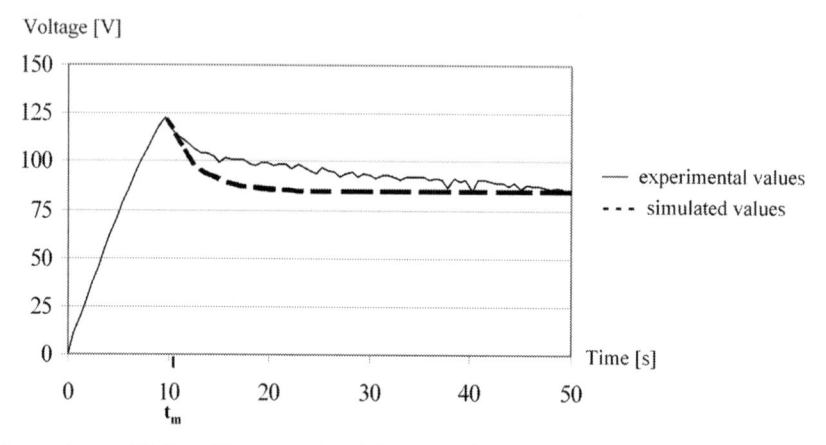

Figure 5: Fitting (for $t \geq t_m$) of the experimental voltage transient (J = 300 A/m^2).

3.3 Electrical transients and materials

3.3.1 Theoretical considerations

While reduction mechanisms were extensively investigated [15,16], only few theoretical studies concern the oxidation, especially for chronopotentiometry on metallic electrode. The more usual case is the oxidation from ion to ion in solution (Red → Ox + ne) on an inert solid metallic electrode M_s, the electrode potential depending on the kinetic of the electronic transfer through a logarithm-type relation. Considering now the anodic dissolution, i.e. the oxidation of a solid metal electrode M_s to ion in solution M^{n+} ($M_s \rightarrow M^{n+} + ne$), the electrode potential is then given by:

$$E = E^0_{M^{n+}/M} + \left(\frac{RT}{nF}\right)\ln\left(\frac{2i\sqrt{t}}{nFS\sqrt{\pi D_{M^{n+}}}}\right) \quad \text{for } C^*_{M^{n+}} = 0 \qquad (5)$$

On the other hand, the case of an anodizing, from a solid metal M_s to a solid oxide MO_s appears more unusual, since this solid-solid oxidation includes also the incorporation of oxygen from the solvent. Moreover, the chemical composition of the anodic layer is in fact complex, including various types of oxi-hydroxides, having a global electric behaviour of a semi-conductor [12]. Consequently, the two previous theoretical cases do not allow to explain the typical appearance of the experimental curves during the anodizing, specially the initial peak after galvanostatic polarisation (Figure 2).

By analogy, these typical "bell shaped" curves should be in fact very similar to the chronopotentiograms obtained during the nucleation phenomena in metal electrodeposition [17,18]. In that case, the initial potential increase is linked to the double-layer charge, while the decrease should be typical of a nucleation phenomenon. The crystallization overvoltage of this second phase of the peak was then expressed by:

$$\eta(t) = \eta_\infty - \eta_\infty \exp(-kt) \qquad (6)$$

where k is a constant and η_∞ is the difference between the electrode potentials respectively at t_∞ and t_m, time corresponding to the maximum value of the potential peak. The theoretical value of η_∞ is usually difficult to obtain because it directly depends of the faradic current i_F, whose value is not exactly known, due to the contribution of the current corresponding to the charge of the double layer. It is interesting to remark now that relation 6 is similar to the previous relation 4, suggesting similar phenomena. However, these phenomena cannot be rigorously compared because the time's scales are greatly different (lower than 0.5s for electrodeposition and higher than 40s for anodizing using low current density). So, it could be probably considered that, in the anodizing case, the initial voltage increase is only due for a weak part to the charge of double layer, and that another phenomenon predominates during the first part of the experimental curves.

3.3.2 Porosity

FEG-SEM plan views (Figure 6(a) and (b)) show that the nanoporosity appears on the sample surface as the function of the current density in the 20-1000 A/m^2 range. At low densities (Figure 6(a)), the surface is quasi-free of porosity while nanopores clearly emerge from the surface at higher density (Figure 6(b)). Other FEG-SEM views (not shown here) reveal similar facts when the anodizing time is increased at constant current density.

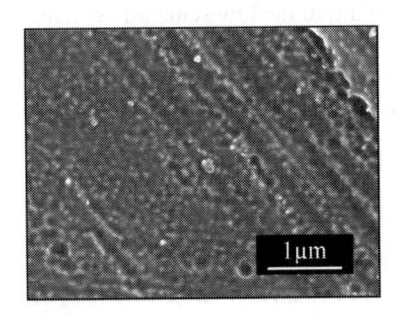

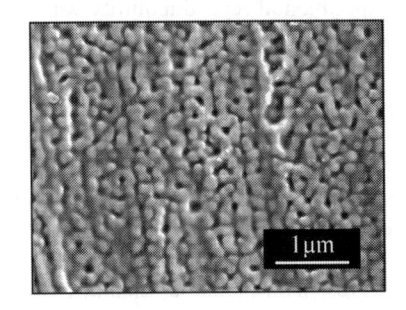

Figure 6: FEG-SEM plan view of sample surface: (a) J=20A/m², (b) J=50A/m².

By analogy with the nucleation phenomena during the metal electrodeposition, the "bell shaped" curves could be interpreted in fact by the initial formation of the highly resistive oxide layer, called the "compact layer" or the "barrier layer", contributing for the main part to the initial voltage increase. The subsequent appearance of the nanopores, inducing then the voltage decrease, induces then preferential conducting points concentrating the current lines distribution, following by analogy the example of the metal initial adatoms during the nucleation phenomena in metal electrodeposition. From this point of view, it is interesting to note that for this anodizing the V_m experimental values obtained with the galvanostatic mode are closed to the previous critical voltage value U_c defining the start of the anodic dissolution in potentiostatic mode.

At the end of the "bell shaped" curve, the steady-state voltage V_{ss} depends simultaneously on both the growth of the anodic layer and its chemical dissolution by the strong acid solution. These explanations clearly show the great influence of the electro-chemical dissolutions in the second part of the curves. So the use of a motionless electrolyte probably induces the deviation of the steady-state voltage V_{ss}, as well as the differences between experimental and fitting curves during the first seconds after the maximum (V_m, t_m).

4 Conclusion

The aim of this study was to develop mathematical relations simulating the electrical transients during the anodizing of highly pure aluminium in phosphoric electrolyte under dc electrical conditions. Electrical transients were recorded during the anodizing carried out in potentiostatic mode (25-150V) or in galvanostatic conditions (20-1000 A/m^2).

The current transients obtained under potentiostatic mode reveal two main types of curve shapes. The first one is a decreasing curve, whereas for higher voltages, the current density increases drastically as a function of the time. A critical voltage value (U_c = 120V) separates the two domains: anodizing for $U < U_c$ and anodic dissolution for higher voltage values.

For the galvanostatic mode, the voltage experimental transients show a "bell shape", characterized by some significant parameters (S_0, t_m, V_m, V_{ss}). The experimental reproducibility was good, according to the low standard deviations apart from the steady-state voltage V_{ss}.

The transient voltage curves were simulated considering two domains (before and after the maximum at t_m) according to the following mathematical relations:

$$V(t) = P_0.t - P_1.t.\exp(\lambda_1.t) \qquad (0 \leq t \leq t_m)$$
$$V(t) = V_{ss} + (V_0 - V_{ss}).\exp(-\lambda_2.t) \qquad (t_m \leq t < \infty)$$

where P_0, P_1, λ_1 and λ_2 are constants for a fixed value of the current density, under the considered anodizing conditions. All the corresponding fittings of voltage transients are in good agreement, especially for the first part of the curves, with the experimental curves for current densities in the 20 - 1000 A/m^2 range.

Then, these computational simulations were correlated with the corresponding experimental FEG-SEM plan-views. By analogy with the nucleation phenomena during the metal electrodeposition, the "bell shaped" curves could be explained by the initial formation of the highly resistive oxide layer, followed by the subsequent appearance of the nanopores, acting then like preferential conducting points concentrating the current lines distribution. The pores formation was in part explained showing that the V_m experimental values obtained with the galvanostatic mode are closed to the previous critical voltage value U_c.

So, this study developed convenient computational simulations in view to simulate and to predict the voltage transients in galvanostatic conditions, allowing finally control of the initial surface state and the nanoporosity characteristics of the anodic film. But now, additional experiments are required to understand the self-nanostructuring phenomenon involved in the preparation of the AAO templates.

References

[1] Masuda H., Fukuda K., Ordered metal nanohole arrays made by a two-step replication of honeycomb structures of anodic alumina, *Science*, **268**, pp. 1466-68, 1995.

[2] Jessensky O., Müller F., Gösele U., Self-organized formation of hexagonal pore arrays in anodic alumina, *Applied Physics Letters*, **72(10)**, pp. 1173-75, 1998.

[3] Inoue S., Chu S-L., Wada K., Li D., Haneda H., New roots to formation of nanostructures on glass surface through anodic oxidation of sputtered aluminum, *Science and Technology of Advanced Materials*, **4**, pp. 269-276, 2003.

[4] Hwang S-K., Lee J., Jeong S-H., Lee P-S., Lee K-H., Fabrication of carbon nanotube emitters in an anodic aluminium oxide nanotemplate on a Si wafer by multi-step anodization, *Nanotechnology*, **16(6)**, pp. 850-858, 2005.

[5] Qin D-H., Zhang H-L., Xu C-L., Xu T., Li H-L., Magnetic domain structure in small diameter magnetic nanowire arrays, *Applied Surface Science*, **239**, pp. 279-284, 2005.

[6] Goad D.G.W., Dignam M.J., Transient Response of the System Al/Al$_2$O$_3$/Electrolyte. Part I. Galvanostatic Transients, *Canadian Journal of Chemistry*, **50(20)**, pp.3259-3266, 1972.

[7] Patermarakis G., Lenas P., Karavassilis C., Papayiannis G., Kinetics of growth of porous anodic Al$_2$O$_3$ films on Al metal, *Electrochimica Acta*, **3**, pp. 709-725, 1991.

[8] Parkhutik V.P., Shershulsky V.I., Theoretical modelling of porous oxide growth on aluminium, *J. Phys D: Appl. Phys.*, **25**, pp. 1258-1263, 1992.

[9] Wu H., Hebert K.R., Electrochemical transients during the initial moments of anodic oxidation of aluminium, *Electrochimica Acta*, **47**, pp. 1373-1383, 2002.

[10] Wu H., Hebert K.R., Reply to comments on « Electrochemical transients during the initial moments of anodic oxidation of aluminum », *Electrochimica Acta*, **48(2)**, pp. 131-133, 2002.

[11] Zamora G., Arurault L., Bes R.S., Energetics of aluminium alloys anodization for porous oxide films elaboration, *Surface Treatment VI – Computer Methods and Experimental Measurements for Surface Treatment Effects*, ed. C.A. Brebbia, J.T.M. de Hosson and S.I. Nishida, WIT Press, Southampton, pp 51-59, 2003.

[12] Zamora G., PhD Thesis, Paul Sabatier University, Toulouse France, 07/01/2005 (in French).

[13] Weast R.C., Handbook of Chemistry and Physics, CRC Press, London, 1975-1976.

[14] Gabe D.R., Density values for anodic films on aluminium and some observations of pore morphology, Trans IMF, **78(6)**, pp. 207-209, 2000.

[15] Delahay P., Mamantov G., Voltammetry at constant current: Review of theoretical principles, *Anal. Chem.*, **27(4)**, pp. 478-83, 1955.

[16] Reinmuth W. H., Chronopotentiometric Potential-Time Curves and Their Interpretation, *Anal.Chem.*, **32(11)**, pp. 1514-17, 1960.

[17] Fleischmann M., Thirsk H.R., Anodic electrocrystallization, *Electrochimica Acta*, **2(1-3)**, pp. 22-49,1960.

[18] Chamelot P., Lafage B., Taxil P., Studies of niobium electrocrystallization phenomena in molten fluorides, *J. Electrochem. Soc.*, **143(5)**, pp. 1570-1576, 1996.

Extraction of a quantitative reaction mechanism from linear sweep voltammograms obtained on a rotating disk electrode

E. Tourwé[1], T. Breugelmans[1,2], R. Pintelon[3] & A. Hubin[1]
[1] *Vrije Universiteit Brussel, Department of Metallurgy, Electrochemistry and Materials Science, Belgium*
[2] *Hogeschool Antwerpen, Department of Industrial Sciences and Technology – Chemistry, Belgium*
[3] *Vrije Universiteit Brussel, Department of Fundamental Electricity and Instrumentation, Belgium*

Abstract

A new methodology to quantitatively determine the mass and charge transfer parameters of an electrochemical reaction is applied successfully in this work for the study of the reduction of ferricyanide to ferrocyanide at a platinum electrode, rotating at 1000 rpm. It is concluded that the reaction can be described by a simple electron transfer and the values of the charge transfer parameters are: α_{ox} = 5.00E-01 ± 8.75E-03 , k_{ox} = 1.39E-08 ± 2.28E-09 and k_{red} = 1.25E+00 ± 2.01E-01.

1 Introduction

The aim of a kinetic study of an electrochemical reaction with a relatively simple mechanism (only mass and charge transfer steps, no adsorption, chemical reactions, etc.) is the determination of its reaction mechanism and the quantification of its kinetic charge transfer parameters (rate constants and transfer coefficients) and mass transfer parameters (diffusion coefficients). Linear sweep voltammetry (LSV) in combination with a rotating disc electrode (RDE) is a powerful technique for providing information on the mechanism and kinetics of an electrochemical reaction. Previously we set-up a statistically founded method to model an electrochemical reaction and to determine its mass and charge transfer parameters quantitatively [1].

The method requires the proposition of an appropriate mechanism for the reaction under study and its mathematical translation into an expression that analytically describes the voltammogram. Powerful parameter estimation algorithms (maximum likelihood combined with Gauss–Newton and Levenberg–Marquardt minimization methods) are used to adjust the values of these model parameters, in order to obtain a good agreement between experimental and modeled data. The values of the model parameters that give rise to the best match, characterize the system quantitatively. Moreover, the method provides error estimates of the obtained parameter values. It is however only after a statistical evaluation of the obtained results, that it is decided whether the model is able to describe the experiments or not.

The simplest electrochemical reactions, which can be found among the different kinds of electrode processes, are those where electrons are exchanged across the interface by flipping oxidation states of transition metal ions in the electrolyte adjacent to the electrode surface [2]. The electrode acts as the source or sink of electrons for the redox reaction and is supposed to be inert. The reduction of ferricyanide to ferrocyanide at a platinum electrode is described in literature [2–5] as an example of such a mechanism, i.e.

$$Fe(CN)_6^{3-} + e^- \leftrightarrow Fe(CN)_6^{4-} \tag{1}$$

Because no complex mechanism is expected to take place, this system is chosen here to evaluate the methodology that was set-up by our group previously [1].

Diverging values for the charge transfer parameters of this reaction are found in literature. The rate constants (defined w.r.t. the overpotential) range from 5×10^{-4} m/s [6,7], over 1×10^{-3} m/s [5], to 2.4×10^{-3} m/s [8]. Beriet and Pletcher [9] found values from $1\ 10^{-3}$ to 2.5×10^{-3} m/s, using both steady state and rapid scan voltammetry. Literature values for α vary from 0.45 to 0.61 [4,7]. This work aims to determine the parameter values in a reliable, statistically founded manner and to provide error estimates on these values.

2 Experimental

2.1 Composition of the electrolyte

The following analytical reagents are used (all Merck p.a.): $K_4[Fe(CN)_6].3H_2O$, $K_3[Fe(CN)_6]$ and KCl. Solutions are made with once-distilled and deionized water. A 1M KCl solution is used as the supporting electrolyte and the concentrations of the electroactive components ferri/ferrocyanide are 0.005M. In that way, a negligible migration flux, constant activity and diffusion coefficients of the electroactive species, a low electrolyte resistance and a uniform current distribution are aimed at.

2.2 Experimental set-up

A typical three electrode set-up is used for the electrochemical experiments [2, 10]. The electrochemical cell contains a Ag/AgCl reference electrode (Schott-Geräte), a platinum rotating disk working electrode and a platinum grid as counter electrode. The RDE is made at our department by embedding a 4 mm diameter polycrystalline platinum rod in an insulating mantle of polyvinlidenefluoride. The electrode is rotated by an RDE control system of Autolab. The rotation speed is set to 1000 rpm. The voltammograms are measured using a high resolution galvanostat/potentiostat PGSTAT30 (Autolab Instruments) of Ecochemie, controlled by the GPES 4.8 software. The scan rate is taken constant at 1mV/s. The step potential is set to 0.00015 V, and this way a maximum number of data points is measured.

All measurements are performed in a 200 ml glass electrolytic cell, thermostatted at $25 \pm 0.5°C$ using a water jacket connected to a thermostat bath (Lauda RE304).

Prior to the measurements, the electrolyte is deoxygenated by bubbling with nitrogen gas (Air Liquide) for 10 minutes, while during the experiment a nitrogen blanket is maintained over the cell. This results in a substantial flattening of the reduction plateau of ferricyanide. Besides, the cell is always shielded from light in order to avoid the following photochemical decomposition of ferrocyanide [11]:

$$Fe(CN)_6^{4-} \stackrel{light}{\leftrightarrow} Fe(CN)_5^{3-} + CN^- \quad (2)$$

$$CN^- + H_2O \leftrightarrow HCN + OH^- \quad (3)$$

2.3 Electrode pretreatment

The reproducibility of the measurements was strongly increased by means of applying the following standardized pretreatment of the electrode surface:
- mechanical polishing of the platinum electrode on a rotating disk (Struers DP10, on cloth), succesively using a diamond paste of 7 μm and of 1 μm (Struers);
- ultrasonic rinsing with deionized water followed by degreasing with chloroform, also in an ultrasonic bath (Elma model T470/H) for 2 minutes;
- before each experiment, the electrode potential was 4 times swept between +0.55 and -0.45V vs Ag/AgCl, at a scan velocity of 10 mV/s and a rotation speed of 1000 rpm.

It is reported in literature [12], that such a pretreatment removes oxide and trace contaminants from the Pt surface, while the O_2 and H_2 evolution reactions are avoided in this potential range.

Each experiment is performed on a freshly prepared electrode.

3 Results and discussion

The method to determine the mechanism and the values of its characteristic parameters is founded on four building blocks [1].

The results of the experimental study These are the current/potential couples defining the polarization curve. The mean of 11 experiments that were performed under identical conditions is used for modeling.

The mathematical expression believed to explain the experimental results This expression is derived from the basic equations that describe what is happening during an electrochemical reaction. It is formulated based on a well-considered model for the studied reaction. It has the following form: *current = function (potential, experimental parameters, model parameters)*, where the *experimental parameters* describe the experimental conditions, like e.g. temperature, rotation speed of the RDE, concentration, ..., and the *model parameters* are the unknown parameters that need to be quantitatively determined, like e.g. rate constants, transfer coefficients, etc.

The fitting procedure In this block the differences between experimental and theoretical data are minimized. Therefore, a weighted least squares cost function V_{WLS} is formulated:

$$V_{WLS} = \frac{1}{2} \sum_{l=1}^{N_{dp}} \frac{(I_m(l) - I(l, \theta))^2}{\sigma_I(l)^2} = \frac{1}{2} \epsilon^T \epsilon \qquad (4)$$

where:

- N_{dp}: the number of data point in the experimental polarization curve
- $I_m(l)$: the mean value of the measured current
- σ_I: standard deviation on I_m, calculated from M repeated measurements
- $\epsilon \in \mathbb{R}^{1 \times N_{dp}}$: the error vector given by

$$\epsilon = \begin{pmatrix} \frac{I_m(1) - I(1, \theta)}{\sigma_I(1)} \\ \vdots \\ \frac{I_m(l) - I(l, \theta)}{\sigma_I(l)} \\ \vdots \end{pmatrix} \qquad (5)$$

- $I(l, \theta)$: the 'model' value of I
- θ: the model parameter vector

The Gauss–Newton and Levenberg–Marquardt method are implemented to minimize this cost function and eventually it provides the parameter values which best describe the data. Moreover, the standard deviation on these parameters is also calculated.

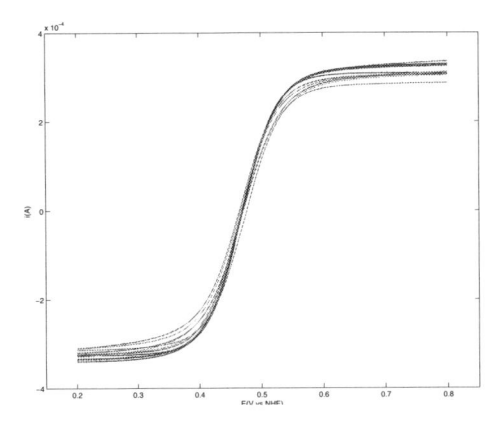

Figure 1: Voltammograms of the reduction/oxidation of 0.005M ferri/ferrocyanide in 1 M KCl, at 1000 rpm.

A statistical evaluation If a statistical evaluation of the fitting results demonstrate a good description of the experiment by the model, a quantitative reaction mechanism is obtained. If, on the other hand, no good agreement between experiment and model is achieved, a new mechanism has to be proposed and the 2 previous steps should be repeated.

The results for each of these blocks are discussed in the following sections.

3.1 Results of the experimental study

The equilibrium potential of a 0.005M ferri/ferrocyanide solution in 1M KCL equalled 0.470 ± 0.001 V/NHE. As advised in [1], 11 polarization curves are measured under identical experimental conditions. The results are shown in figure 1.

From figure 1 it is seen that the current depends on the electrode potential for overvoltages of about ± 100 mV. At higher overvoltages, a limiting current is reached.

Figure 2 shows the mean experimental voltammogram and its 95% confidence interval ±2σ, with σ the standard deviation of the current, calculated from figure 1.

3.2 The mathematical expression believed to explain the experimental results

For reaction (1) a mathematical expression that describes the polarization curves is derived, taking into account mass and charge transfer. The basic equations which are used, were described previously [1].

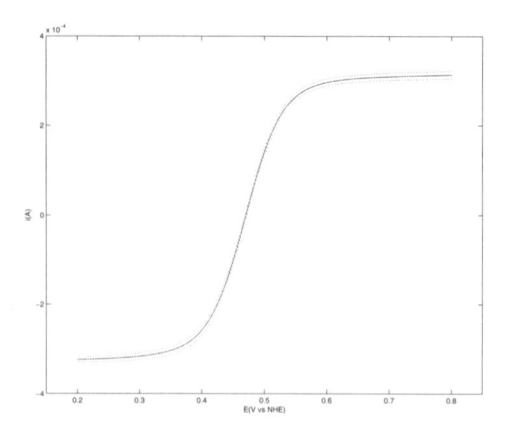

178 Simulation of Electrochemical Processes II

Figure 2: Mean voltammogram and its 95% confidence interval of the reduction/oxidation of 0.005M ferri/ferrocyanide in 1 M KCl, at 2000 rpm.

The following expression for the current as a function of potential is obtained:

$$i = \frac{nFS(K_{ox}c^*_{red} - K_{red}c^*_{ox})}{1 + nFS\left(\frac{K_{ox}c^*_{red}}{i_{lim,ox}} - \frac{K_{red}c^*_{ox}}{i_{lim,red}}\right)} \qquad (6)$$

with:
- i: the current(in A)
- n: the number of electrons exchanged in the reaction
- F: Faraday's constant (96485 C/mol)
- K_{ox}: the potential dependent rate constant for the oxidation half reaction, given by: $K_{ox} = k_{ox} \exp \frac{\alpha_{ox} nFE}{RT}$
- K_{red}: the potential dependent rate constant for the reduction half reaction, given by: $K_{red} = k_{red} \exp \frac{-\alpha_{red} nFE}{RT}$
- E: the potential (in V vs NHE)
- R: ideal gas constant (8.32 J/molK)
- T: the absolute temperature (in K)
- c^*_{red}: the bulk concentration of the reducing agent (ferrocyanide) (in mol/m^3)
- c^*_{ox}: the bulk concentration of the oxidizing agent (ferricyanide) (in mol/m^3)
- $i_{lim,ox}$: the oxidation limiting current (in A)
- $i_{lim,red}$: the reduction limiting current (in A)
- k_{ox}: the rate constant for the oxidation half reaction (in m/s)
- k_{red}: the rate constant for the reduction half reaction (in m/s)
- α_{ox}: the transfer coefficient in the sense of the oxidation
- α_{red}: the transfer coefficient in the sense of the oxidation

3.3 The equations needed for the fitting procedure

Assuming that the sum of α_{ox} and α_{red} equals one, a model for the current (equation 6) with 3 unknown parameters is obtained, viz. α_{ox}, k_{ox} en k_{red}.

Next to this expression for the current, the minimization algorithms require expressions for the derivatives of the current w.r.t. the unknown parameters in the Jacobian matrix. These are given by:

$$\frac{\delta i}{\delta \alpha_{ox}} = \frac{n^2 SEF^2 i_{lim,ox}^2 i_{lim,red}^2 c_{red}^* \exp\frac{\alpha_{ox}EFn}{RT} k_{ox}}{R(c_{red}^* \exp\frac{\alpha_{ox}EFn}{RT} F i_{lim,red} k_{ox} nS + i_{lim,ox}(i_{lim,red} - c_{ox}^* \exp\frac{(-1+\alpha_{ox})EFn}{RT} F k_{red} nS))^2 T}$$

$$- \frac{n^2 SEF^2 i_{lim,ox}^2 i_{lim,red}^2 c_{ox}^* \exp\frac{(-1+\alpha_{ox})EFn}{RT} k_{red}}{R(c_{red}^* \exp\frac{\alpha_{ox}EFn}{RT} F i_{lim,red} k_{ox} nS + i_{lim,ox}(i_{lim,red} - c_{ox}^* \exp\frac{(-1+\alpha_{ox})EFn}{RT} F k_{red} nS))^2 T}$$

(7)

$$\frac{\delta i}{\delta k_{ox}} =$$

$$\frac{c_{red}^* \exp\frac{\alpha_{ox}EFn}{RT} F i_{lim,ox} i_{lim,red} nS(-i_{lim,ox} i_{lim,red})}{(c_{red}^* \exp\frac{\alpha_{ox}EFn}{RT} F i_{lim,red} k_{ox} nS + i_{lim,ox}(i_{lim,red} - c_{ox}^* \exp\frac{(-1+\alpha_{ox})EFn}{RT} F k_{red} nS))^2}$$

$$+ \frac{c_{red}^* \exp\frac{\alpha_{ox}EFn}{RT} F i_{lim,ox} i_{lim,red} nS c_{ox}^* \exp(-1+\alpha_{ox})EFnRTF(i_{lim,ox} - i_{lim,red}) k_{red} nS}{(c_{red}^* \exp\frac{\alpha_{ox}EFn}{RT} F i_{lim,red} k_{ox} nS + i_{lim,ox}(i_{lim,red} - c_{ox}^* \exp\frac{(-1+\alpha_{ox})EFn}{RT} F k_{red} nS))^2}$$

(8)

$$\frac{\delta i}{\delta k_{red}} =$$

$$\frac{c_{ox}^* \exp\frac{(-1+\alpha_{ox})EFn}{RT} F i_{lim,ox} i_{lim,red} nS(-i_{lim,ox} i_{lim,red})}{(c_{red}^* \exp\frac{\alpha_{ox}EFn}{RT} F i_{lim,red} k_{ox} nS + i_{lim,ox}(i_{lim,red} - c_{ox}^* \exp\frac{(-1+\alpha_{ox})EFn}{RT} F k_{red} nS))^2}$$

$$+ \frac{c_{ox}^* \exp\frac{(-1+\alpha_{ox})EFn}{RT} F i_{lim,ox} i_{lim,red} nS c_{red}^* \exp\frac{\alpha_{ox}EFn}{RT} F(i_{lim,ox} - i_{lim,red}) k_{ox} nS}{(c_{red}^* \exp\frac{\alpha_{ox}EFn}{RT} F i_{lim,red} k_{ox} nS + i_{lim,ox}(i_{lim,red} - c_{ox}^* \exp\frac{(-1+\alpha_{ox})EFn}{RT} F k_{red} nS))^2}$$

(9)

3.4 Fitting results and statistical evaluation

To initiate the minimization procedure starting values are needed for the unknown model parameters and the values found in [13] are used for this purpose. The values for the other parameters (like temperature, bulk concentrations, etc.) can be found in the experimental section. Values for the limiting currents can be easily derived from figure 2. The oxidation and reduction limiting currents are fixed to 3.1×10^{-4} A and -3.3×10^{-4} A respectively.

At this point everything is ready to start the fitting procedure. The theoretical expression for the current (equation 6) will be fitted to the mean experimental polarization curve of figure 2, using the method described previously in [1].

The fitting results are illustrated in figure 3. Part (a) of this figure shows a comparison between the experimental polarization curve and a modeled curve, calculated with the best-fit-parameters. An exceptional agreement is obtained

(a)

parameter	estimated value	rel. std
α_{ox}	6.16E-01	3.87E-03
k_{ox}	1.52E-09	4.05E-02
k_{red}	1.40E-01	3.94E-02

(b)

cost	condition number
9.53E+02	8.78E+01

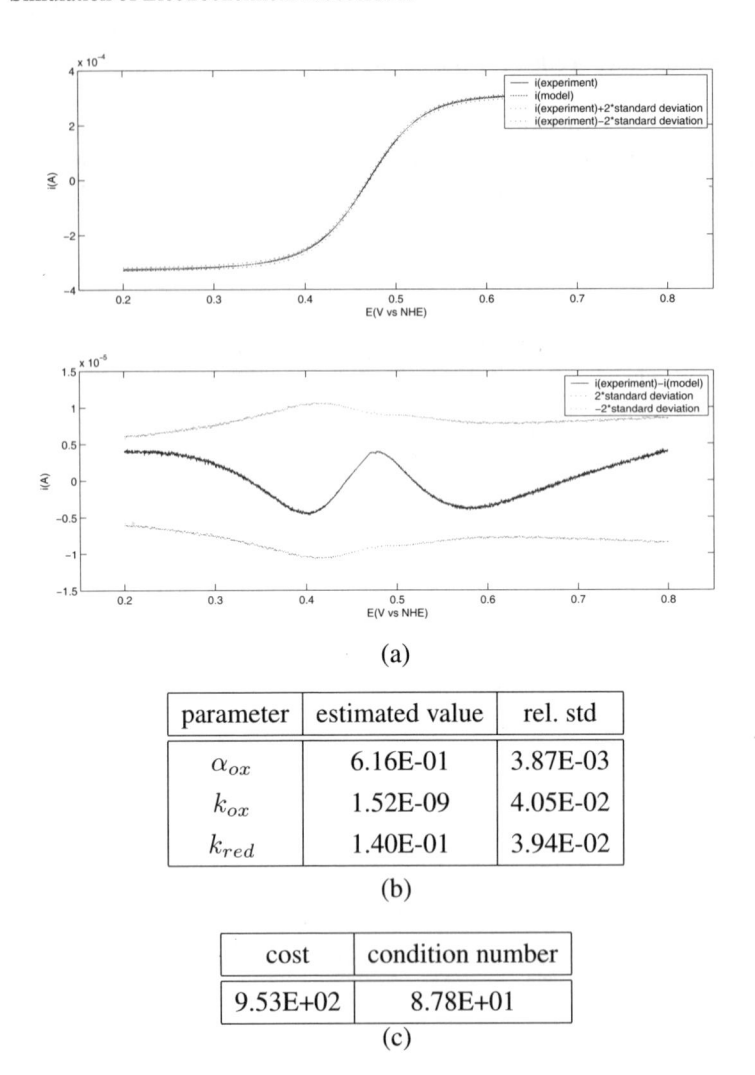

(c)

Figure 3: Results of the fitting of equation 6 to the mean polarization curve at 1000 rpm (3 parameter model): (a) comparison of the mean experiment with the model. (b) best-fit parameter vector and its relative standard deviation and (c) cost function and condition number.

and both curves are quasi indistinguishable. Also, the difference between the experimental and modeled data is plotted. It is observed that this difference lies in the confidence band, which is defined by ± two times the standard deviation on the current, calculated from figure 1. When performing multiple experiments, 95% of the experiments are expected to fall in this interval. It is therefore concluded that the model is able to describe the experiment appropriately. This is also evidenced

Table 1: Best-fit parameter vector and its relative standard deviation, obtained when fitting the 5 parameter model to the mean polarization curve at 1000 rpm.

parameter	estimated value	rel. std
α_{ox}	5.00E-01	8.76E-03
k_{ox}	1.39E-08	8.19E-02
k_{red}	1.25E+00	8.07E-02
$i_{lim,ox}$	3.11E-04	5.79E-04
$i_{lim,red}$	-3.22E-04	6.12E-04

by the low value of the cost function. The best-fit values for the the model parameters and their *relative* standard deviation are shown in part (b) of figure 3.

3.5 A 5 parameter model

The oxidation and reduction limiting currents are in fact also model parameter that need to be estimated. They replace the more obvious, intrinsic model parameters, viz. the diffusion coefficients of the oxidizing and reducing species. Previously the limiting currents were not considered as model parameters because good estimates are available from the experimental curves. However, it is preferred now to include them as model parameters because this way better estimates can be obtained. Consequently, a 5 parameter model is obtained and the expression for the Jacobian is adapted accordingly.

Again, an excellent match between modeled and experimental data is observed (figures not shown). The value of the cost function decreases from 9.53E+02 for the 3 parameter model to 1.42E+02. This indicates an even better agreement between model and experiment. The values for the best-fit-parameters and their relative standard deviation are shown in table 1. They differ slightly from those obtained for the 3 parameter model. As in the latter model the values for the limiting currents are not determined as accurately as in the 5 parameter model, the results of the 5 parameter model are considered as the best estimates.

It has to remarked that by performing these experiments at several rotation speeds of the RDE, the diffusion coefficients of the ferri- and ferrocyanide species can be calculated by the Koutecky-Levich method. As very accurate values for the limiting current are provided by this methodology, the diffusion coefficients will also be estimated accurately.

If the values for the rate constants are calculated w.r.t. the overpotential η (e.g. $K_{ox} = k'_{ox} \exp \frac{\alpha_{ox} nF\eta}{RT}$) instead of w.r.t. the potential one obtains the following values: k'_{ox} = 1.9E-04 and k'_{red} = 1.9E-04. These are of the same order of magnitude as the literature values [4–7, 9].

4 Conclusions

A methodology to quantitatively determine the mass and charge transfer parameters of an electrochemical reaction that was proposed previously in [1] is validated and applied successfully in this work for the study of the redox couple ferri/ferrocyanide. It is concluded that the reaction mechanism is given by reaction (1) and the values of the charge transfer parameters are: α_{ox} = 5.00E-01 ± 8.75E-03, k_{ox} = 1.39E-08 ± 2.28E-09 and k_{red} = 1.25E+00 ± 2.01E-01.

Acknowledgements

Els Tourwé thanks the Flemish Institute for support of Scientific-Technological Research in Industry (I.W.T.).

References

[1] Tourwé, E., Pintelon, R. & Hubin, A., *Journal of Electroanalytical Chemistry*, **594(1)**, pp. 50–58, 2006.
[2] Bamford, C. & Compton, R., *Electrode Kinetics: Principles and Methodology*, volume 26 of *Comprehensive Chemical Kinetics*. Elsevier Science Publishers, 1986.
[3] Iwasita, T., Schmickelr, W., Hermann, J. & Vogel, U. *Journal of Electrochemical Society*, **130**, p. 2026, 1983.
[4] Angell, D. & Dickinson, T. *Journal of Electroanalytical Chemistry*, **35**, p. 55, 1972.
[5] Bruce, P., Lisowska-Oleksiak, A., Los, P. & Vincent, C. *Journal of Electroanalytical Chemistry*, **367**, p. 279, 1994.
[6] Jahn, D. & Vielstich, W. *Journal of Electrochemical Society*, **109**, p. 849, 1962.
[7] Tanaka, N. & Tamamushi, R. *Electrochimica Acta*, **9**, p. 963, 1964.
[8] Daum, P. & Enke, C. *Analytical Chemistry*, **41**, p. 653, 1969.
[9] Beriet, C. & Pletcher, D. *Journal of Electroanalytical Chemistry*, **361**, p. 93, 1993.
[10] Diard, J.P., Le Gorrec, B. & Montella, C., *Cinétique électrochimique*. Hermann, 1996.
[11] Eisenberg, M., Tobias, C. & Wilke, C. *Journal of Electrochemical Society*, **101**, p. 306, 1954.
[12] Robertson, B., Tribollet, B. & Deslouis, C. *Journal of Electrochemical Society*, **135**, p. 2279, 1988.
[13] Vandeputte, S., *Frequency response analysis of electrochemical systems at a rotating disk electrode under sinusoidal potentials or flow modulation*. Ph.D. thesis, Vrije Universiteit Brussel, Brussel, Belgium, 1996.

IRDE and RDE electrochemical cells evaluation: comparison of electron and mass transfer

H. Van Parys[1], E. Tourwé[1], M. Depauw[1], T. Breugelmans[1,2], J. Deconinck[3] & A. Hubin[1]
[1]*Vrije Universiteit Brussel, Department of Metallurgy, Electrochemistry and Materials Science, Brussels, Belgium*
[2]*Hogeschool Antwerpen, Department of Industrial Sciences and Technology-Chemistry, Antwerp, Belgium*
[3]*Vrije Universiteit Brussel, Computational Electrochemistry Group, Department of Electrical Engineering and Power Electronics, Brussels, Belgium*

Abstract

An inverted rotating disk electrode reactor (IRDE) is constructed to facilitate the study of electrochemically formed gas bubbles. The hydrodynamic and mass transfer characteristics of this new cell design are validated by means of an electrochemical reaction with known characteristics, i.e. the ferri/ferro cyanide redox reaction. This validation is done both qualitatively and quantitatively. The qualitative validation consists of comparing the entire polarization curve obtained in the IRDE with the one obtained in the classical RDE set-up under the same experimental conditions. To quantitatively validate the cell another approach was pursued. A previously developed statistically founded method to model an electrochemical system and to quantitatively determine its mass and charge transfer parameters is used. It is found that the mass and charge characteristics in both configurations agree very well.
Keywords: inverted rotating disk electrode, IRDE, ferri/ferro cyanide, modeling.

1 Introduction

The study of electrochemically formed gas bubbles on an electrode surface and their influence on mass transfer are of great industrial importance. The formation of these gas bubbles can have several effects on the electrode behaviour, which have to be taken into consideration when dealing with gas evolution reactions. For example if gas evolution takes place on the electrode surface, not only will it consume part of the current, it will also have an important effect on the local process parameters and current distribution. Although indispensable for reactor design and optimization of new and existing industrial processes, no general model has yet been developed to describe the influence of gas bubbles on mass transport. The rotating disk electrode configuration is most often used in deterministic studies of the reaction kinetics. Due to the controlled convection obtained by rotating the disk electrode, it is possible to distinguish between mass and charge transport control in surface reactions. But the RDE has one drawback when dealing with gas evolution reactions, namely the formed gas bubbles tend to stick to the electrode disk and shield the active electrode surface in this way. The inverted RDE configuration can provide a solution to this problem, since the electrode surface is facing upwards instead of downwards. This way the generated gas bubbles can detach either by buoyancy or be swept away by the rotational movement of the electrode. So the gas bubbles can rise freely to the surface and are no longer shielding the electrode. Although the change in position of the working electrode in the inverted RDE from the top to the bottom of the electrochemical cell seems to be a minor change in the experimental set-up, it has to be validated whether the mass and charge transfer equations governing the classical RDE can be transferred to the inverted RDE configuration. Therefore, before going to electrochemical reactions involving gas evolution, a system needs to be chosen which has well-known mass transfer characteristics and where no influence of the gas bubbles on the flow exists. Since the ferri-/ferro cyanide redox system is extensively studied [1-5], it is chosen as a model system to validate the inverted rotating disk electrode configuration (IRDE). In literature [6-9] the mass transfer characteristics of the IRDE are tested by measuring limiting currents in the ferricyanide reduction system. Subsequently the diffusion coefficient is calculated from the Koutecky-Levich plot and is compared to the ones available in literature. Since mass transfer also plays its role in the part of the curve where charge transfer is dominant, a more rigorous validation technique consists in considering the whole polarization curve. Moreover, the validation will not only be done in a qualitative way by comparing the polarization curves obtained in the classical rotating disk set-up versus the ones measured in the inverted one, but also in a quantitative way. A new statistically founded method to model an electrochemical system and to quantitatively determine its mass and charge transfer parameters is recently developed and will be used in this work [10]. We believe this is, compared to the technique of qualitatively comparing part of the experimental data, a much more powerful tool to validate an electrochemical cell.

In this work an inverted rotating disk electrode is constructed and is validated in both a qualitative and a quantitative way. The aim is to determine the charge and mass transfer parameters of the ferri/ferro redox system obtained in both the classical and the inverted RDE set-up and to compare them. In this way it can be decided on a reliable and statistically founded basis whether the inverted RDE obeys the same hydrodynamic and mass transport conditions as the classical RDE set-up, providing a suitable instrument for the study of electrochemical reactions taking place with simultaneous gas evolution.

2 Experimental

Electrochemical measurements are performed with a three electrode configuration using a PGSTAT100® potentiostat from Ecochemie®, controlled by the GPES 4.8 software. The counter electrode is a large platinum grid and the reference electrode is a saturated Ag/AgCl electrode (Schott-Geräte). The working electrode is a rotating disk electrode of platinum, made by embedding a platinum rod (Alfa Aesar, 99.99% purity) in an insulating PVDF cylinder. The rotation speed of the working electrode is kept constant at 1000rpm. The polarization curves are obtained at a rate of 1mVs^{-1} to reach quasi steady-state measurements at each point of the curve. The step potential is fixed at 0.00015V in order to measure the maximal amount of data points the software allows.

For the preparation of the solutions once-distilled and demineralized water and analytical reagents (all Merck p.a.) are used. Equimolar solutions of potassium ferro- and ferricyanide 0.005M $K_3Fe(CN)_6$ en $K_4Fe(CN)_6$ are used. A 1M KCl solution is used as supporting electrolyte. In this way the migration flux can be neglected and the activity constants and diffusion coefficients of the electro-active components Fe^{2+} and Fe^{3+} can be considered to remain constant during the oxidation/reduction process. Prior to each measurement the electrolyte is bubbled with nitrogen gas for 5 minutes in order to deoxygenate the electrolyte. In this way the oxygen reduction is reduced and doesn't interfere with the reduction of ferricyanide resulting in a flattened reduction plateau. The electrochemical cell is shielded from light in order to avoid the following photochemical decomposition of ferrocyanide [9], i.e.

$$Fe(CN)_6^{4-} \xleftrightarrow{light} Fe(CN)_5^{3-} + CN^-$$
$$CN^- + H_2O \leftrightarrow HCN + OH^-$$
(1)

All measurements are thermostatically controlled at 25 ± 0.1°C by means of a waterjacket around the cell, of which the temperature is controlled by a thermostat (Lauda RE304).

To improve the reproducibility of the measurements the following standardized pretreatment of the electrode surface is applied [9]:
- Mechanical polishing of the platinum electrode on a rotating disk (Struers DP10, on cloth), successively using a diamond paste of 7μm and 1 μm (Struers)

- Ultrasonic rinsing with deionized water followed by degreasing with chloroform, both in an ultrasonic bath (Elma® model T470/H) for 3 minutes
- Before each experiment, the electrode potential was swept 3 times between +0.55V and -0.45V vs. Ag/AgCl sat., at a scan velocity of 0.01V/s and a rotation speed of 1000rpm. It is reported in literature [11], that such a pretreatment removes oxide and trace contaminants from the Pt surface, while the O_2 and H_2 evolution reactions are avoided in this potential range.

Before each experiment this electrode pretreatment procedure is repeated.

3 Results and discussion

3.1 Cell design of the IRDE

As in the classical RDE set-up, the inverted configuration consists of a cylindrical vessel with a rotating working electrode, now placed at the bottom of the cell, a counter electrode positioned at a fixed point at the top of the vessel and the reference electrode also placed at the top of the cylindrical cell (see Figure 1). A construction of O-rings keeps the electrolyte from leaking at the bottom of the rotating electrode. The main difference between the two configurations is that in the inverted RDE set-up the working electrode is protruding into the solution what can lead to the creation of additional vortices in the flow introducing a different flow pattern compared to the classical RDE.

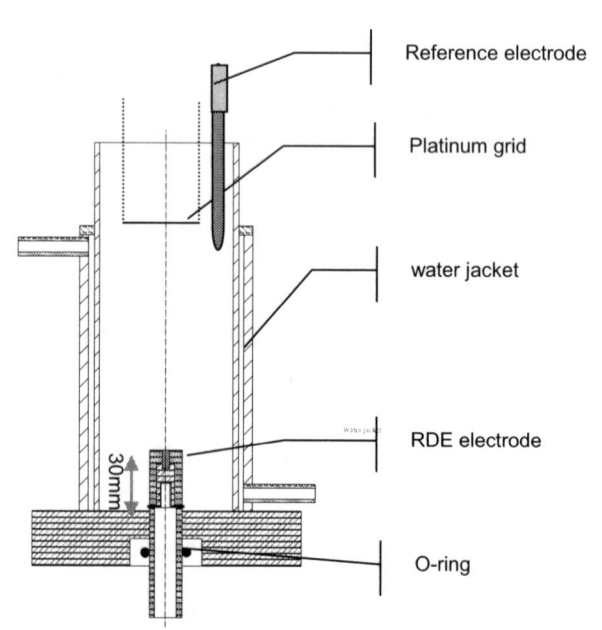

Figure 1: Sketch of the IRDE cell (front view), not a scale drawing.

3.2 Validation of the inverted RDE configuration

The reduction of ferricyanide to ferrocyanide at a platinum electrode is an example of a relatively simple electrode process in which an electron is exchanged across the electrode/electrolyte interface and in which no complex mechanism is expected to take place, i.e.

$$Fe(CN)_6^{3-} + e^- \leftrightarrow Fe(CN)_6^{4-} \qquad (2)$$

As advised in [10] at least 11 polarization curves with the same experimental conditions are recorded in the inverted RDE. Exactly the same experiments are performed in the RDE set-up in the papers of E. Tourwé [12,13]. The latter experimental series are considered to be reference data to compare our data to.

3.2.1 Qualitative validation of the mass transfer characteristics in IRDE

The mean polarization curve is calculated from the 11 polarization curves by calculating the mean of the current at each data point. The 95% confidence interval is obtained in a similar way, namely by calculating the standard deviation of the mean current at each data point. The 95% confidence interval is then given by +/- 2 times the standard deviation. This interval gives the range of current values that is expected to include 95% of all experimental data. The mean polarization curves and their respective confidence intervals obtained in both the classical and the inverted RDE configuration are compared in Figure 2. Note that the confidence interval for the IRDE is very small and therefore difficult to observe in figure 2. In the inset, a magnification of an arbitrary part of the polarization curve is shown. It is seen that the 95% confidence interval obtained in the inverted RDE falls within the one obtained in the classical RDE. So on an experimental basis it can be concluded that the same polarization curves are measured in both configurations and that they compare well. Remark that the confidence interval of the polarization curve obtained in the IRDE is smaller that the one obtained in the RDE. This could indicate that the IRDE is a more accurate tool to measure reaction kinetics than the RDE. Yet, this still has to be established by an extended set of data under different experimental conditions, such as rotating speed, temperature, etc.

3.2.2 Quantitative validation of the mass transfer characteristics in IRDE

To determine the kinetic behavior of the system under study, the methodology described in the papers written by E. Tourwé [10,12] is used. Therefore an appropriate reaction mechanism has to be proposed. This mechanism is then translated in a mathematical expression that analytically describes the polarization curve. In a next step powerful parameter estimation algorithms (maximum likelihood combined with Gauss–Newton and Levenberg Marquardt minimization methods) are used to adjust the values of these model parameters, in order to obtain a good agreement between experimental and modeled data. When a good match between the experimental and simulated data is achieved, the model can be retained and the kinetic parameters of the system are quantified. However it is only after a statistical evaluation of the fit that it is decided whether the model describes the system well.

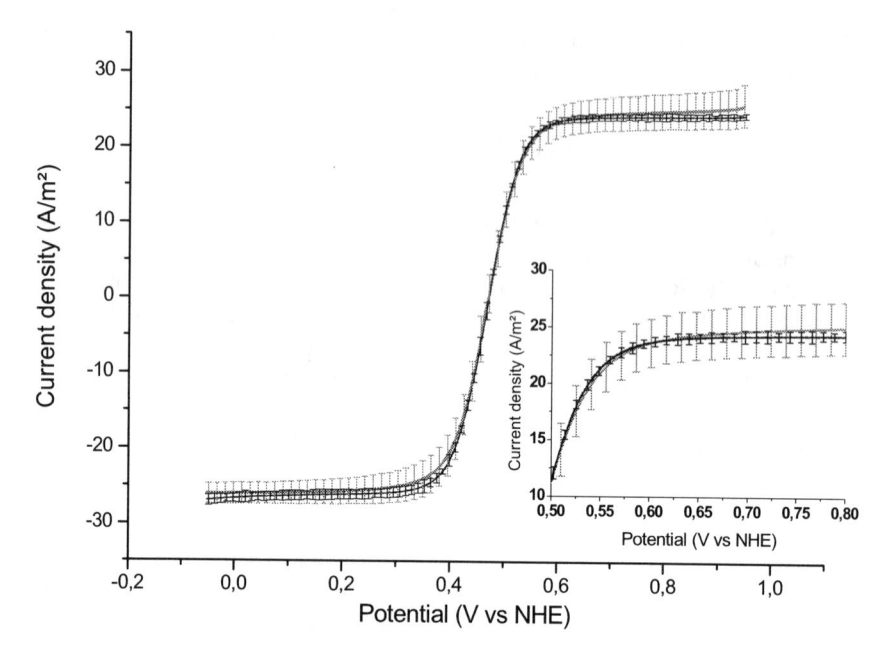

Figure 2: Mean voltammogram of the reduction/oxidation of 0,005M ferri/ferrocyanide in 1M KCl obtained in the classic RDE set-up (grey) and in the inverted RDE set-up (black) and their respective 95% confidence interval, at 1000 rpm.

The mathematical expression believed to describe the current density-potential relationship of the electrochemical reaction (see eqn (2)) is the following set of equations. These expressions take into account mass and charge transfer.

$$red \leftrightarrow ox + n\,e^-$$

$$j = \frac{nF(K_{ox}c_{red}^* - K_{red}c_{ox}^*)}{\frac{1}{S} + nF(\frac{K_{ox}c_{red}^*}{i_{\lim,ox}} - \frac{K_{red}c_{ox}^*}{i_{\lim,red}})} \tag{3}$$

with: $K_{ox} = k_{ox} \exp\dfrac{\alpha_{ox} nFE}{RT}$ (4)

$K_{red} = k_{red} \exp\dfrac{-\alpha_{red} nFE}{RT}$ (5)

j: the current density (A/m^2)
n: the number of electrons exchanged, here 1
F: Faraday's constant (96485 C/mol)
S: electrode surface (m^2)
K_{ox}: the potential dependent rate constant for the oxidation half reaction
K_{red}: the potential dependent rate constant for the reduction half reaction
E: the potential (V vs. NHE)
R: ideal gas constant (8,32 J.mol^{-1}.K^{-1})
T: the absolute temperature (K)
c_{red}^*: the bulk concentration of the reducing agent (mol/m^3), in this case Fe(CN)$_6^{4-}$
c_{ox}^*: the bulk concentration of the oxidizing agent (mol/m^3), in this case Fe(CN)$_6^{3-}$
$i_{lim, ox}$: the oxidation limiting current (A)
$i_{lim, red}$: the reduction limiting current (A)
k_{ox}: the rate constant for the oxidation half reaction (m/s)
k_{red}: the rate constant for the reduction half reaction (m/s)
α_{ox}: the transfer coefficient for the oxidation
α_{red}: the transfer coefficient for the reduction

    The theoretical expression will be fitted to the mean polarization curve obtained in the inverted RDE configuration (see Figure 2), using the fitting algorithm. In Figure 3 the fitting results are represented. Part a shows the fitted and the experimental polarization curve and the 95% confidence interval of the experiments. As can be seen, the fitted plot concurs very well with the experimental polarization curve. In part b of Figure 3 the residual plot is shown in which the difference between the calculated and the experimental data is represented as a function of the potential. This graph gives a clear visual representation of the goodness-of-fit. If the residuals fall within the 95% confidence interval, a good fit is obtained. The graph shows that the residuals indeed fall within the 95% confidence interval. So we consider this to be a good match and the model put forward in eqn (2) is able to describe the kinetics of the ferri/ferro redox reaction in the IRDE.

    In table 1 the quantified model parameters are represented and compared to these obtained in the classical RDE [12,13]. The diffusion coefficients are not calculated here because in this paper only one rotation speed is considered. In order to calculate the diffusion coefficients in a good way, a data set taken at different rotation speeds is needed. Table 1 shows that the parameters agree well. We have to point out that the 95% confidence interval for the respective parameters obtained in the RDE and IRDE configuration don't overlap, although they lie very close to one another. This doesn't necessarily mean that the values are inconsistent. The values are so close that it shows that these parameters are very sensitive to the starting condition of the working electrode which cannot be fully controlled e.g. the presence of dislocations, scratches, etc. Moreover the parameters agree very well with the values found in literature. Literature values for α_{ox} vary between 0.45 and 0.61 [3,4]. When the rate constants of table 1 are calculated (see eqn (6)) taking into account the equilibrium constant E_0=0.47V/NHE the following values are obtained: k'_{ox} = 1.98E-04 m/s and k'_{red} = 1.94 E-04 m/s.

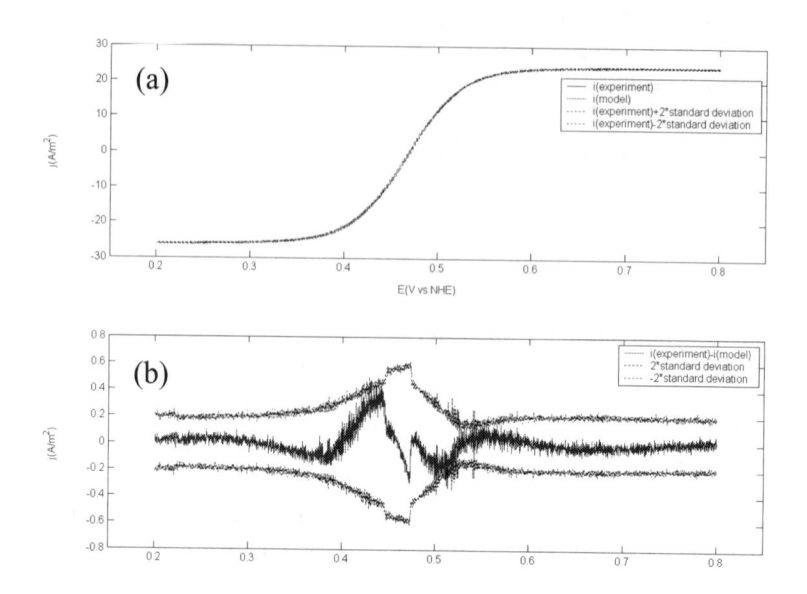

Figure 3: Comparison of the mean experiment obtained in the IRDE with the model, at 1000rpm: (a) the calculated and the simulated polarization curve together with the 95% confidence interval, (b) the difference between the experimental and simulated polarization curve at each potential together with the 95% confidence interval.

Table 1: Kinetic parameters of the ferri/ferrocyanide redox reaction with the respective absolute standard deviation.

	Classical RDE	*Inverted RDE*
α_{ox}	*5.00E-1* ± *8.76 E-03*	*5,37E-01* ± *3.87e-03*
k_{ox}	*1.39E-08* ± *2.28 E-09*	*1,06E-08* ± *7.70E-10*
k_{red}	*1.25E+00* ± *2.01E-01*	*9,33E-01* ± *6.66E-02*
$j_{lim,,ox}$	*2.47E+01* ± *1.57E-02*	*2,42E+01* ± *7.65E-03*
$j_{lim,,red}$	*-2.56E+01* ± *1.43E-02*	*-2,61E+01* ± *1.14E-02*

$$k'_{ox/red} = k_{ox/red} \exp\frac{\alpha_{ox} n F E_0}{RT} \tag{6}$$

These are of the same order of magnitude as the ones reported in literature, which range from 5E-04 m/s until 1E-03 m/s [3-5,14-15].

4 Conclusions

It is shown that the IRDE obeys the same hydrodynamic conditions as the classical RDE set-up. It has the same advantages of well-defined mass transport conditions. In future work, the setup will be used for the study of the kinetics of gas evolution reactions taking advantage of the fact that the formed gas bubbles can freely rise to the surface.

Secondly it is shown that the fitting algorithm is a very powerful and sensible tool to validate the mass transfer characteristics of a new reactor, such as the IRDE. The validation was done making use of the ferri/ferro cyanide reference system. The values of the charge transfer parameters are: α_{ox}= 5,37E-01 ± 3.87e-03, k_{ox} = 1,06E-08 ± 7.70E-10 m/s and k_{red} = 9,33E-01 ± 6.66E-02 m/s.

Acknowledgements

This research is funded by the Flemish Institute for support of Scientific-Technological Research in Industry (IWT SBO contract number 040092, project acronym: Mutech and post-doc position of E. Tourwé).

References

[1] Bamford, C. & Compton R., Electrode Kinetics: Principles and Methodology, volume 26 of Comprehensive Chemical Kinetics, Elsevier Science Publishers, 1986
[2] Iwasita, T., Schmickelr, W., Hermann, J. & Vogel, U. *Journal of Electrochemical Society*, 130, p. 2026, 1983
[3] Angell, D. & Dickinson, T. *Journal of Electroanalytical Chemistry*, 35, p.55, 1972
[4] Bruce, P., Lisowska-Oleksiak, A., Los, P. & Vincent, C. *Journal of Electroanalytical Chemistry*, 367, p.279, 1994
[5] Beriet, C. & Pletcher, D. *Journal of Electroanalytical Chemistry*, 361, p.93, 1993
[6] Zdunek, A.D. & Selman, J.R. *Journal of the Electrochemical Society*, 139 (9), p.2549, 1992
[7] Bressers, P.M.M.C. & Kelly, J.J. *Journal of the Electrochemical Society*, 142 (7), p. L114, 1995
[8] Bradley, P.E. & Landolt, D. *Journal of the Electrochemical Society*, 144 (6), p. L145, 1997
[9] Vandeputte S., PhD Thesis, Vrije Universiteit Brussel, Brussels, 1996
[10] Tourwé, E., Breugelmans, T., Pintelon, R. & Hubin, A. *Journal of Electroanalytical Chemistry*, 594(1), p. 50, 2006
[11] Robertson, B., Tribollet, B. & Deslouis, C. *Journal of Electrochemical Society*, 135, p. 2279, 1988
[12] Tourwé, E., Pintelon, R. & Hubin, A. *Journal of Electroanalytical Chemistry, accepted*
[13] Tourwé, E., Pintelon, R. & Hubin, A. *submitted to this journal*

[14] Jahn, D., Vielstich, W. *Journal of Electrochemical Society,* 109, p.849, 1962
[15] Tanaka, N., Tamamushi, R. *Electrochimica Acta,* 9, p. 963, 1964

Experimental study and modelling of heat transfer during anodizing in a wall-jet set-up

T. Aerts[1], G. Nelissen[2], J. Deconinck[2], I. De Graeve[1] & H. Terryn[1]
[1]*Vrije Universiteit Brussel, Department of Metallurgy, Electrochemistry and Material Sciences, Brussel, Belgium*
[2]*Vrije Universiteit Brussel, Computational Electrochemistry Group, Department of Electrical Engineering and Power Electronics, Brussel, Belgium*

Abstract

The anodizing of aluminium is an electrochemical surface treatment yielding the formation of an alumina film, the characteristics of the formed oxide strongly depending on the considered anodizing conditions. Heat transfer has an important influence on the anodizing process, which can be explained by considering the production of heat near the aluminium anode, combined with the significant influence of the local electrode temperature on the process of oxide formation. The influences of temperature and heat transfer on the growth of the anodic oxide film during anodizing of high purity Al are studied on a laboratory scale in a wall-jet electrode reactor. The impinging jet configuration of the reactor creates a non-uniformly accessible electrode with variable convection as a function of the radial position on the anode. The influence of the resulting non-uniform heat transfer on the local temperature of the electrode is monitored by local temperature measurements on the backside of the aluminium anode, whereas its influence on local film growth is evaluated by means of FEG-SEM surface and cross sectional analyses. A comparison between the simulated and experimentally acquired data is presented. The controlled and known electrolyte flow in the wall-jet reactor enable numerical simulations of the convection which supply additional information on the encountered conditions of heat transfer. The anodizing process itself is simulated using a model based on the high field theory.

Keywords: aluminium, anodizing, heat transfer, wall-jet reactor, modelling.

1 Introduction

Anodizing of aluminium is a well-known electrochemical process during which an anodic oxide layer is formed on the aluminium anode. The properties of this alumina layer strongly depend on the considered anodizing conditions. When performed in acid electrolytes (e.g. solutions of sulphuric, phosphoric or oxalic acid) it leads to the formation of a porous oxide, which improves, among other properties, corrosion resistance, wear resistance, hardness, compatibility with adhesives and paint of the underlying aluminium. Anodizing in sulphuric acid electrolytes leads to the formation of a porous oxide film with pore diameters and barrier layers generally in the range of 10 to 30nm [1]. Products of sulphuric acid anodized aluminium are almost universally used for decorative and protective purposes [1].

Despite the fact that anodizing of aluminium is a widely used and relatively old process, depending on the convection conditions and anodizing geometry industrial anodizers are still confronted with non-uniform oxide thicknesses. This is undesired and might lead to a reduced corrosion protection, local differences in appearance, etc. To avoid these problems reactor optimisation is often based on a trial-and-error approach. Hence from the viewpoint of reactor design simulations, which accurately predict the evolution and outcome of the anodizing process, would certainly be an improvement.

To enable such simulations a model, which correctly describes the anodizing process, is required. The difficulty in the formulation of this model can mainly be attributed to the high sensitivity of the anodizing process to temperature. In contrast to most electrochemical systems in which mass transfer is the determining factor, anodizing of aluminium is a process in which heat transfer needs to be considered. This is due to the fact that on one hand heat is produced during anodizing, the main source being the Joule heating due to ionic current passing through the highly resistive oxide layer on the aluminium electrode. On the other hand the local electrode temperature plays an important role in the process of oxide formation [2,3]. Inadequate removal of the produced heat results in non-uniform films with locally thicker oxide layers [2-4] or in the extreme case even in burning [1-3,5], a local phenomenon which leads to the formation of a region with a very thick but highly degraded oxide layer [1]. Hence, for the modelling of anodizing of aluminium a correct description of the complex relation between current density and (electrode) temperature, in combination with an accurate calculation of the conditions of heat transfer, is a key issue.

For the validation of a model for anodizing the possibility to compare simulated and experimentally measured data is essential. Inevitably these data need to be acquired during anodizing in a set-up with known and controlled convection. In this paper a wall-jet electrode reactor is considered for this purpose. The flow pattern in the wall-jet configuration is due to a submerged fluid jet impinging perpendicularly on the working electrode and spreading out radially, with the fluid outside the jet being at rest [6]. For laminar as well as for turbulent flow the characteristics of the convection are described in literature

[6-9]. The convective and thermal transfer coefficients are the highest at the centre of the electrode, where the fresh electrolyte impinges on the surface, and decline towards the border of the electrode [7,10,11]. Additionally, information on the influence of the varying local transfer coefficients across the electrode surface on the local electrode temperature is obtained by measurement of local electrode temperatures at different radial positions [2,3,5].

In this paper the use the wall-jet reactor as a tool for the validation of a numerical model for the anodizing of aluminium is described. Experimentally anodizing experiments have been performed under conditions of varying current density and convection. Acquired global and local parameters have been compared with numerically simulated ones.

2 Experimental

Disk shaped high purity aluminium samples (99.99% Al sheet 0.3mm), with a diameter of 55mm were used, the diameter of the active surface of the anode being 40mm. Prior to anodizing samples were alkaline etched in 60g/l NaOH solution at 60°C for 60s, followed by a desmutting treatment consisting out immersion in a 1:1 concentrated HNO_3:H_2O solution at room temperature during 90s.

The configuration of the wall-jet electrode cell is illustrated in Figure 1. A circular nozzle with diameter 2.0mm was used, the distance between the nozzle exit and the working electrode H was 27mm. A cylindrical aluminium rod was used as a counter electrode (CE). The distance in height between the CE and the WE was 80mm; the distance in height between the RE and the WE was 150mm. Local temperatures were measured at the backside of the WE by five thermocouples (type T), embedded in the sample holder at 5 different radial positions, these being -15mm, -10mm, 0mm, +5mm and +15mm from the centre of the anode respectively.

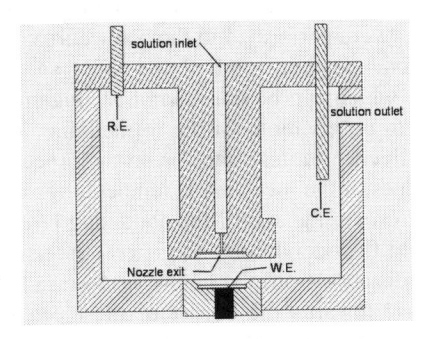

Figure 1: Schematic visualisation of the wall-jet electrode cell [5].

As anodizing electrolyte a 145g/l H_2SO_4 + 5g/l $Al_2(SO_4)_3 \cdot 18H_2O$ solution at 45.0°C was used. Thermostatic control of the electrolyte temperature (±0.1°C) was ensured by a Lauda RP845 thermostat that controlled the flow of H_2O

through a glass heat exchanger immersed in the 45l electrolyte reservoir and which was equipped with an external PT-100 temperature probe.

Anodizing was performed under galvanostatic conditions, the current being supplied by a Delta Elektronica SM 300-20 power source. Two different current densities were applied: 4A/dm² and 8A/dm². A constant charge density of 720C/dm² was considered, which corresponded to an anodizing time of 300s and 150s at 4A/dm² and 8A/dm² respectively. Three different convection conditions were considered: the first at a low flow rate only slightly higher than natural convection, the second within the laminar regime and the third in the turbulent regime. The corresponding Reynolds numbers (based on the nozzle diameter) Re_a were 200, 800 and 5000 respectively. For each experimental condition at least 3 measurements were performed. The evolution of the potential of the working electrode (WE) versus a Ag/AgCl reference electrode (RE) (+222mV vs. NHE at 25°C) was recorded with a National Instruments M-6220 DAQ card and using National Instruments VI-Logger software.

FE-SEM observations were performed using a Jeol JSM-7000F FE-SEM. To avoid charging effects due to the non-conductive properties of the oxide, the surfaces of the samples for FE-SEM analysis were covered with a 1.5nm Pt/Pd layer, applied by a Cressington 208 HR sputter coater equipped with a Cressington MTM-20 thickness controller. Information on the oxide thickness of the porous films at different radial positions on the anode was acquired by FE-SEM cross-section images of the oxides at different radial positions. On these images the thickness of the observed cross-sectioned oxide was measured using the Jeol SMile View software.

3 Anodizing experiments

3.1 Results

Under conditions of varying convection and applied current density several anodizing experiments were performed in the wall-jet reactor. Systematically the evolutions of the anode potential and of the local anode temperatures were recorded. These data, which can be considered as fingerprints of the anodizing process, combined with observations of the formed anodic film, can be used to gain insight on the influence of heat transfer on the process of oxide formation. In Figures 2 mean values of the local anode temperatures after an applied charge density of 1200C/dm² (i.e. at the end of the anodizing process) are presented for the different considered flow conditions. In general the temperature of the anode increases with radial distance from the centre up to a radial position of 10mm, followed by a decrease towards the border of the electrode. This temperature difference between the centre of the anode and a radial position of 10mm decreases with increasing Re_a. At a higher current density of 8A/dm² the differences in local temperatures between the centre and a radial distance of 10mm increase. Here local temperature differences are still observed for Re_a=5000 whereas practically a uniform temperature is recorded at this flow rate at 4A/dm². For all considered anodizing conditions the temperature decreases from 10mm towards 15mm from the centre. Upon applying a current density of 4A/dm² the overall anode temperature increases with increasing Re_a. This effect is observed on a significantly reduced scale when 8A/dm² is applied.

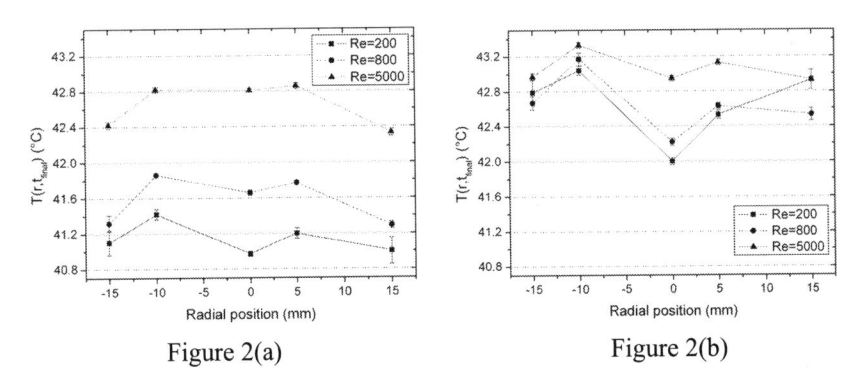

Figure 2(a)

Figure 2(b)

Figure 2: Local anode temperatures after an applied charge density of 1200C/dm²; (a) applied current density 4A/dm²; (b) applied current density 8A/dm².

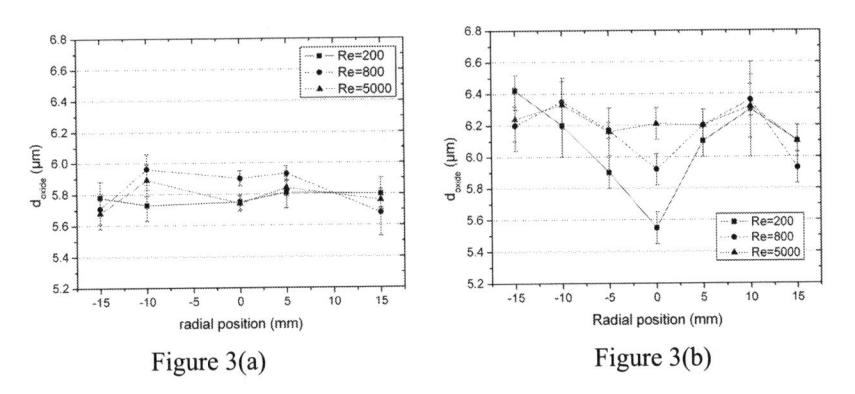

Figure 3(a)

Figure 3(b)

Figure 3: Oxide thickness distribution along electrodes anodized under an applied current density of (a) 4A/dm² and (b) 8A/dm².

The thicknesses of the formed anodic oxide films, determined by cross-sectional FE-SEM analyses at different positions on the anodes, are displayed in Figure 3. The distribution of the oxide thickness along the anodes on the samples anodized at 4A/dm² displays slightly the same trend as the corresponding local electrode temperatures. The oxide thickness increases from the centre of the anode up to a radial distance of 10mm, followed by a decrease towards the edge of the active surface. Though, the observed differences in thickness are relatively moderate and, in contrast to the local temperatures, the local oxide thickness is not directly linked to the flow rate. However, when a current density of 8A/dm² is applied, the convection does significantly influence the local oxide thickness.

The tendency of an increasing thickness from the centre towards a position of 10mm, followed by a decrease towards the edge is present under all considered convection conditions and becomes more pronounced when the electrolyte flow is reduced.

3.2 Discussion

The moderate differences in local temperatures when anodizing at 4A/dm^2 indicate that under this condition the amount of heat produced at the anode by the anodizing process can still sufficiently be removed by the convection. Indeed, the observed pronounced influence of the considered flow condition on the overall anode temperature confirms that under these conditions the main heat input is not due to the anodizing reaction but due to heat transfer created by the impinging jet of warm electrolyte. The relatively uniform oxide thicknesses on these samples are consistent with the small differences in local electrode temperatures. As known from literature [2-4] differences in local oxide thickness can directly be linked to differences in local electrode temperatures during anodizing. The differences in overall anode temperatures observed when varying convection are not reflected in different oxide thicknesses. Due to their small scale, in combination with sufficient heat removal by convection and the galvanostatic anodizing conditions, the encountered differences in overall temperature might moderately affect the microstructure of the formed anodic film [12] but will not lead to different overall oxides thicknesses.

At 8A/dm^2 the effect of varying convection conditions on the overall anode temperature is strongly reduced. Additionally the heat locally produced at the anode by the anodizing process has increased to a level where convective heat removal is no longer sufficient to avoid the occurrence of differences in local temperatures. In this case the local heat generation by anodizing has become more important than heat input due to convection. As indicated above, the different local temperatures are in correspondence with observed variations of local oxide thickness: higher local temperatures will locally induce the growth of thicker films [2-4].

In the next paragraph anodizing experiments under these conditions will be simulated and numerically and experimentally acquired data will be compared.

4 Numerical simulations

4.1 Model

An overview of the governing equations and boundary conditions used to describe the electrochemical system is presented in [13]. Concerning the fluid dynamics the velocity field $\vec{v}$ and pressure p are calculated in each point of the reactor by solving the Navier-Stokes equation in combination with the conservation of mass for an electrolyte, which is assumed to be incompressible. Additionally, a low-Re k-ω turbulence model [14] is used to calculate the turbulent viscosity. When the fluid flow field is known the temperature distribution in the reactor and along the anode are determined. The local heat dissipation in the fluid and the aluminium due to Ohmic losses are taken into account in the model, but are negligible. On the other hand, for the heat balance of the electrode the heat generated due to anodizing process will be a more important heat source. Furthermore, in calculation of the temperature of the anode the heat loss from the reactor to the surrounding was taken into account. Neglecting concentration gradients in the electrolyte, the electrical potential distribution in the reactor is given by the Laplace equation. This is also the case for the aluminium anode. These equations are solved with the appropriate boundary conditions, of which the electrode boundary conditions for the electrochemical reactions are the most complex. On the cathode still a simple linear behaviour is assumed to describe the hydrogen evolution, though the boundary condition describing the relation between temperature, overpotential and local current density for the anodizing process is much less straightforward. Based on theoretical considerations, the following temperature dependent Butler-Volmer overpotential relation was suggested:

$$J = D.T^m .e^{\frac{B}{T}} \left(e^{\frac{\alpha zF\eta}{RT}} - e^{\frac{-(1-\alpha)zF\eta}{RT}} \right)$$

with $D = 9.86E-6$, $B = -2600$, $\alpha = 0.0116$, $z = 3$ and $m = 0.5$. The values of these parameters were fitted from experimental data in an experimental set-up with limited convection [16].

In the scope of this paper the calculated temperature and local current density distributions along the anode at the end of the anodizing process will be considered. A direct comparison between simulated and measured local temperatures is possible, whereas to come to a similar comparison between simulated and measured values, the local oxide thickness distribution is calculated from the numerically acquired local current density. Based on Faraday's Law, and assuming a constant growth rate, the following expression can be found for the evolution of the oxide thickness d_{ox} with anodizing time:

$$d_{ox} = \varepsilon \frac{Mj\Delta t}{zF\rho} \quad (1)$$

with ε the anodizing efficiency, M the molecular weight and ρ the density of the formed of Al_2O_3. ε was determined based on experimentally observed average oxide thicknesses. The influence of sulphate incorporation into the alumina film on ε was not taken into account [2]. Under the considered conditions the efficiency was found to vary with the applied current density but to be independent of the considered flow condition. Respective values of 0.85 and 0.90 for applied current densities of 4A/dm² and 8A/dm² were obtained.

4.2 Results

Due to the axis-symmetry of the wall-jet reactor the equations mentioned above are solved axis-symmetrically to obtain a full 3D result. A comparison between the simulated and experimentally measured local electrode temperatures from the centre towards the border of the anode is presented in Figure 4. As can be observed in all cases too high overall temperatures are calculated. Also an inverted influence of the flow rate on the overall temperatures is found. Whereas experimentally the overall temperature increased with increasing Re_a, the simulations predict an increasing overall temperature with decreasing Re_a. Concerning the local temperatures, for an applied current density of 4A/dm² the calculated distributions underestimate the radial increase of the temperature up to 10mm from the centre in the case of Re_a=200 and 800. On the other hand, besides a difference in overall temperature there is a good correspondence between the experimental and calculated temperature evolution at Re_a=5000. This is also the case when a current density of 8A/dm² is considered. At 8A/dm² in general the radial increase of the local temperature up to a distance of approximately 10mm from the centre, followed by a decrease towards the border of the electrode, is numerically also obtained. In contrast to 4A/dm² the simulations indicate an overestimation of the radial temperature increase in the case of Re_a=200 and 800.

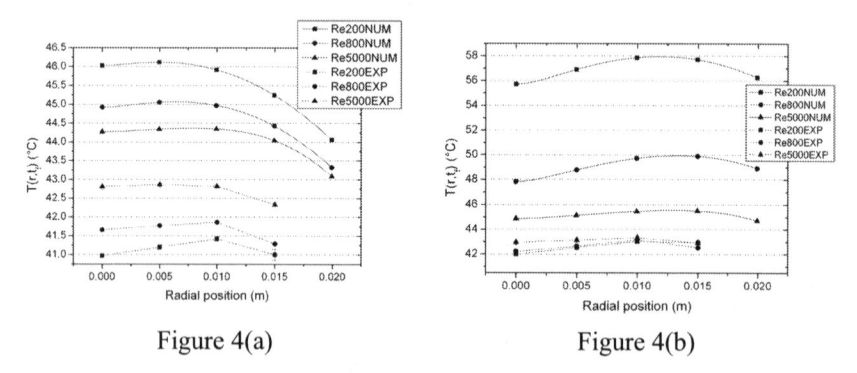

Figure 4(a) Figure 4(b)

Figure 4: Simulated and experimental local anode temperatures after an applied charge density of 1200C/dm². "NUM" and "EXP" respectively indicate numeric and experimental values. Applied current density (a) 4A/dm² and (b) 8A/dm².

Figure 5 displays a comparison between the experimentally determined and numerically calculated evolutions of oxide thickness along the anodes. For both current densities the large variations observed at the centre and at the border of the electrode are due to edge effects occurring during the calculation of the current density and will be neglected in the following discussion. Considering a current density of 4A/dm² the outcome of the numerical simulations is a uniform oxide thickness, whereas experimentally small variations in oxide thickness, in correspondence with variations in local temperature, are observed. Despite the numerical overestimation of the local temperature differences in the case of an applied current density of 8A/dm² and Re_a values of 200 and 800, the model also predicts a too uniform oxide thickness evolution in comparison to the experiments under these conditions. Furthermore, in the case of Re_a=200 at 8A/dm² a higher overall current density and thus overall oxide thickness is calculated due to the significantly overestimated overall temperature.

Figure 5(a)   Figure 5(b)

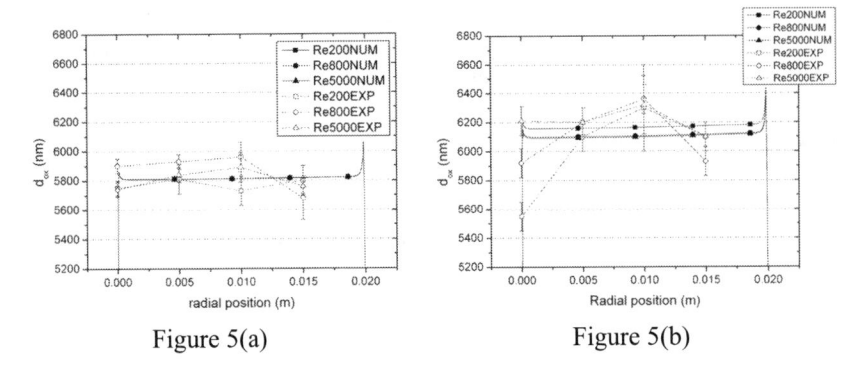

Figure 5: Simulated and experimental evolution of oxide thickness along electrodes. "NUM" and "EXP" respectively indicate numeric and experimental values. Applied current density (a) 4A/dm² and (b) 8A/dm².

4.3 Discussion

The generally overestimated overall temperatures and the numerically obtained trend of an increasing overall temperature with decreasing Re_a point out that the simulations overestimate the local heat generation of the anodizing process, or that the heat transfer due to the impinging jet is underestimated, or a combination of both. The overestimation of the radial temperature increase in the case of Re_a=200 and 800 at 8A/dm² can also be explained by considering these effects. On the other hand the underestimated radial temperature increases in the case of Re_a=200 and 800 at 4A/dm² indicate an overestimated heat loss at the edge of the anode. This is most pronounced under these conditions since in this case the heat input from the anodizing process as well as for the impinging jet of warm electrolyte are the most reduced.

For both considered current densities numerically too uniform evolutions of the local current density and local oxide thickness are obtained. Hence, in the boundary condition describing the anodizing process the local current density is not sensitive enough towards variations in *local* temperature.

Improvement of numerical simulations will involve the use an updated boundary condition for the anodizing process, better describing the interaction between local current density, local temperature and overpotential. Not only more insight on the point of the anodizing process is necessary, a more advanced description of the heat loss from the reactor to the surrounding is necessary.

5 Conclusions

The use of the wall-jet reactor as a tool for the experimental investigation, as well as for the numerical simulation of the influence of heat transfer on the anodizing of aluminium has been demonstrated. The influence of a variation in convection and applied current density on the local electrode temperatures and oxide thickness was investigated. Experimentally acquired data were compared with numerical calculated data. To improve accordance between experiments and simulations modifications to and refinement of the used model are necessary.

Acknowledgements

The authors acknowledge the support from the Instituut voor de aanmoediging van innovatie door Wetenschap & Technologie in Vlaanderen (IWT, contract nr. SBO 040092).

References

[1] P.G. Sheasby, R. Pinner, The Surface Treatment and Finishing of Aluminium and its Alloys, 6th Edition, ASM International, USA/ Finishing Publications Ltd, UK, 2001.
[2] I. De Graeve, H. Terryn, G.E. Thompson, *J. Electrochem. Soc.* **150**, p.B158, 2003.
[3] I. De Graeve, H. Terryn, G.E. Thompson, *J. Appl. Electrochem.* **32**, p.73, 2002.
[4] A.J. Bosch, P. Boerstoel, Th. Zuidwijk, A. Hovestad, A. Plomp, J.A. van de Heuvel, *ATB Metallurgie*, **43**, p.65, 2003.
[5] T. Aerts, I. De Graeve, H. Terryn, *ATB Metallurgie*, in press.
[6] M.B. Glauert, *J. Fluid Mech.* **1**, p.625, 1956.
[7] H. Marin, in *Advances in Heat Transfer*, Vol.13, ed. by J.P. Hartnett and T.F. Irvine Jr., Academic Press, New York, 1977, p.1-60.
[8] W.J. Albery, C.M.A. Brett, *J. Electroanal. Chem.* **148**, p.211, 1983.
[9] D-T. Chin, C-H. Tsang, *J. Electrochem. Soc.* **125**, p.1461, 1978.
[10] D.J. Bizzak, M.K. Chyu, *Int. J. Heat Mass Transf.* **38**, p.267, 1995.
[11] B. Ellison, B.W. Webb, *Int. J. Heat Mass Transf*.**37**, p.1207, 1994.
[12] T. Aerts, Th. Dimogerontakis, I. De Graeve, J. Fransaer, H. Terryn, *Surf. Coat. Technol.*, in press.
[13] G. Nelissen, B. Van Den Bossche, A. Van Theemsche, J. Deconinck, *Proc. AESF SUR/FIN*, p.386, 2001.
[14] D.C. Wilcox, Turbulence modelling for CFD, DCW Industries 2nd edition, 1998.
[15] G. Nelissen, A. Van Theemsche, C. Dan, B. Van den Bossche, J. Deconinck, *J. Electroanal. Chem.*, **563**, p.213, 2004.
[16] B. Van den Bossche, J. Deconinck, *Proc. BEM 14 Conference*, 1992.

Scanning Vibrating Electrode Technique as an application for measuring corrosion activity of carbon steel welded pipelines

A. Abdurrahim[1] & R. Akid[2]
[1]*Petroleum Research Centre, Corrosion Research Unit, Tripoli, Libya*
[2]*Materials and Engineering Research Institute, Sheffield Hallam University, Sheffield, UK*

Abstract

The scanning vibrating electrode technique, also known as SVET, is a relatively new technique which offers the opportunity to obtain information concerning corrosion activity on a small scale. However, its utilisation to investigate the corrosion of welds has been limited. SVET experiments were carried out on specimens of different welded sections within two different electrolytes, e.g., 0.35% NaCl solution alone and buffered 0.35% NaCl solution with CO_2 saturations. SVET was used during these investigations to evaluate preferential corrosion susceptibilities of weldments.
Keywords: SVET, localised corrosion, weldments.

1 Introduction

The scanning vibrating electrode technique (SVET) is an electrochemical method, which is able to resolve and quantify highly localised corrosion currents occurring at the metal-electrolyte interface [1]. It consists of a platinum microelectrode tip at the end of insulated wire thinned down to a fine point, positioned close to the surface to be scanned. A schematic of the SVET arrangement is given in Figure 1. The SVET microtip electrode is vibrated mechanically at a constant amplitude and frequency using a simple electromagnetic or piezoelectric driver [2,3]. The SVET has found a wide range of applications in the study of localised corrosion phenomena [4,5].

2 Experimental work

2.1 Introduction

Corrosion measurements were performed using welded carbon steel pipeline materials designated types Y8R27 (St-1) and Y8R24 (St-2). The preferential corrosion of the welded joints within chloride solution alone and buffered solution saturated with CO_2 is the result of galvanic differences between the PP, HAZ and WM. The most active region suffers accelerated attack, with a reduction of the corrosion over more noble surfaces. Corrosion tests were conducted within different environments at ambient temperature. The pH for both environments was measured and found to be 6.3 ± 0.3 and 6.2 ± 0.1 within chloride solution alone and chloride solution with CO_2 saturation respectively.

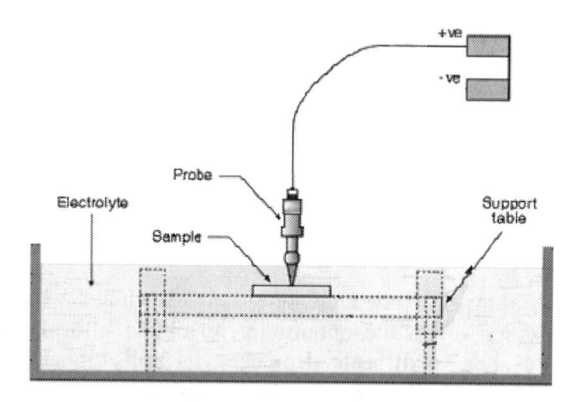

Figure 1: Schematic diagram shows SVET circuit used for measuring corrosion activity of the welded steel samples.

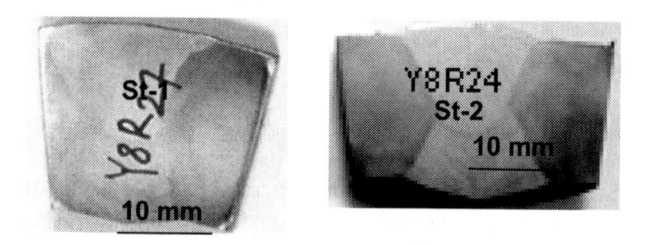

Figure 2: Schematic illustrations of the welded specimens used in the SVET tests.

3 Results and discussion

3.1 SVET measurements

All SVET measurements were carried out within different environments using a vibrating platinum microelectrode tip, mechanically scanned at a small fixed distance from the corroding surface. SVET map scans were recorded at room temperature over a period of 1, 2, 3 and 4 hours.

Due to the fact that SVET resolution and output signal is dependent upon the conductivity of the solution, low conductivity solution were used for this analysis.

It should be noted that resolution and output signal decrease with increasing conductivity [6].

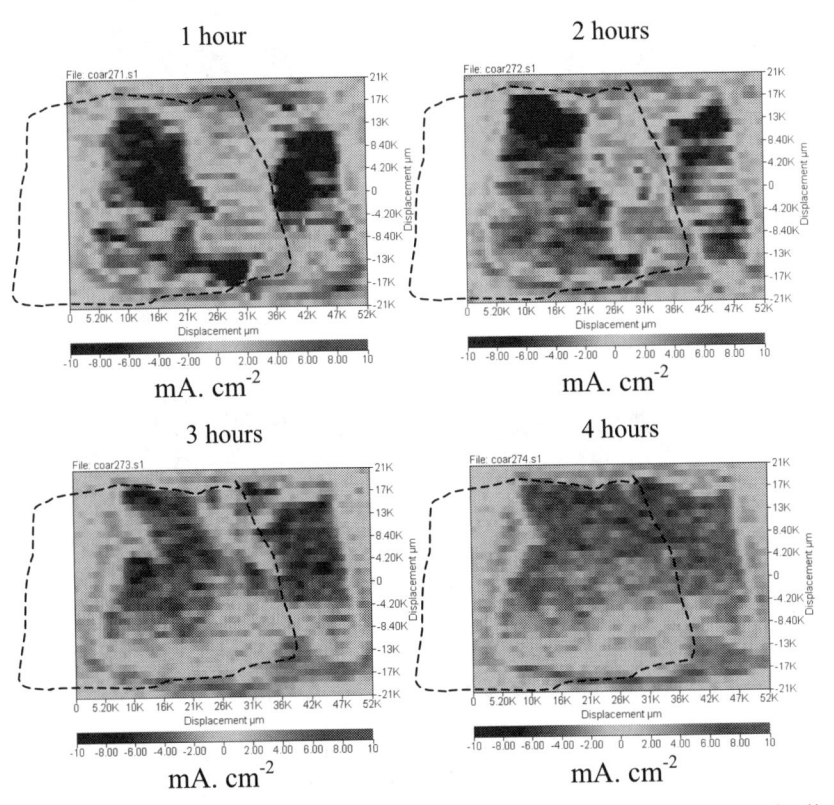

Figure 3: Showing the SVET area map scan on welded carbon steel pipeline designated of St-1 within aqueous chloride solution for different time scans.

3.2 SVET measurements within chloride solution

Figures 3-4 present the area map scans of steels 1-2 respectively in sodium chloride solution. The colours at the more positive end of the palette, i.e., red, represents relatively cathodic potentials, whereas those at the negative end of the palette, i.e., blue, represents anodic areas. The SVET tests were conducted on the welded sections for periods of time between one and four hours due to the aggressiveness of the electrolyte, which induced highly localised attack.

The localised activity for steel-1 seemed to be concentrated on the PP and HAZ during the first two hours and being uniform over the following 3-4 hours.

Finally, the SVET maps showing the localised activity of the steel-2 was severe in the WM and HAZ whilst less activity observed in the PP during first two hours and thereafter reduced over the following 3-4 hours in the WM.

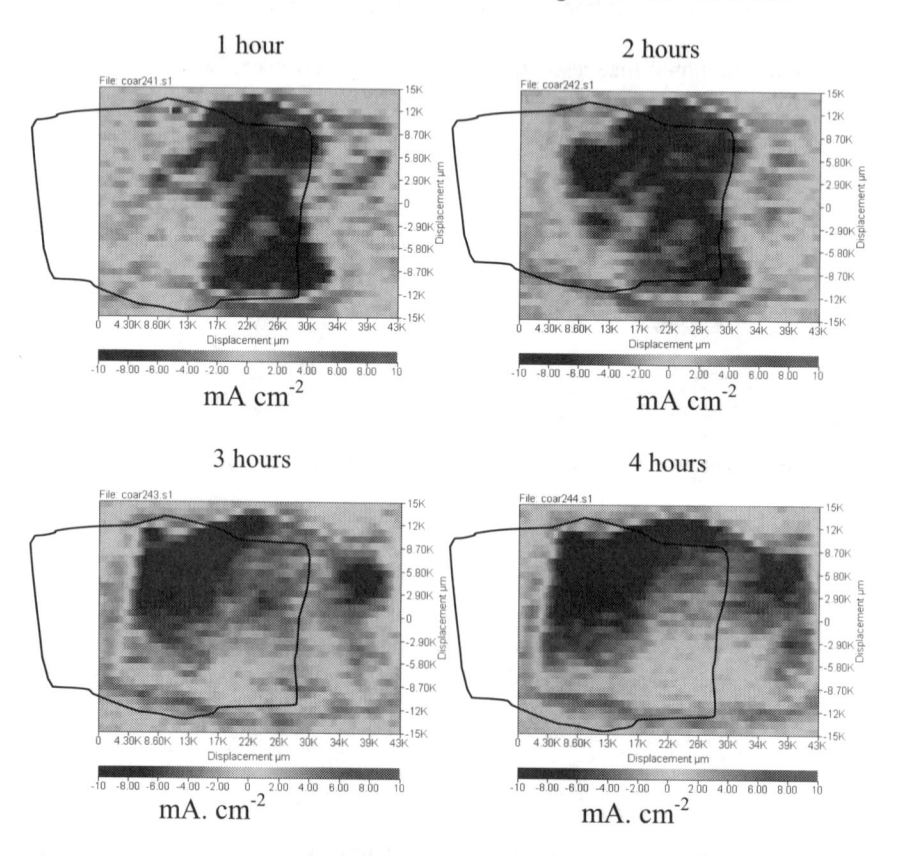

Figure 4: Showing the SVET area map scan on welded carbon steel pipeline designated of St-2 within chloride solution alone for different time scans.

3.3 SVET measurements within CO_2 saturations

Figures 5-6 represent the area map scans for the different microstructures of welded steels within chloride solutions saturated with CO_2. The localised attack could be seen in Figure 5, the area maps of the sites with high localised activity produced of the St-1 on the HAZ and to some extent at the PP and less than that on the WM during 1-2 hours and over the following 3-4 hours there is less activity on the WM and more uniform corrosion at the HAZ and PP. Figure 6 shows area map scans for steel-2, where localised activity is being uniformly during 1-4 hours for all the steel microstructures.

SVET was found to be a sensitive and reliable technique that can be used for rapid and reliable evaluation of weldments. Results of the map scans over weldments show that, although the activity varied from time to time, the most pronounced attack seemed to be concentrated in the ferrite and pearlite microstructures and least activity for some steels occurred in the bainite and martensite microstructures.

Results from the corrosion tests showed a clear significant difference in the corrosion behaviours between the weld zone and parent plate microstructures, i.e., some location for welded steel microstructures are more harmed or attacked than others.

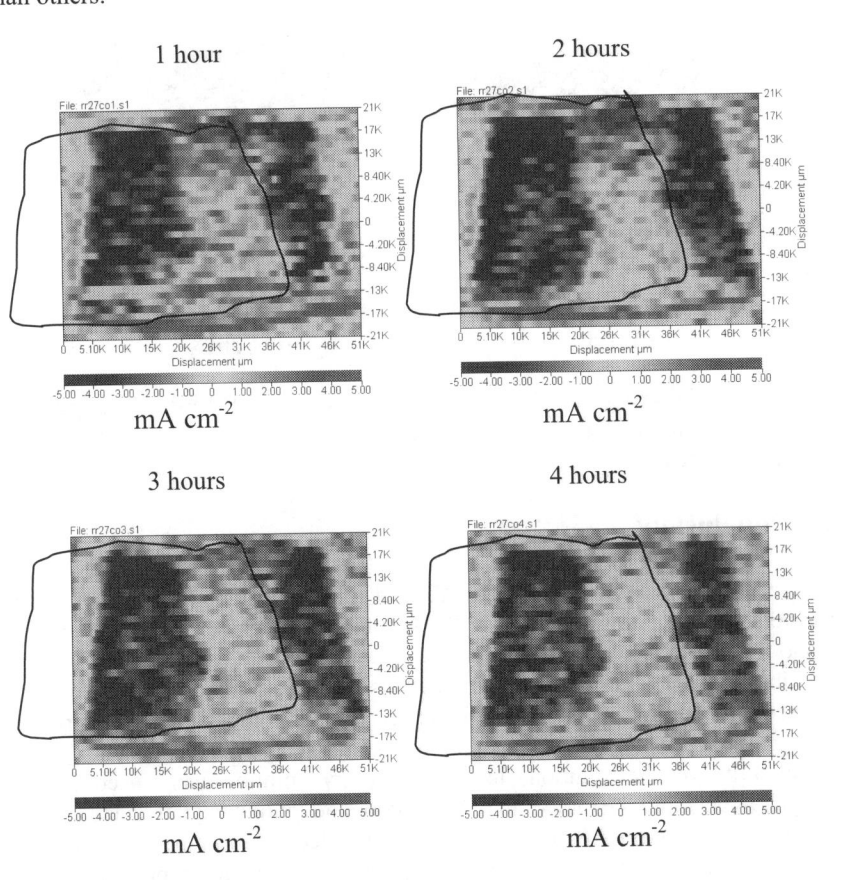

Figure 5: Showing the SVET area map scan on welded joints designated of St-1 within chloride solution saturated with CO_2 for different time scans.

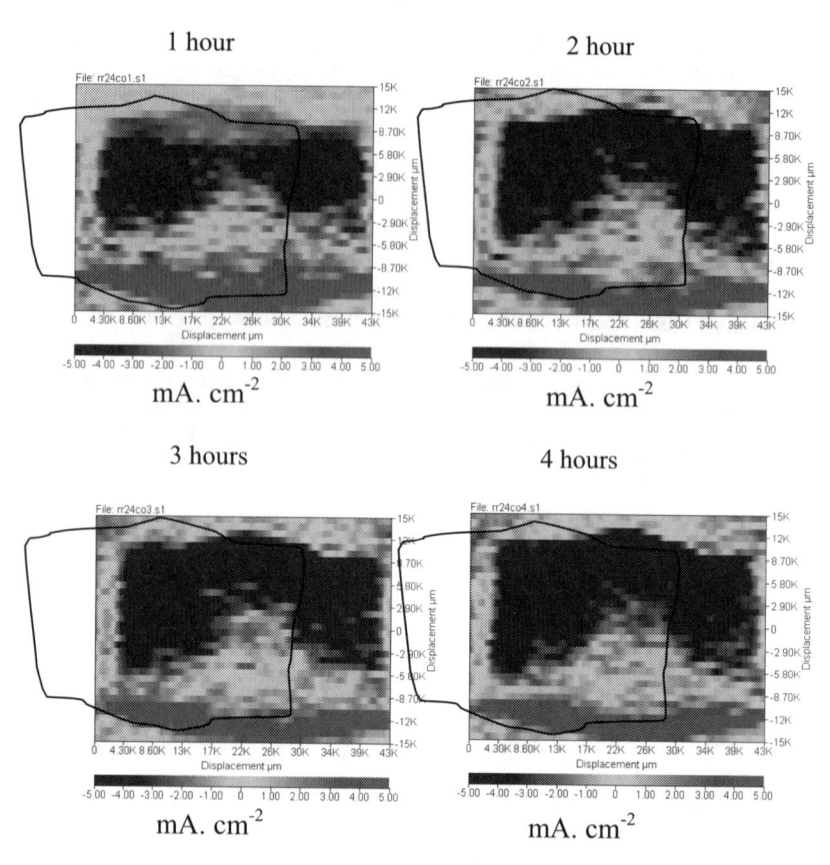

208 Simulation of Electrochemical Processes II

1 hour   2 hour

3 hours   4 hours

Figure 6: Showing the SVET area map scan on welded joints designated of St-2 within chloride solution saturated with CO_2 for different time scans.

4 Conclusions

The scanning microtip electrode is vibrated relative to the scanned surface and registers an alternating potential at the vibration frequency, this potential being proportional to electrical field strength or potential gradient in the direction of the vibration. The alternating potential arises from the oscillation of the microtip in the potential field generated ohmically by the ionic current flux passing through the electrolyte. SVET resolution and output signal is dependent upon the conductivity of the solution, low conductivity solution were used for this analysis. It should be noted that resolution and output signal decrease with increasing conductivity. The scanning vibrating electrode technique was found to be a sensitive and reliable technique that can be used for rapid and reliable evaluation of weldments.

Results of the map scans over weldments show that, although the activity varied from time to time, the most pronounced attack seemed to be higher localised activity in the weld metal and/or the heat affected zone after duration of time and least activity showed in the parent plate, moreover to be concentrated in the bainite and martensite microstructures and least activity for some steels occurred in the pearlite and ferrite microstructure. It can be concluded that the corrosion measurements showed a clear significant differences in the corrosion behaviours between the weld zone and parent plate microstructures, i.e., some location for welded steel microstructures are more harmed or attacked than others.

References

[1] H. S. Isaacs and B. Vyas, Electrochemical Corrosion Testing, ASTM STP 727, 1981, pp. 3.
[2] H.S. Isaacs and Y. Ishikawa, Electroch. Technique, ed. R. Baboian, Houston, TX. NACE, 1986, pp. 17.
[3] H.S. Isaacs, Corrosion Science Vol. 28, 1988, pp547.
[4] M. J. Franklin et al., Corrosion. Sci., Vol. 32, No. 9, 1991, pp. 945.
[5] T. Shibata et al., Corrosion Engineering, Vol. 39, 1990, pp. 331.
[6] R. Akid and M. Garma, Electyrochimica Acta, Vol. 49, 2004, pp. 2871.

Section 4
Interference and signature control

Predicting corrosion related signatures

R. Adey & J. M. W. Baynham
C M BEASY Ltd, Ashurst Lodge, Southampton, UK

Abstract

Computer simulation has been widely used to predict the corrosion related electric and electromagnetic signatures of naval vessels. The modelling strategy has varied from simple dipole type models to detailed boundary element models of the vessel and its environment. For the dipole models users have had to choose the location and strength of the dipoles based upon experience, using range data or data from similar vessels. Whereas the boundary element model enables the user to define the actual geometry of the vessel, the electrochemical properties of the materials and the properties of the environment as data to the model and obtain predictions of the electric and magnetic signatures.
   This paper describes the simulation tools developed as part of the BEASY software to predict electric and magnetic signatures. Comparison is made between results of the boundary element detailed models and the dipole type models.

1 Introduction

The design of a cathodic protection system is of interest to defence organisations not only to ensure the integrity of the ship but also because of the electric fields generated in the sea water by the Cathodic Protection (CP) system. These fields are known as the signature of the ship.
   Electromagnetic signatures are playing an important role in the detection of naval vessels and in the fusing of intelligent mines. The static electric signature is the electric field associated with the DC corrosion or cathodic protection current that flows through the seawater around a vessel. This is sometimes referred to as the Underwater Electrical Potential or UEP. The corrosion related magnetic (CRM) field is the coupled magnetic field caused by the corrosion related electric currents flowing in the seawater between the anodes and the ship hull.

It is important to note that UEP and CRM signatures exist even in the absence of a cathodic protection system. They are caused by the galvanic potential differences between the metallic structures in contact with the seawater. For example, the relative position in the electrochemical table of steel and bronze provides a sufficient driving potential to create an electric field.

In order to control the signatures and to preserve the integrity of a vessel it is essential to be able to predict the impact of the design and operation of the CP system on the electric fields.

Computational models have been widely used to predict the electromagnetic fields associated with vessels due to on board systems and ferromagnetic aspects. The software BEASY [1] has been widely used to predict the performance of cathodic protection systems by modelling the coupled electric fields and electrochemistry for complete ships and other structures. Other authors have used simple dipole models to make predictions. In this paper an integrated approach is presented to enable the performance of the CP system, the corrosion related electric field and the corrosion related magnetic field to be predicted.

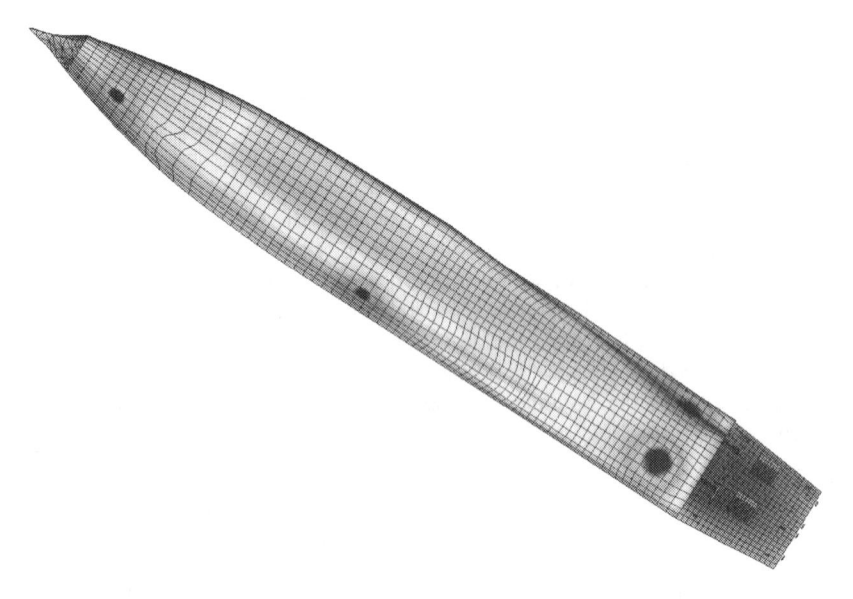

Figure 1: BEM model of a ship (note: only elements on the wetted surface of the ship).

1.1 Theoretical aspects

The ability of the boundary element method (BEM) to provide excellent solutions to problems associated with corrosion simulation has been discussed in [2]. The BEM requires the user to describe only the boundary or surface of the ship to be modelled, thus simplifying the modelling process. Other benefits are described in [2]. Figure 1 shows an example of a BEM model of a ship hull, including propeller shafts, rudders, and propellors.

To represent a CP problem the computer model must simulate the IR drop through the electrolyte and the electrochemical electrode kinetics on the metallic surfaces. BEM solution of the Laplacian in the sea water, combined with representation of the electrode kinetics associated with different surface types, has been shown to provide accurate results for the complex current fields in the sea water [3 to 7] and hence the UEP signature.

The CRM signature generated by the "volume currents" flowing in the seawater *may* be found by solving the vector potential equation:

$$\nabla^2 \underline{A} = -\mu \underline{J}$$

where $\underline{A}$ = vector potential
  $\underline{J}$ = Vector of current density components

but this approach is computationally expensive.

Recently an alternative approach has been used, in which the CRM is obtained by means of an integration over the boundaries of the volume through which the volume currents are flowing. This approach, introduced by Allan [8], is significantly less expensive, and is used in the examples shown in this paper.

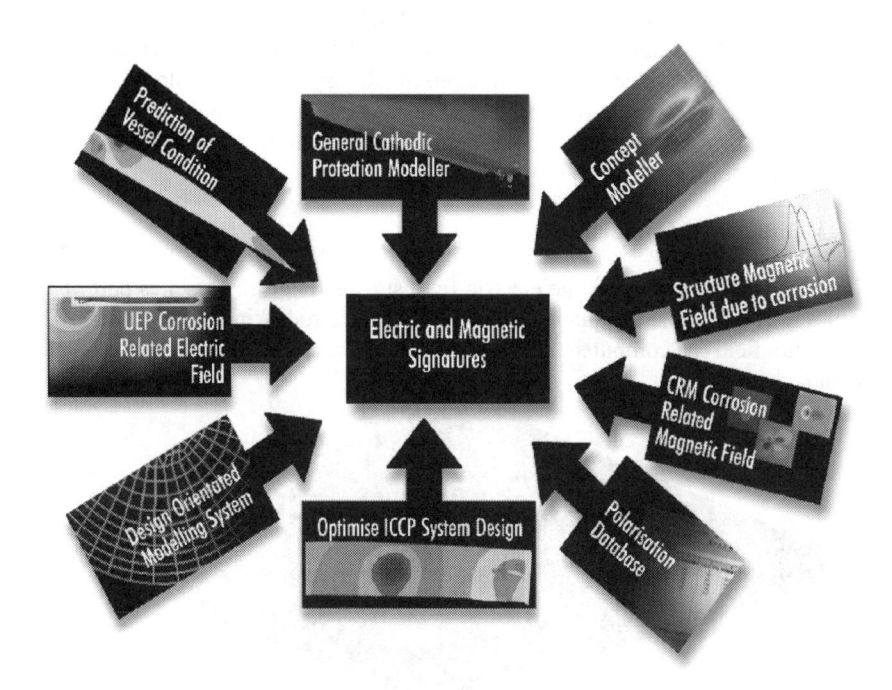

Figure 2: View of the simulation tools available to compute the components which make up a vessels signature.

1.2 Signatures

The corrosion related signature of a vessel is made up of two components, the electric field and the magnetic field. The electric field is calculated directly as part of the BEM solution of the corrosion model. The magnetic field due to the currents flowing through the sea water can be computed once the electric field is known.

However this is only part of the magnetic signature as there is also current flowing through the metal structure of the vessel back to the power supply for an ICCP system or back to the anode. The magnetic field due to this current flow through the vessel must be added to the magnetic field caused by the volume currents to obtain the total electromagnetic field.

2 Signature simulation tools

The signature tools available within BEASY are as follows.
- Detailed BEM modeller capable of predicting the corrosion related electric and magnetic fields generated by the currents flowing through the sea water and sea bed.
- Structure magnetic field modeller capable of predicting the magnetic field generated by the currents flowing through the ships structure.
- Dipole modeller capable of predicting the electric and electromagnetic fields generated by a vessel approximated by a set of dipoles
- Characterisation tool capable of generating a set of dipoles to match an electric and magnetic field obtained from ranging or the detailed BEM model.

2.1 Detailed BEM modeller

The BEM model computes the electric field surrounding the vessel as part of the solution of the Laplace equation and the metal electrode kinetics equations. The magnetic field is computed using the known current densities to perform a boundary integration.

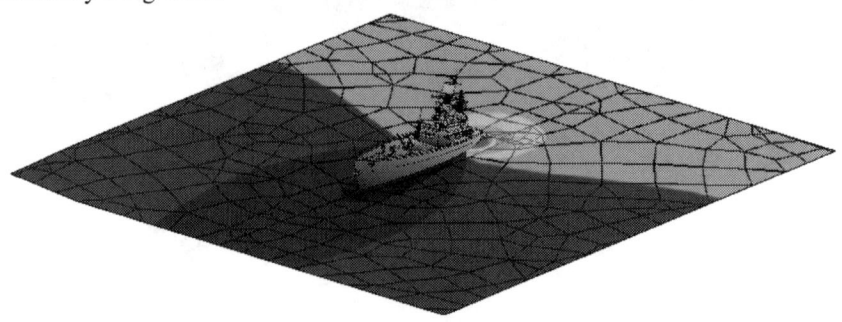

Figure 3: The BEM modeller can predict the electric field on the BEM mesh, and the electric and magnetic fields at internal point positions.

2.2 Structure magnetic field model

The internal structure of the vessel is idealised as a series of wires and the current flowing in each wire is determined from the current densities (on the surface of the vessel) which have been obtained using the detailed BEM model.

The user simply selects the structure and the software automatically calculates the current flowing. The magnetic fields generated by each wire are determined using Biot-Savart.

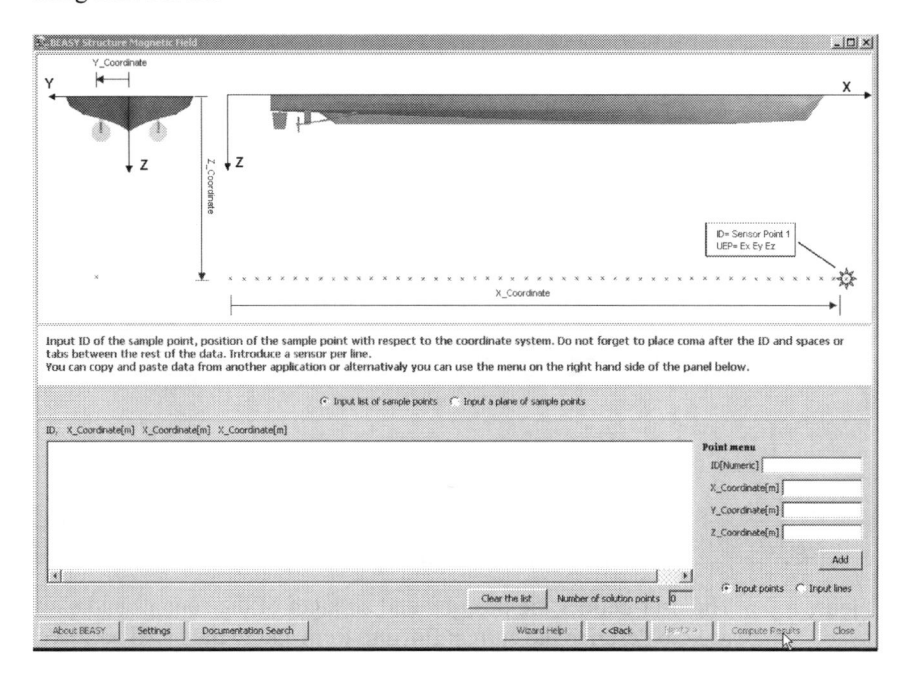

Figure 4:   View of the Structure Magnetic Field software.

2.3 Dipole simulation tool

In contrast to the modelling tools described above (which use the detailed current distributions caused by the real hull shape), the dipole simulation tool requires the user to specify the current flowing in each dipole, and the position of each dipole source and sink.

The dipole model allows multiple horizontal dipoles, and uses a two-layer solution (in which the depth of layer 1 can be selected by the user) to calculate the resulting electric and magnetic fields.

2.4 Characterisation tool

The characterisation tool provides a way of automatically obtaining a dipole distribution which reproduces a known electric signature. The signature may

218 Simulation of Electrochemical Processes II

have come from the detailed BEM model, or could also come from experimental measurements.

The user defines the overall dimensions of the ship, and can define the depth of the top conducting layer. The software uses an optimisation technique to create the set of dipoles.

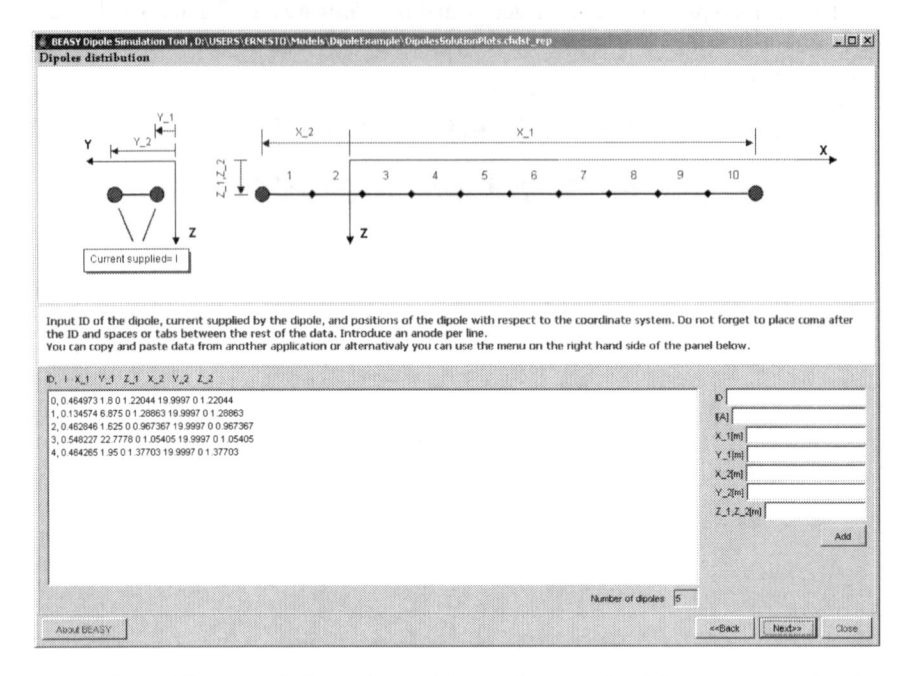

Figure 5:   The user defines the position and strength of the sources and sinks used to describe the current flowing from and to the metal surfaces.

3 Comparison

The objective of this example is to use the various tools described above to determine the magnetic field generated by current flowing into and out of the ends of a semi-cylindrical "hull" floating at the surface of an electrolyte with conductivity of 4 S/m. The hull (shown in Figure 7) is at the surface of a volume of electrolyte with depth 1000 metres, and extending 1000 metres horizontally.

Firstly a detailed BEM model of the hull was used to determine the volume currents flowing in the electrolyte, and the currents flowing through the surface of the hull. The currents were used to determine the magnetic field caused by the volume currents, and the magnetic field caused by the currents flowing along the return path through the hull.

Next a dipole solution was used to determine the electric and magnetic fields, for a dipole with the same current, and with the source and sink separated by the same distance as the ends of the hull, i.e. at +/- 50 metres, and at approximately the "centroid" of the anode and cathode areas.

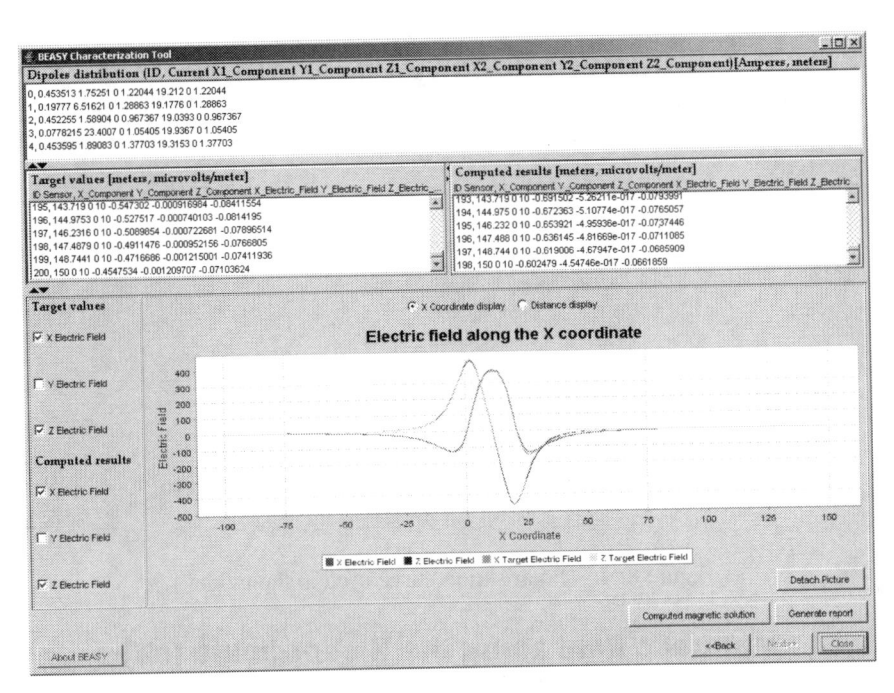

Figure 6: Comparison between electric field generated by the equivalent dipoles and the actual signature.

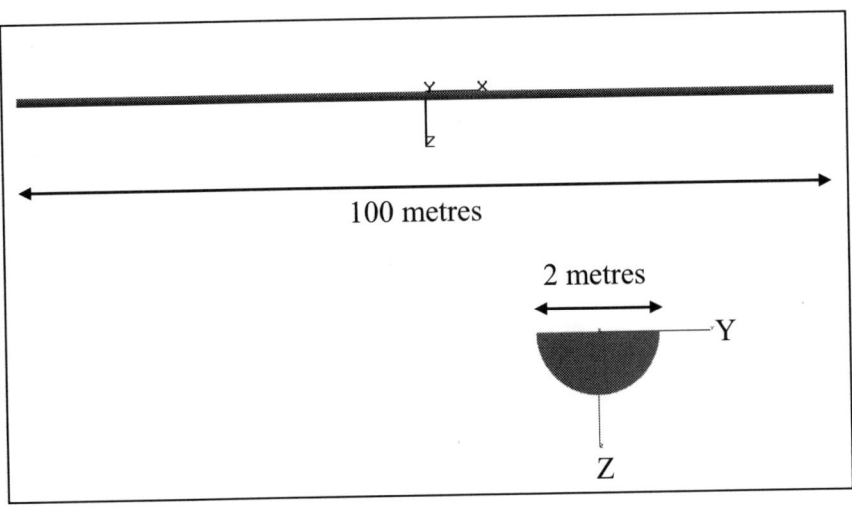

Figure 7: The "hull" used in the comparison.

One end of the hull was given an anodic current of 20 Amperes, and at the other end a polarization curve for steel (Figure 8) was used.

Results were obtained along a line of 401 internal points placed under the hull from position (-200,0,20) to position (200,0,20).

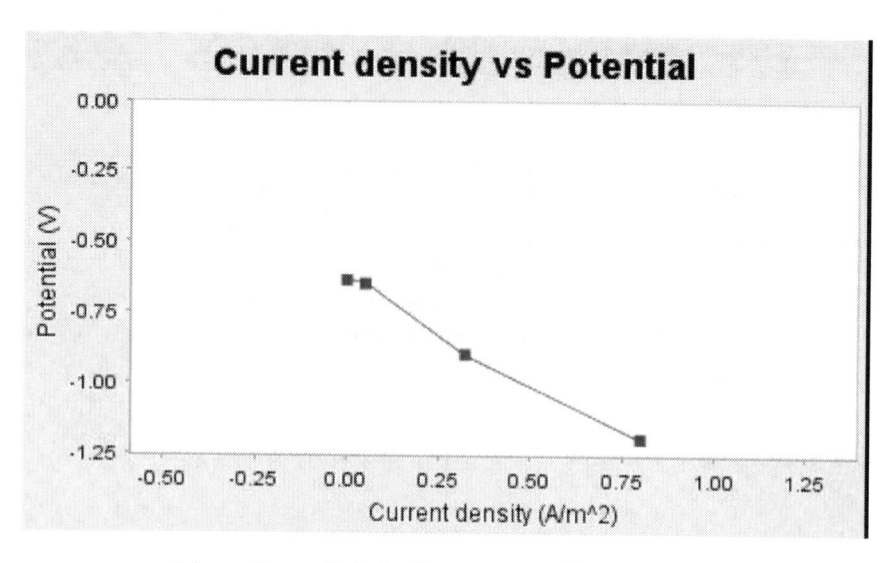

220 Simulation of Electrochemical Processes II

Figure 8: Polarisation curve used in the model.

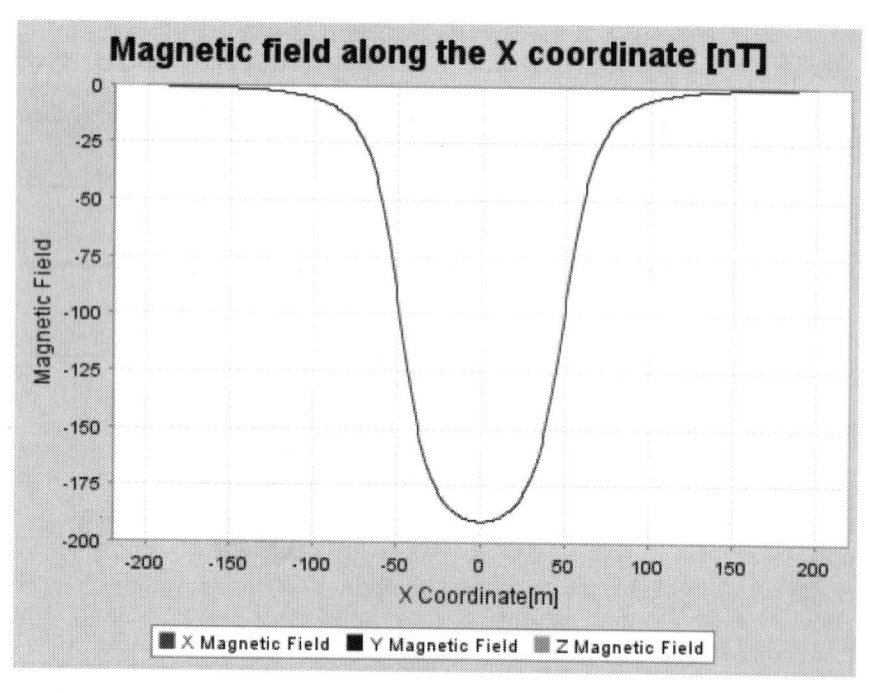

Figure 9: Magnetic field predicted by the structure magnetic field tool.

Return Path Currents

The magnetic field due to the currents flowing in the return path through the structure are shown in Figure 9. In this model and for the result positions chosen, the vertical and axial components are zero.

Comparison of the transverse magnetic field along the line of sampling points is shown in Figure 10. It can be seen that the sum of the volume and structural components is very similar to the result from the dipole. Immediately underneath the middle of the hull, the dipole solution is 239.75 nanoTesla, and the sum of the results from volume and structural currents is 241 nanoTesla.

The combined magnetic field (volume currents plus return path currents) and the magnetic field from the dipole agree to within about 0.5%, making the curves "Vol+struct" and "dipole" in Figure 10 indistinguishable in the figure.

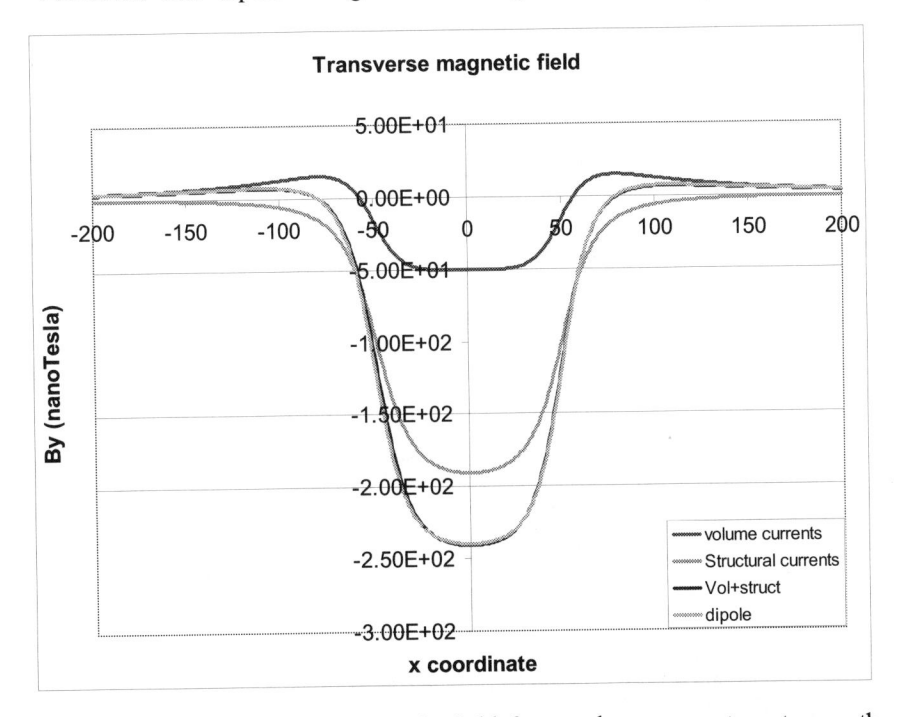

Figure 10: Comparison of magnetic field from volume currents, return path currents, and the dipole. "Vol+struct" and "dipole" curves overlap.

Equivalent Dipole model

Finally, a dipole was computed for this structure by using the Characterisation tool to match the electric signature obtained with the detailed BEM model.

The characterization tool matched the electric field as shown in Figure 11, and came up with an "equivalent dipole" with source and sink at x=+/- 48 metres, y=0, and z=0.5 metres.

Comparison for the equivalent dipole

The sum of the magnetic field caused by the volume currents and the magnetic field caused by the return path currents through the structure are compared with the magnetic field due to the calculated *equivalent dipole* in Figure 12. Immediately underneath the middle of the hull, the solution from the *equivalent dipole* is 238.3 nanoTesla.

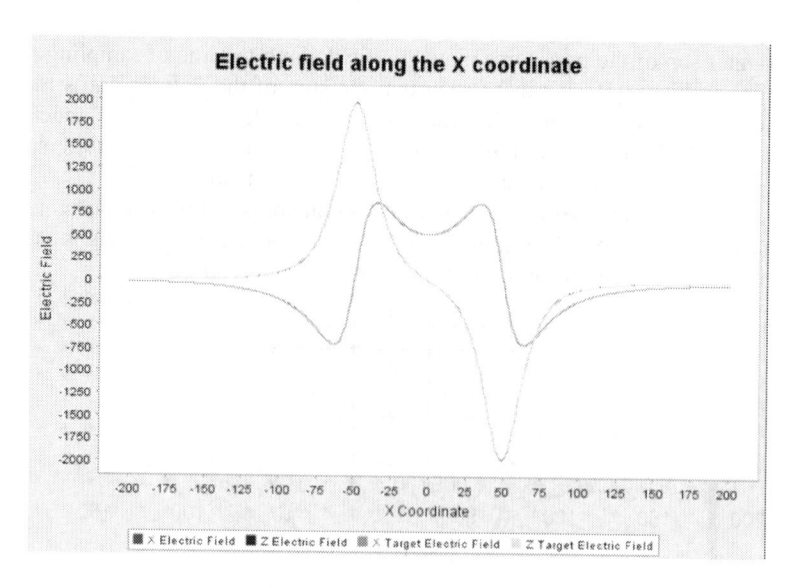

Figure 11: Comparison between the electric field computed with the BEM model and the *equivalent dipole* model.

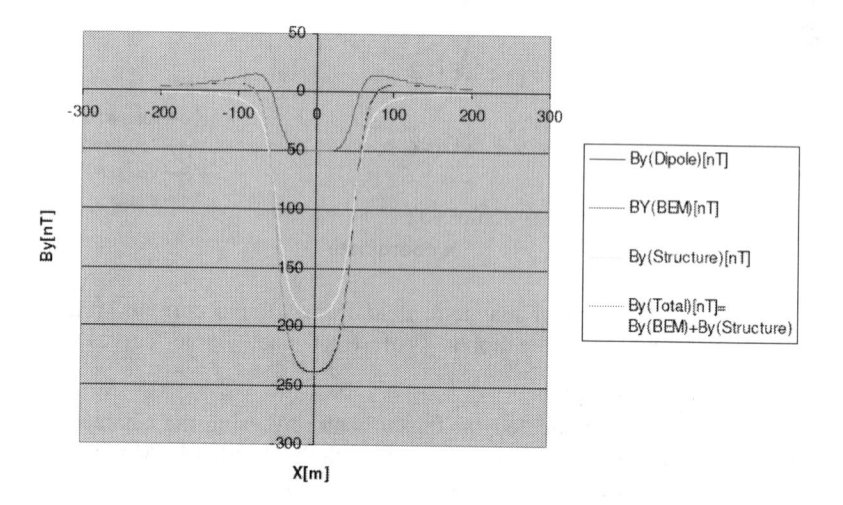

Figure 12: Comparison of magnetic field from volume currents, return path currents, and the *equivalent dipole*.

The magnetic field from the equivalent dipole determined by the characterization tool is within about 1% of the field from the combined structural and volume currents.

4 Conclusions

It has been demonstrated that for a very simple "hull" shape calculation of magnetic fields caused by volume currents and separate calculation of magnetic fields caused by currents in the structure gives nearly the same combined result as the magnetic field calculated using a dipole. This validates the method of obtaining magnetic fields by means of detailed BEM combined with structural models.

Clearly the "detailed BEM combined with structural model" method offers most benefit when the volume current field is complicated by factors such as non-planar boundaries between the sea layer and the seabed layer(s), complex hull shape and so on.

A method has been shown of determining the equivalent dipole(s) that reproduce some target UEP (along a line of sampling points). The "target UEP" could come from a detailed BEM model of a hull, or from experimental measurements made on a range.

It has been shown that the contribution of the volume currents to the total magnetic field is significant (around 21% for the simple case studied), and so it is clear that accurate representation of the volume currents is important.

References

[1] BEASY User Guide, Computational Mechanics BEASY, Southampton, 2001.
[2] Adey R A, Niku S M, "Application of BEM in Corrosion Engineering", Chapter in Topics in Boundary Element Research Ed, C A Brebbia, Springer Verlag, 1990.
[3] Adey R A, Baynham J M W, "Design and Optimisation of Cathodic Protection Systems Using Computer Simulation", NACE Corrosion Conference, Orlando USA, 2000.
[4] Adey R A, Pei Yuan Hang, "Computer Simulation as an aid to Corrosion Control and Reduction", NACE Corrosion Conference, San Antonio Texas USA, 1999.
[5] T. Bailey, A. Parker, R. Twelvetrees, M. Turner, S.J. Davidson. Advanced Signature Control Systems. Ultra Electronics Magnetic Division. United Kingdom. 2000.
[6] E Santana Diaz, R Adey, J Baynham, Y H Pei, Optimisation of ICCP systems to minimise electric signatures. Marelec 2001.
[7] E Santana Diaz, R Adey, B Tims. A Complete Underwater Electric and Magnetic Signature Scenario Using Computational Modeling. Marelec 2006.
[8] Allan, P.J., Ph.D. Thesis, A Study of the Electromagnetic Signatures of Marine Vessels using the Boundary Element Method, University of Glasgow, 2004.

Fast solution techniques for corrosion and signatures modelling

A. J. Keddie[1], M. D. Pocock[1] & V. G. DeGiorgi[2]
[1]*Frazer-Nash Consultancy Ltd., Dorking, Surrey, UK*
[2]*Naval Research Laboratory, Washington, DC, USA*

Abstract

The boundary element method (BEM) involves field nodes interacting with each other. Solving the ensuing matrix equations often requires an iterative solver to be used at a cost that scales with the second power of the number of nodes per iteration. This limits the size of the problem that can be solved. The fast multipole method (FMM), introduces hubs to reduce the number of direct interactions between field nodes in the BEM. The cost of calculating the matrix-vector multiplication using the FMM scales linearly with problem size. This paper contains a brief mathematical description of the FMM for Laplace's equation in which a Taylor series expansion is used to model Green's function. The computational performance of the FMM applied to modelling an impressed current cathodic protection (ICCP) system of a naval vessel is then investigated and the results compared to those of a commercial BEM solver and experimental (physical scale model) results. For this relatively small example model, it is shown that the cost benefit of the FMM is eight times greater than that of the commercial solver. Greater savings will be obtained on larger models. The results confirm that larger, more detailed, corrosion problems can be solved faster using the FMM. It is also shown that the capabilities of the FMM offer the choice between reduced processing time and enhanced accuracy. This provides the user with the opportunity to sacrifice accuracy in order to run less computationally expensive problems, for example during parametric studies.
Keywords: boundary element methods, fast multipole method, ICCP system.

1 Introduction

ICCP systems exploit the electrochemical nature of corrosion and the establishment of cathode-anode regions on the ship. Cathodes (exposed metal regions on the hull, rudder and propeller) are protected regions and anodes are either discrete components of sacrificial metal or electron source points positioned along the underwater hull at specific locations. For a ship at sea, the wetted surface area of the hull and the appendages such as bilge keel, rudder and propeller, are the cathodes, which require CP.

A typical ICCP system consists of non-sacrificial noble anodes connected to power supplies, reference cells to monitor hull potential state and a controller to adjust the current output of the anodes. By ensuring that the on-board power supply provides controllable anodic current to inert anodes, the ICCP system inhibits corrosion of a material that would otherwise act as an anode by forcing it to behave cathodically. The ideal system is designed with anodes located so current is evenly distributed to ensure that a uniform voltage is maintained for all points on the underwater hull. In reality, cathodes created by paint damage, components made of non-similar materials, geometric features and openings in the hull result in a varied profile.

In the evaluation of material or system performance an appropriate material characterisation is necessary. In the case of electrochemical corrosion, the sensitivity of the electrical current-electrical potential relationship to environmental and electrolyte characteristics must be considered when defining an appropriate characterisation. The accuracy of corrosion prediction, whether based on analytical, experimental or computational analysis, will depend on how well the system defined in the analysis matches both the real structure and the environment surrounding the structure.

Polarisation, which for present purposes may be characterised as the relationship between electrical current and electrical potential, is an observed behaviour resulting from the combined effects of multiple oxidation and reduction reactions and is usually non-linear. This often complex behaviour between electrical current and potential severely complicates the process of corrosion prediction and sensitivity studies can be used as a method for determining the suitability of polarisation relationships.

2 Ship geometry

The geometry used in this study is the same as that used by DeGiorgi et al [1]. It is representative of navy ship hulls with one propeller and one rudder located along the port-starboard centerline. The ship is outfitted with a 2 zone ICCP system. Anode placement, power supply sizing, and reference cell locations are defined based on established design rules.

In [1], the authors used the boundary element code FNREMUS [2] to compare the performance of a 2 zone ICCP system for four zonal operational combinations. The focus of this study will be for the case of normal operating condition in which both forward and aft zones are fully functional. In [1], the authors compared computational results for this case with those obtained using physical scale modelling (PSM) for the same problem. PSM is a physics based experimental process used to design and evaluate ICCP systems. Details of this approach can be found in [3].

The authors have duplicated the geometry, ICCP system, material configuration, material properties and loading conditions defined in [1]. The ship hull used is representative of the type of ship that can be found in the US Navy but is not a duplicate of any active ship. The ICCP system is a conceptual system similar to that which may be found on US Navy ships. The ship hull boundary element mesh (Figure 1) is identical to that found in [1]. The commercial code PATRAN [4] was used to generate the ship hull mesh. Only the portion below the design waterline is included in the model. Results are calculated for the minimum damage (3% bare material concentrated in the docking blocks and propeller) static flow condition. The intent is not to create a new analysis of the design but to compare solver CPU capabilities on a standard published problem. The ship model is submerged in a large container of seawater representing infinite expanse of open ocean. The electrolyte is modeled as scaled (1/96) seawater with a conductivity of 5.69×10^{-2} Siemens/m. Material properties are identical to those used in [1]. Further details on material property definitions can be found in [5]. A reference cell set value of -0.85 V Ag/AgCl electrode is used in the analysis.

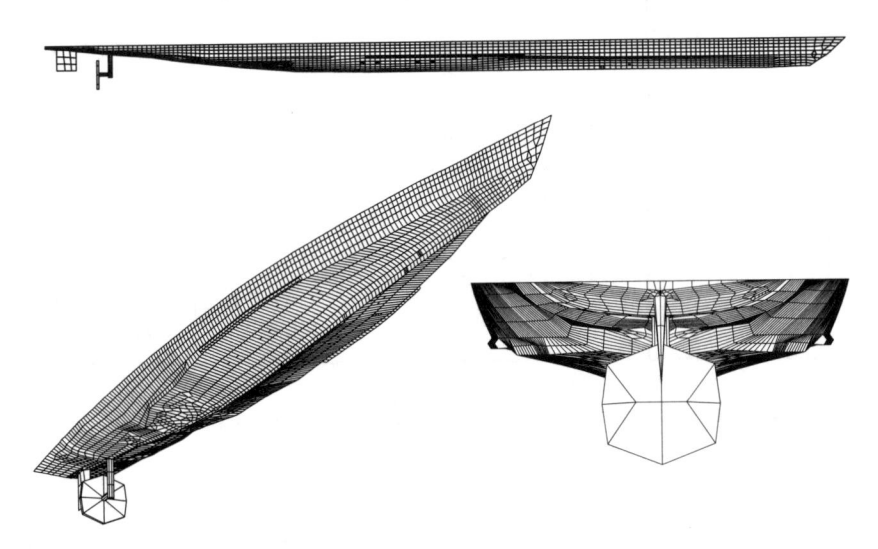

Figure 1: Boundary element mesh.

3 ICCP controller model

The solution algorithm is an iterative process in which the anode current for each iteration is determined from the difference between the reference cell's set potential and achieved potential at the previous iteration. The relationship between potential and current can be a simple linear gain, a Proportional-Integral-Derivative (PID) type control or a more complex algorithm, and can take account of maximum supply currents or other physical constraints of the ICCP system. At each iteration the currents at all anodes are adjusted to minimize the difference between set and achieved potential at each reference cell. This involves employing another iterative process around the (usually) non-linear material polarisation relationship.

The surface potential on a vessel has traditionally been calculated at each iteration by employing the boundary element method (BEM) to numerically solve Laplace's equation. Boundary element techniques require solving a system of equations for which the computational cost scales with the second power of the problem size. This becomes prohibitively large as the number of nodes for a particular problem increases. As will be discussed in the following section, the fast multipole method (FMM) reduces this computational and storage cost by approximating the integrals within the BEM.

Table 1: ICCP system details-scale geometry dimensions, [1].

	Frame	Below Waterline (cm)	Off Centerline (cm)	Symmetry
Forward Zone Anodes	49.9	2.3	1.8	Port & Starboard
	189.1	2.8	4.6	Port & Starboard
Forward Zone Reference Cells	103.7	3.3	1.0	Port & Starboard
Aft Zone Anodes	279.4	2.8	4.3	Port & Starboard
	368.8	0.7	3.8	Port
	379.5	0.5	3.8	Starboard
Aft Zone Reference Cell	397.0	0.5	0.2	Starboard

4 Numerical methods for solving Laplace's equation

4.1 Boundary element modelling and cost scalings

Within each iteration of the FNREMUS code it is necessary to solve Laplace's equation. Laplace's equation, which models the electric field, is expressed as

$$\nabla^2 \phi(\mathbf{y}) = 0 \qquad (1)$$

for some location $\mathbf{y}$.

Green's function for Laplace's equation is:

$$G(\mathbf{x}, \mathbf{y}) = \frac{1}{4\pi |\mathbf{x} - \mathbf{y}|} \qquad (2)$$

The Boundary Integral form for Laplace's equation can be expressed as:

$$\frac{1}{2}\phi(\mathbf{x}) = \int_{\partial\Omega}\left(G(\mathbf{x},\mathbf{y})\frac{\partial\phi}{\partial n'} - \phi\frac{\partial}{\partial n'}(G(\mathbf{x},\mathbf{y}))\right)ds' \qquad (3)$$

where ds' indicates that we are integrating over the surface with respect to $\mathbf{y}$.

Discretising (3) and evaluating the integrals using Gaussian quadrature, gives the following system of equations:

$$2\pi\phi_i + \sum_{j=1}^{N} a_{ij}\phi_j - \sum_{j=1}^{N} b_{ij}\left(\frac{\partial\phi}{\partial n'}\right)_j = 0 \qquad (4)$$

where, N represents the number of nodes in the problem after discretising. In matrix form this gives:

$$\begin{bmatrix} A \end{bmatrix}\begin{bmatrix} \phi \end{bmatrix} = \begin{bmatrix} B \end{bmatrix}\begin{bmatrix} \phi' \end{bmatrix} \qquad (5)$$

where $\phi' = \partial\phi/\partial n'$.

The cost of storing and evaluating (5), using an indirect solver, scales with the second power of the size of the problem. A problem that doubles in size will therefore approximately quadruple in memory requirement and computational time. This becomes prohibitively large as the size of N approaches 10^5.

4.2 Fast multipole method and cost scalings

One of the methods that has promised the most in recent years, with respect to reducing the operation count and storage requirements of the BEM is the fast multipole method. The FMM was initially introduced by Rokhlin [6] as a fast solution method for the two-dimensional Laplace equation. The FMM can be thought of as a method for separating the kernel of the BEM into a product of functions of its dependant variables, i.e. if we can write

$$G(\mathbf{x},\mathbf{y}) = g_1(\mathbf{x})g_2(\mathbf{y}) \qquad (6)$$

In this case we may re-express (3), after moving the g_1 term outside of the integral, as

$$\frac{1}{2}\phi(\mathbf{x}) = g_1(\mathbf{x})\int_{\partial\Omega}\left(g_2(\mathbf{y})\frac{\partial\phi}{\partial n'} - \phi\frac{\partial}{\partial n'}(g_2(\mathbf{y}))\right)ds' \qquad (7)$$

We see that in (7), g_1 and g_2 take on N different values. The calculation of the integral will therefore require just order N operations, and the evaluation of g_1 at all locations $\mathbf{x}$, a further order N operations.

Most kernels of interest cannot be expressed directly in such a simple form as (7) and the FMM can be thought of as a way of achieving such a decomposition, by in part, expanding the kernel as an infinite series. Expanding the function $F(\mathbf{x},\mathbf{y})$, around a reference point, $\mathbf{z}$, using a Taylor series expansion gives:

$$F(\mathbf{x},\mathbf{y}) = \sum_{\beta=0}^{\infty}\frac{1}{\beta!}((\mathbf{z}-\mathbf{y})\cdot\nabla)^{\beta} F(\mathbf{x},\mathbf{z}) \qquad (8)$$

which, for simplicity, can be expressed as a truncated series of products of functions that depend on $\mathbf{x}$, $\mathbf{y}$ and $\mathbf{z}$:

$$F(\mathbf{x},\mathbf{y}) \approx \sum_{\beta=0}^{p} f_1^{(\beta)}(\mathbf{z},\mathbf{y}) f_2^{(\beta)}(\mathbf{x},\mathbf{z}) \qquad (9)$$

Expressing Green's function in the form of (9) and substituting into (3) gives

$$\frac{1}{2}\phi(\mathbf{x}) = \sum_{\beta=0}^{p} g_1^{(\beta)}(\mathbf{x},\mathbf{z}) \int_{\partial\Omega} \left(g_2^{(\beta)}(\mathbf{z},\mathbf{y}) \frac{\partial\phi}{\partial n'} - \phi \frac{\partial}{\partial n'} \left(g_2^{(\beta)}(\mathbf{z},\mathbf{y}) \right) \right) ds' \qquad (10)$$

As with many series expansions, the rate of convergence is of utmost importance. Such expansions converge efficiently only for small ranges of the dependant variables. Indeed for the Taylor series expansion, and hence (10), to converge, the condition that $|\mathbf{z}-\mathbf{y}|<|\mathbf{x}-\mathbf{z}|$ must be satisfied. As such, it is not possible to evaluate (10), for all locations $\mathbf{x}$ and $\mathbf{y}$, by choosing a single value of $\mathbf{z}$.

To overcome this limitation, we introduce a hierarchical decomposition of the spatial domain to carry out grouping of nodes. This is achieved by decomposing the domain into a finite number of groups, $\partial\Omega_l$, each containing a centre point $\mathbf{z}_l$, such that (10) can be expressed as

$$\frac{1}{2}\phi(\mathbf{x}) = \sum_{\beta=0}^{p} \sum_{l} g_1^{(\beta)}(\mathbf{x},\mathbf{z}_l) \int_{\partial\Omega_l} \left(g_2^{(\beta)}(\mathbf{z}_l,\mathbf{y}) \frac{\partial\phi}{\partial n'} - \phi \frac{\partial}{\partial n'} \left(g_2^{(\beta)}(\mathbf{z}_l,\mathbf{y}) \right) \right) ds' \qquad (11)$$

Evaluation of (11) can be performed for nearly every $\mathbf{x}$ and $\mathbf{y}$ location. This is achieved by only allowing collocation nodes, $\mathbf{x}$, to interact with relatively distant groups of integration nodes, $\mathbf{y}$, with centre $\mathbf{z}_l$. The FMM goes on to decrease the computational cost by performing additional expansions. For example, the functions g_1 and g_2, which, like (9) depend on two vectors, can also be expressed as Taylor series expansions. By doing so, the number of direct interactions between nodes is decreased. This is illustrated in Figure 2 for the BEM and FMM.

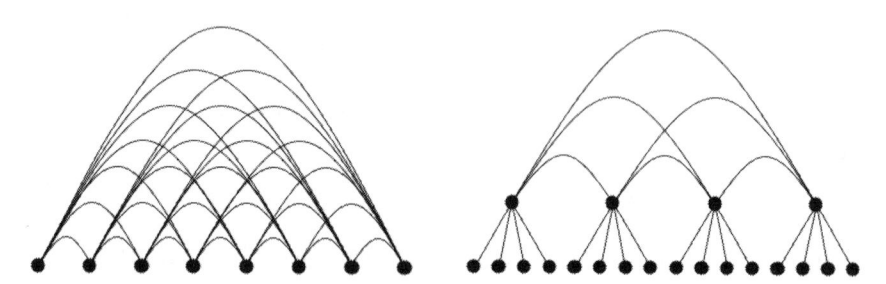

Figure 2: Diagram showing the interaction of nodes in the BEM (left) and FMM (right). Hubs are established in the FMM to reduce the number of nodal links and hence the computational cost.

While performing additional expansions produces a complex grouping structure, doing so will result in the computational and storage cost of the FMM to scale approximately with the size of the problem, N. Clearly, the reduction in cost, will allow larger problems to be solved faster. Indeed, even for the modestly sized surface ship mesh, described in Section 2, it would be expected that significant savings in CPU time would be achieved using the FMM. This offers the time-constrained user the ability to perform more detailed sensitivity studies than would otherwise be possible. Additionally, the CPU-time required to solve a problem can be further decreased by reducing the number of expansion terms, p, used. This is likely to cause a slight reduction in the accuracy of the overall solution since the relevant function in (9) will be approximated using fewer terms. However, this is a useful tool in the initial stages of sensitivity analyses and allows the user the chance to perform preliminary studies in relatively quick time before performing more accurate and detailed analyses.

5 Results

In this section computational results using two Laplace solvers within FNREMUS will be studied. These are:
- BEASY-Thermal, a commercial boundary element solver; and
- the FMM code as described in Section 4.

These will henceforth be referred to as FNREMUS-BEASY and FNREMUS-FMM, respectively. Experimental results, obtained using physical scale modelling (PSM), will also be used to validate the computational results. Results calculated using both PSM and FNREMUS-BEASY were calculated as part of the paper by DeGiorgi et al [1].

In order to compare its performance and accuracy, the results calculated using FNREMUS-FMM were obtained using different numbers of expansion terms. While increasing the number of terms in the Taylor series expansion will generally increase the accuracy of the solution, the benefits of doing so must be offset against the additional computational cost of calculating the extra expansion terms.

Key results obtained using PSM and computational methods are shown in Table 2. As discussed by DeGiorgi et al [1] variations can be seen as indications of the non-uniqueness of the multiple anode-multiple cathode-variable input current problem. This issue is not discussed here since the objective of this work is the comparison of solver performance. Of importance here is the comparison between the two calculated results. In both FNREMUS solvers, the current from the anodes will equal the sum of the current from the propellers and docking block. It can be seen from FNREMUS-BEASY results that the total anodic and cathodic currents are equal to 1.973mA. While the total anodic current shows good agreement with the PSM results, the individual (fore and aft) currents show differences. This is the same pattern as seen in [1]. The FNREMUS-FMM results in Table 2 show that there is a total anodic and cathodic current of between 2.011mA and 2.009mA depending on the number of expansion terms used. The difference between FNREMUS-FMM and PSM anode currents is greater than the differences calculated for FNREMUS-BEASY and PSM. However, there is a smaller difference between the total cathodic currents for FNREMUS-FMM and PSM. Indeed, the FNREMUS-FMM values of anodic/cathodic currents are closer to the average of the PSM anodic/cathodic currents than FNREMUS-BEASY. As with the FNREMUS-BEASY case, the fore and aft currents show differences between PSM and FNREMUS-FMM. This could again be attributed to multiple combinations of zonal input values that result in similar overall system performance parameters.

Table 2:   Corrosion results from ship hull calculated using Physical Scale Modelling (PSM), FNREMUS-BEASY and FNREMUS-FMM.

	PSM	FNREMUS-BEASY	FNREMUS-FMM p=2	FNREMUS-FMM p=3	FNREMUS-FMM p=4	FNREMUS-FMM p=5
Fore anodes (mA)	-1.21	-0.161	-0.143	-0.183	-0.182	-0.204
Aft anodes (mA)	-0.76	-1.812	-1.868	-1.828	-1.828	-1.805
Total	-1.97	-1.973	-2.011	-2.011	-2.010	-2.009
Docking blocks (mA)	0.89	0.889	0.915	0.915	0.914	0.913
Propeller (mA)	1.15	1.084	1.097	1.096	1.096	1.096
Total	2.04	1.973	2.011	2.011	2.010	2.009
Forward reference electrode (mV)	-850	-851	-851	-851	-851	-851
Aft reference electrode (mV)	-850	-850	-849	-849	-849	-849

Figure 3 shows the CPU times required to run the model using FNREMUS-BEASY and FNREMUS-FMM with increasing expansion terms. The total CPU time taken to run FNREMUS-BEASY is approximately 24,000 seconds. This is at least twice as long at it takes to run FNREMUS-FMM for any of the cases shown. The time taken to solve the problem using FNREMUS-FMM with two expansion terms is approximately 3,000 seconds or eight times quicker than FNREMUS-BEASY. As larger problems are solved, the difference in time taken to solve them using both FNREMUS solvers will be exacerbated due to the cost scalings of each method.

It can be seen that the currents in Table 2 don't appear to converge towards a single value as the number of expansion terms increases. This is most likely a result of the FNREMUS ICCP algorithm which employs an iterative solution technique around the non-linear material polarisation relationship. The currents shown in Table 2 are the first iterative values within the specified tolerance.

6   Summary

In this work the fast multipole method has been successfully applied to the computational modelling of electrochemical corrosion on a surface ship. The ship chosen is one similar to US Navy ships with a system similar to that which can be found on these ships. The problem chosen is representative of the type of design problem which may be encountered and does have a basis in reality. Cost savings estimated in this work would therefore translate to real cost savings. The CPU-time required to solve this problem was more than eight times faster than a commercial boundary element when solved using two expansion terms and twice as fast when using five terms. Furthermore, the results obtained using the FMM corrosion solver demonstrate the same type of agreement and variation with PSM as shown earlier with conventional BEM solvers [1]. This indicates that there are no additional errors introduced into the modelling process by the use of the fast multipole method.

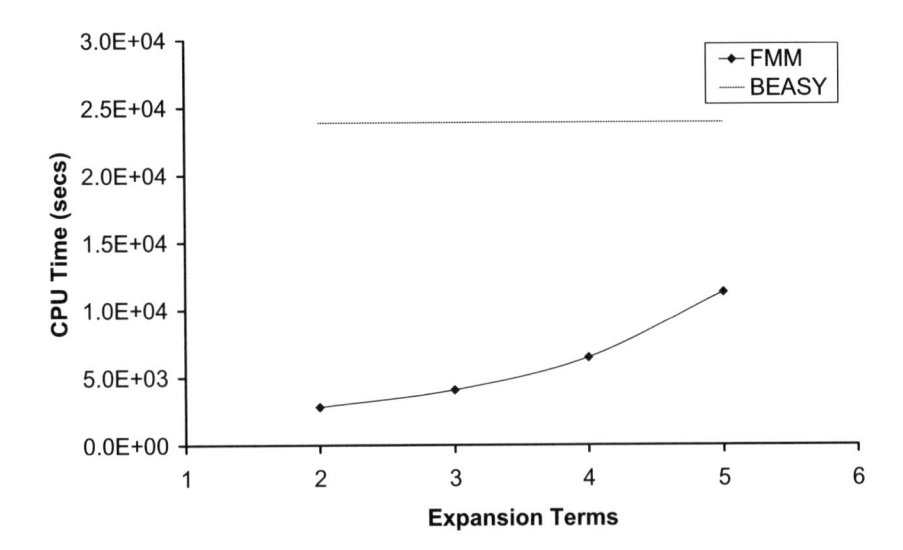

Figure 3: CPU time required to run FNREMUS using different Laplace solvers on a Frigate model with ICCP system. FNREMUS-BEASY was only run once but has been shown here for comparison with FNREMUS-FMM.

Based on the comparison of runtimes generated by analysis of a realistic ship problem, it has been determined that incorporation of the fast multipole method into BE code would greatly enhance the ability to use BEM for large problem solutions. This could readily lead to more use by those interested in the solution of electric fields on- and off-board of large complex structures such as modern military ships.

References

[1] DeGiorgi, V.G., Pocock, M.D., Wimmer, S.A., & Hogan, E.A., "Zonal ICCP System Control Interactions," *Simulation of Electrochemical Processes*, 15-24, 2005.
[2] Frazer-Nash Consultancy, FNREMUS Detailed Modeller User Guide, FNC 5421/21133R, Issue 1, 2000.
[3] DeGiorgi, V.G., Hogan, E., Lucas, K.E. & Wimmer, S.A., "Shipboard Impressed Current Cathodic Protection System," Chapter 2, *Modeling of Cathodic Protection Systems*, WIT Press, 13-44 2006.

[4]　MSC Software Corp., MSC Patran 2001 User's Manual, 2001.
[5]　DeGiorgi, V.G., "Evaluation of Perfect Paint Assumptions in Modeling of Cathodic Protection Systems," *Engineering Analysis with Boundary Elements*, 26/5, 435-445, 2002.
[6]　Rokhlin, V., "Rapid Solution of Integral Equations of Classical Potential Theory," *Journal of Computational Physics*, 1985. **60**: 187-207.

Corrosion related electromagnetic signatures measurements and modelling on a 1:40th scaled model

L. Demilier[1], C. Durand[1], C. Rannou[2], E. Hogan[3], M. Krupa[3], A. M. Grolleau[1], J. Blanc[4] & A. Guibert[1]
[1]*DCN, Paris, France*
[2]*DGA GESMA, Brest Armées, France*
[3]*NRL, Washington DC, USA*
[4]*DGA CTA, Arcueil, France*

Abstract

The Corrosion Related Magnetic Field (CRM) is a part of the static magnetic signature of ships that cannot be precisely quantified for steel hull ships because of the ferromagnetic signature that is added to the CRM in magnetic measurements. For this reason, the CRM is generally evaluated by computation.

Most of CRM models are based on analytical formulas of the magnetic field due to electric dipoles which decrease more slowly with distance than the magnetic signature due to ferromagnetism (spatial decay law in $1/r^2$ instead of $1/r^3$). This is the reason for the following assumption usually met in the technical literature: CRM is dominates the ferromagnetic signature at long distance. But theses models don't take into account a significant part of the CRM source: the currents that flow through the metallic structures of the ship. As a result, conclusion on the distance for which the CRM is dominating the ferromagnetic or the distance for which the CRM remains important can be corrupted.

CRM measurements have been carried out on a ship mock-up to validate a new Finite Element software for CRM that takes into account all the CRM sources: currents in the sea and currents flowing through the metallic structures of the ship. This paper describes the measurements of low levels of CRM on the mock-up, and the first results of its validation with measurements.

Keywords: cathodic protection; electromagnetic silencing; CRM; UEP; MINE.

1 Introduction

During the Second World War II, intelligent mines have been developed and extensively used. Theses weapons contained several types of sensors, allowing them to detect the proximity of the vessel and wait for the best moment for triggering (see Figure1). They might be equipped either with an acoustic sensor, a pressure sensor and a magnetic sensor or the combination of the three.

A mine explosion at a closed distance of the vessel can cause severe damages to the hull and might sink it rapidly in certain cases.

Some recent conflicts have demonstrated that the mine threat cannot be neglected and is of major concern for all navies because such weapons are very cost effective and their deployments are not risky compared to other weapons.

Another threat can use Electromagnetic indiscretion to detect a target. The MAD (Magnetic Anomaly Detector) is a highly sensitive magnetometer (see figure 2) installed on aircraft and used for submarine detection at high distance.

Figure 1: (a) A picture of a WWII bottom magnetic mine, (b) an underwater mine explosion.

Figure 2: A maritime patrol aircraft P3-Orion equipped with a MAD.

Usually, a way to represent the risk of a vessel to interrupt its mission is illustrated with equation (1):

$$R \approx 1 - \left[1 - \frac{(p \times f)}{L}\right]^n \quad (1)$$

where:
- R: the Risk that the vessel mission could be stopped
- p: the probability of the vessel to be detected
- f: the dangerous front, distance for which the mine explosion will provoke subsequent damages to the vessel
- L: the minefield width
- n: the number of mines

It is possible to have an influence on the risk, R. For example, by using a dedicated naval force to hunt or sweep the mine at sea, n could be decreased. It is also possible to design the vessel sufficiently strong to resist a mine explosion. If the vessel is designed discreet enough, the probability that the mine sensor can detect the vessel could be limited. Usually, not only one solution is used, but a combination depending on operational and economical parameters.

The role of the DCN Electromagnetic Silencing group is to study and eventually minimise the indiscretion used by mine or aircraft to detect a potential target. Nowadays, mines can be micro-controlled and can exploit all indiscretions generated by a target with high sensitivity. The corrosion related electromagnetic signatures come from different sources but are always dealing with an aspect of galvanic corrosion. The most noted warship galvanic couple is the NAB propeller coupled with the steel hull.

The purpose of this paper is to illustrate how we have proceeded to evaluate, measure and model corrosion related electromagnetic signatures.

2 Main challenge

2.1 Mechanisms

Usually made from steel, the vessel hull is submitted to corrosion in sea water. The galvanic couple created between the dissimilar metals in electrical contact with the sea water induce some currents in the sea water, which result in the destruction (oxidation) of the less noble metal (called the anode). To fight against such destruction, the solution consists of providing some DC currents (by means of sacrificial anodes or impressed current cathodic protection (ICCP) anodes) in the sea water, on surfaces to be protected. In doing so, the potential of the surface to be protected is brought to a potential referred to as an "immunity potential" where corrosion is minimized.

The solution adopted by DCN for ship corrosion protection is a combination of an efficient coating and a cathodic protection system that protects the ship as the coating degrades over time. Unfortunately, these anti-corrosion currents create an electric and a magnetic indiscretion, which can be used by mines for triggering (see Figure 3).

2.2 Difficulties

The main difficulties in studying and modelling the corrosion signatures lie in acquiring accurate experimental data as well as in identifying and implementing a numerical method allowing us to evaluate theses signatures.

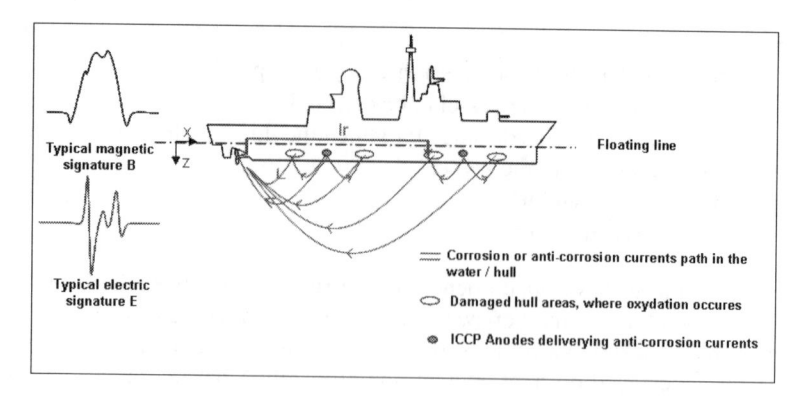

238 Simulation of Electrochemical Processes II

Figure 3: Illustration of the current distribution in a cathodic protection problem.

In order to have the best control as possible on all parameters involved in the corrosion and signatures problems, it has been decided to use a physical scaled model.

Measurements took place at the U.S. Naval Research Laboratory, Center for Corrosion Science and Engineering, Marine Corrosion Facility in Key West, Florida. This laboratory, focused on corrosion studies, incorporates and electrically isolated, 208 cubic meters test tank, instrumented for corrosion and electric field measurements.

Electric signature measurements require low noise sensors and a high quality electronics amplifier. The natural ambient electrical noise is quite low. As a result, it is possible to obtain excellent measurements. For the magnetic signature measurements, the problem is complex because several parameters are to be taken into account in order to obtain exploitable measurements.

- The ambient geomagnetic field is high, several thousand times more than that of the target magnetic field strengths to be measured.
- The ambient lab magnetic field is perturbed (metallic structure, periodic passage of trucks, etc.).
- Presence of ferromagnetic sheets (damage area) on the model.
- The CRM signature is very low due to the scale effect (40 times less than expected for full scaled vessel)

For all of these reasons, direct CRM measurements were not possible. It was therefore decided to acquire differential magnetic measurements with all the contributors with and without the ICCP system ON. In doing so, the magnetic contribution of the ICCP system was extracted from measurements analyzing the field difference. The validity of these differential measurements were considered satisfactory because it was possible to set the ICCP system in the same configuration and status during all the acquisition of measurements by maintaining control on the current density and potential along the model.

3 Physical scale modelling

3.1 1:40th scaled model

A 1:40th scale frigate model was built and equipped with an equivalently scaled ICCP system, discreetly wired damage areas, representing surface coating damage areas, and numerous hull Silver- Silver Chloride (Ag/AgCl) reference cells utilized in the monitoring of the surface hull potential fields and for ICCP system control (see Figure 4).

The discreet damage areas were wired such that each individual steel or NAB electrode could be switched in or out of circuit to enable several different damage scenarios. More importantly, the ability to electrically isolate each electrode was imperative as to not impede the overall CRM measurements by introducing small galvanic currents that would be present if any two electrodes were wired to one switch and switched out of circuit as a pair.

The entire electrical circuit and wire pathways of discreet model components were implemented on the model with accuracy and followed two different, isolated separate circuits (see Figure 5). Both circuits were designed that they could be then switched on or off in order to evaluate the importance of the structure current return path on the magnetic signature comportment.

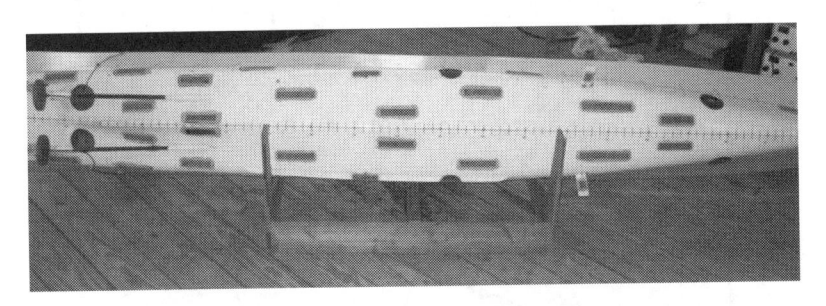

Figure 4: 1/40th scale model mock-up with hull coating damage areas.

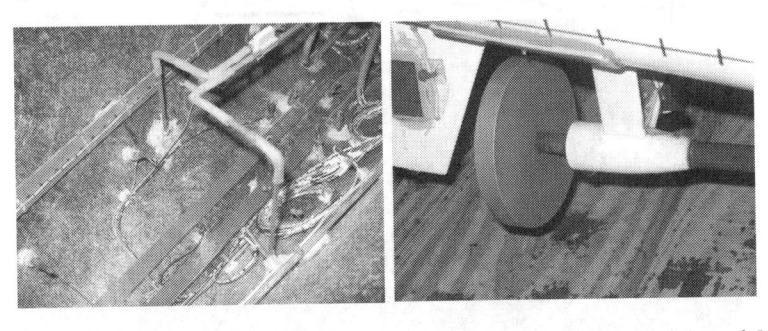

Figure 5: Pictures of the detailed wiring of anode cables (right) and NiAl Bronze Prop disk (left).

3.2 Measurements

Measurements have been separated into two different parts, each one with its own technique:
- The Electric signature (UEP) measurement. For this signature, a 2-axis electric sensor (Silver/silver chloride electrode) has been used. Acquisition has been made by moving the sensor under the model and scanning electric signature during the displacement.
- The Magnetic signature (CRM) measurement. Three 3-axis magnetometers (Bartington® MAG03MSS-L70 MSS) have been installed on a rigid non magnetic mast. In the case of magnetic measurement, it is highly difficult to move the magnetic sensor in a very important magnetic ambience. As a result, acquisition has been realised with sensors stopped. Sensors displacement was then realised before new measurement.

The configuration of the measurement is illustrated in Figure 6. Measurements were taken on three longitudinal lines, at different three different distances from the model.

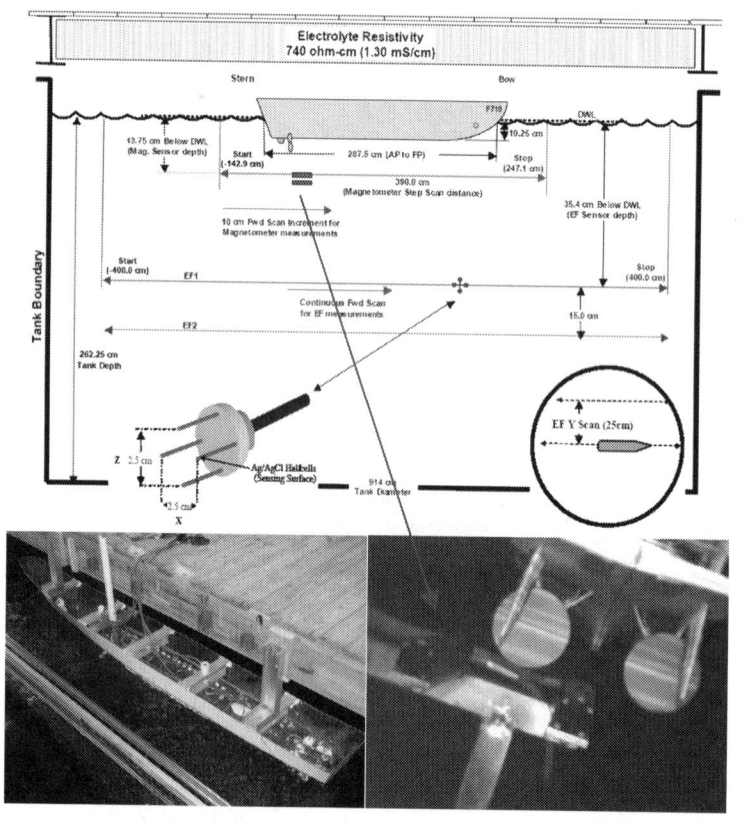

Figure 6: Schematic and pictures of testing tank setup, mock-up in test and close-up of magnetometers under stern.

3.3 Modelling

The main target of this study was to measure the corrosion related signatures in order to have at our disposal a set of experimental data but also to identify and validate the tools by which to evaluate theses signatures on any different operational scenario.

The 1:40th scale model has been modelled using finite element software (see section 6). Additionally, both of the complete electrical circuit pathways, inside the model, have been described and meshed, in order to evaluate their impact on the signature (see Figures 7 and 8).

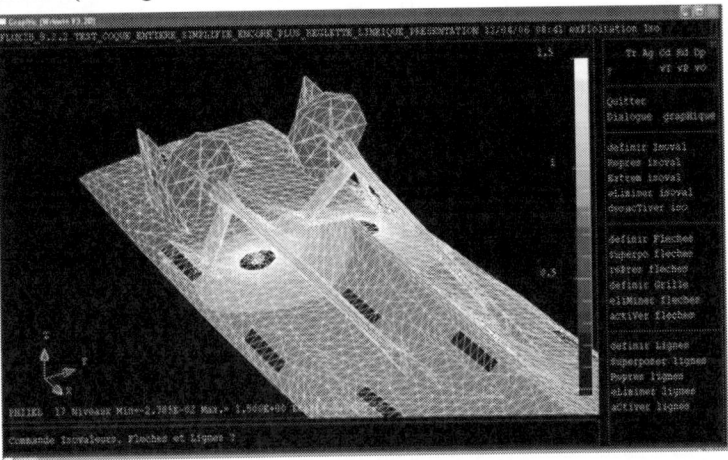

Figure 7:   Aft part of the f.e.m scale model with anode and damage areas.

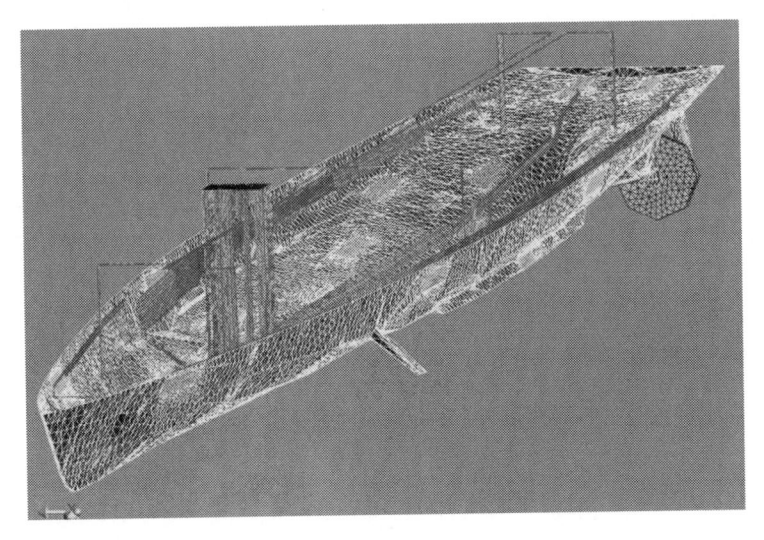

Figure 8:   Internal electrical pathways description and meshing.

4 Comparison

4.1 Electric signatures

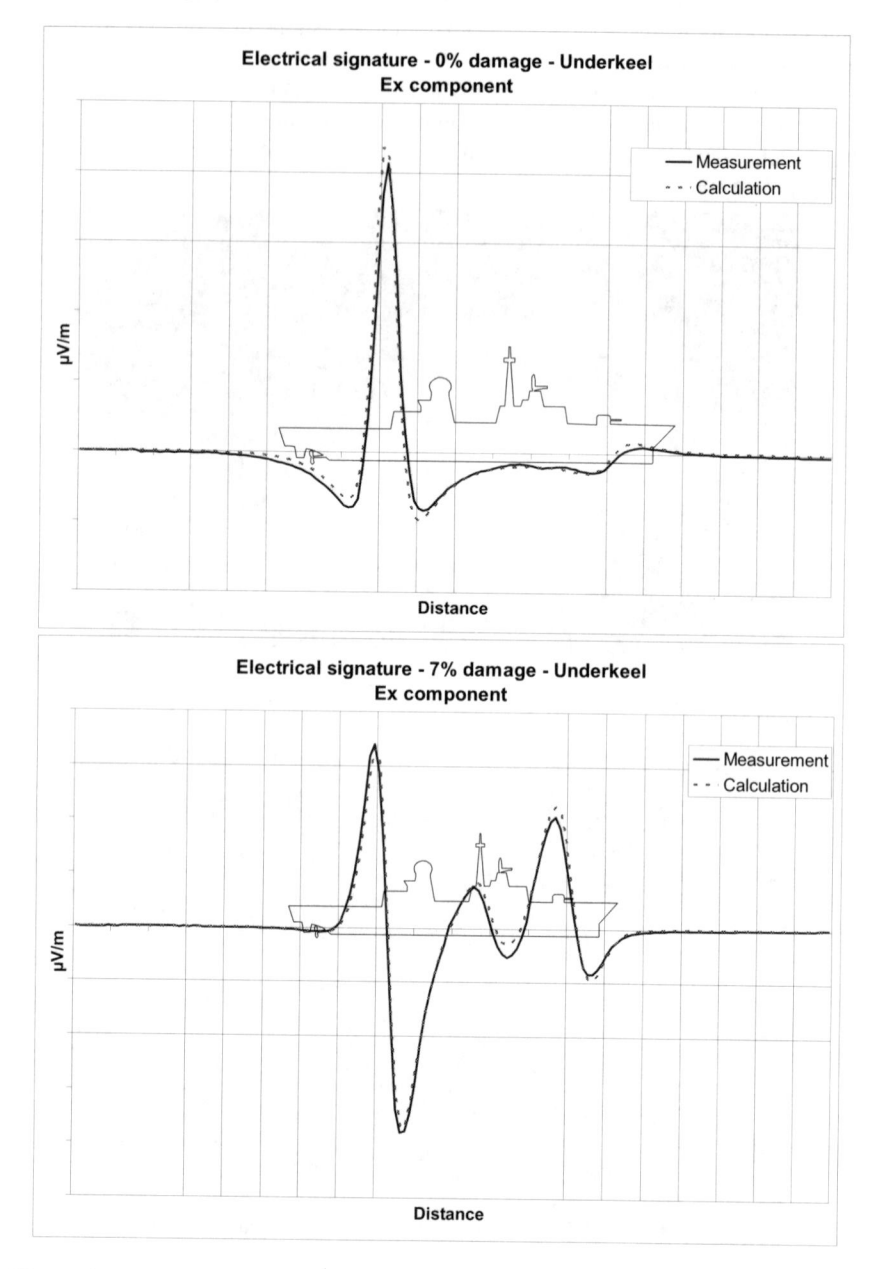

Electrical signature - 0% damage - Underkeel Ex component

Electrical signature - 7% damage - Underkeel Ex component

Figure 9: Comparison between UEP calculation and measurement for two different damage scenarios (0% and 7%).

4.2 Magnetic signatures

The CRM measurements shows clearly that the way the DC current is flowing back through the structure (large loop electrical path or small loop electrical path) has a major role in the CRM magnitude (see Figure 10) and certainly on its spatial comportment (decay law in $1/d^2$ or $1/d^3$ or an intermediate) according to the measurement distance.

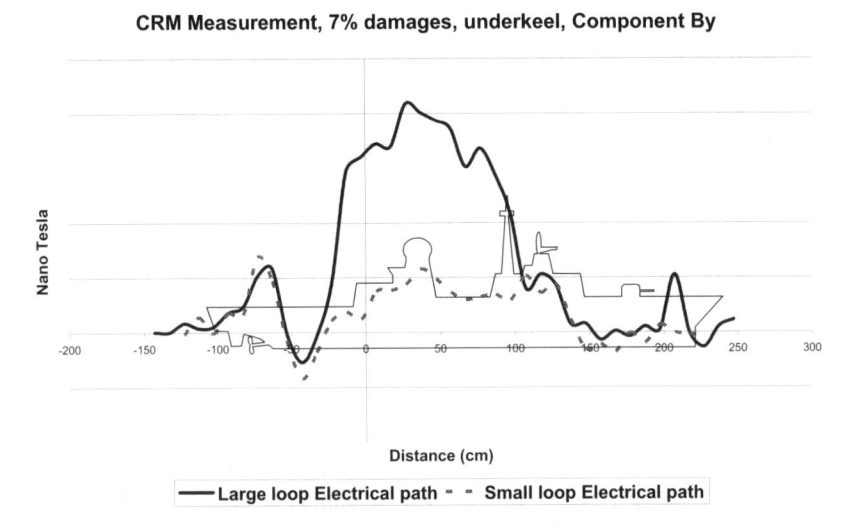

Figure 10: CRM measurement for 2 electrical different wires.

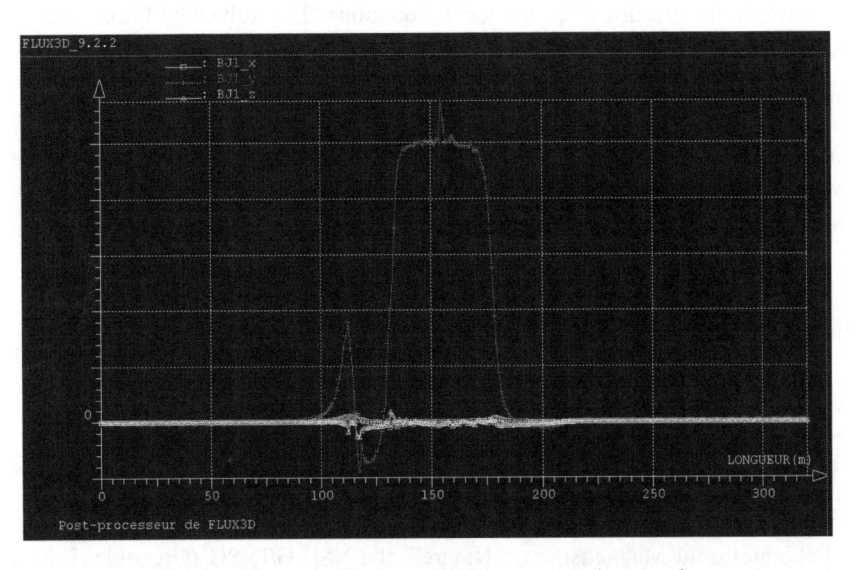

Figure 11: Preliminary CRM computation result.

The validation of CRM computation is not yet completed (see Figure 11). The first preliminary modelling result shows that calculations seem to be coherent with the measurement. The study is still in progress.

5 Conclusions and further studies

Calculations of the electric signatures have shown excellent correlations with measurements. Calculations of the CRM signatures of the 1:40th scaled model are still in progress. Preliminary results have shown interesting results but deeper investigations shall be undertaken. Once the calculations are validated, it is intended to study different operational configurations:
- Influence of a ferromagnetic hull on the CRM signature comportment
- Identification of the CRM decay law related to the electrical configuration
- Study of the CRM signature on non magnetic hull vessel (GRP, Titanium, stainless steel, carbon, etc.)

Then, it should be possible to determine the importance of the CRM signature related to the other magnetic sources and define in which conditions and distance, the CRM signature becomes dominant.

6 Finite element software

Calculations were done with Flux® software. The Electrolysis module allows handling cathodic protection problems for complex structures, such as warship and calculating the related electromagnetic signatures (UEP and CRM). Its unique feature is to take into account the whole current distribution (in the water and also in the metallic ship structure). Equations (2, 3) solved by Flux® are:

$$div\left([\sigma]\vec{grad}(V)\right) = 0. \qquad (2)$$

in the water and the metallic structure.

Solving this equation using the finite element method gives distributions of the electric potential V, of the electric field E ($E=-gradV$) and the current density vector j ($j=-\sigma\, gradV$). Then, Flux® computes the magnetic field H using Biot and Savart law:

$$\vec{H}(\vec{r}) = \frac{1}{4\pi}\iiint \vec{j}.dv \wedge \frac{\vec{r}}{r^3}. \qquad (3)$$

in the metallic structure and in the water.

Acknowledgements

The Authors wish to thank the French MoD – DGA (Direction Générale pour l'Armement) for founded this project.

Special thanks for Mr Arnaud Guibert, currently in PhD for the DGA, in the "Laboratoire du Magnétisme du Navire" of ENSIEG/INPG (Grenoble, France) for his tenacity in the modelling of the scaled frigate.

Section 5
Stress corrosion, cracking and corrosion fatigue

Modelling environmentally assisted cracking in pipeline steels

A. Plumtree & S. B. Lambert
*Department of Mechanical and Mechatronics Engineering,
University of Waterloo, Ontario, Canada*

Abstract

An investigation has been conducted on the environmentally assisted cracking (EAC) of ferritic-pearlitic pipeline steels in contact with simulated groundwater. The loading and environmental conditions were similar to those for buried natural gas pipelines in service. An anaerobic, dilute near-neutral pH solution was used in conjunction with the open-circuit potential for this system. The intent of this work was to determine and model the growth rate of environmentally assisted cracks in the form of transgranular stress corrosion cracks (TGSCC) that have been observed following field investigations. Combinations of low frequency cycling and high stress ratio R (=minimum load/maximum load), can produce transgranular fracture and a quantitative relationship between these two parameters has been developed for the conditions under which TGSCC takes place. The recorded crack growth rates were similar to those in the field and a superposition model was applied to the experimental data, giving good agreement between the observed and predicted single crack growth rates. Applying the superposition model to operating natural gas pipeline data indicated that more realistic predictions of crack growth would result by considering the interaction of multiple cracks, as observed in the field colonies.
Keywords: crack growth, near-neutral pH solution, pipeline steel, superposition model, transgranular stress corrosion cracking.

1 Introduction

Pipeline steels with a ferritic-pearlitic microstructure are highly susceptible to environmentally assisted cracking (EAC) known as a transgranular stress corrosion cracking (TGSCC) in dilute near-neutral (pH ~ 6.5) solutions [1]. This form of EAC which appears to be insensitive to temperature has been observed under the disbonded region of the pipeline coating where cathodic protection is not effective. Hydrogen assisted cracking due to the cathodic reaction at the crack tip and subsequent entry into the steel has been identified as the active mechanism [1,2]. TGSCC is termed "near neutral SCC" since it is uniquely different from intergranular SCC that is observed in concentrated carbonate/bicarbonate electrolytes of high pH (~10) at slightly elevated temperatures [3]. However, it is likely that a general four stage model proposed by Parkins [4], originally for the high pH case, is applicable to near-neutral SCC. It is expected that the first stage involves the breakdown of the pipe coating. This is followed by the initiation, growth and coalescence of microcracks forming a colony of very small individual macro-cracks. During the third stage, these grow and coalesce. If sufficient cracks coalesce to form a dominant, long shallow crack, relatively rapid crack growth will lead to failure. In most cases, however, cracks become dormant throughout the life-time of the pipeline [4].

The purpose of this work is to study the morphology and growth of single EAC cracks and apply a superposition model [5] to describe the later stages of crack growth using laboratory testing techniques that duplicate pipeline operating conditions. This model forms the basis for predicting the remaining life when the natural gas pipeline stress histories are analyzed.

2 Experimental procedure

Tests were carried out by Zhang et al [5] on a section of low alloy pipeline steel (API X60) 510 mm in diameter with a wall thickness of 6.35 mm. The section had been removed from service by Trans Canada Pipelines Ltd. in 1988 when some TGSCC colonies were discovered after 16 years of operation. The chemical composition of this steel was 0.22% C, 1.36% Mn, 0.34% Si, 0.018% S, 0.009% P, 0.025% Ni, 0.08% Cr, 0.01%Mo, 0.04% Cu, 0.22% Nb with traces of V, Ti, B and Al. The microstructure consisted of fine ferrite-pearlite with an average ferrite grain size of ~ 14µm. The specified minimum yield strength (SMYS) was 414 MPa, and the measured yield stress varied between 421 MPa and 470 MPa.

After conducting dye-penetrant tests to verify that TGSCC was absent, rings 10.4 mm wide were cut from the hoop orientation of the pipe and then sectioned into four 360 mm long samples which would be the base for edge crack specimens. Longitudinal welds were excluded. Inner and outer surfaces of each specimen were degreased with trichlorethane and then cleaned using a water-based alkaline solution to avoid contaminating the test electrolyte.

Using a V-shaped cutter, three notches 1.5 mm deep and 0.35 mm wide were machined in the top surface of each edge crack specimen. In this manner, more data from each test specimen could be obtained. These notches were cut normal to the specimen axis and positioned 50 mm apart after it was determined by finite-element analysis that the notches would respond independently to the applied load.

Thomas et al [6] prepared wider (40 mm) surface cracked specimens with surface notches 3 mm long and 1.5 mm deep. For these surface crack samples, as well as the edge crack samples, three or four-point bending test rigs were used to grow a fatigue crack from each starter notch to a final depth of about 3 mm, measured using the potential drop technique. Care was taken to ensure that the maximum crack tip loading was lower than that used in subsequent EAC tests.

After fatigue precracking, the specimens were carefully hand-polished to a 0.06-μm alumina finish, and the precracks were opened slightly using a small calibrated loading rig to clearly identify each crack tip location. When open, K_{max} for each crack was lower than that used during the fatigue precracking stage. A pair of microhardness indentations was made adjacent to the tip of each precrack to serve as reference points for further measurements. This was carried out on both sides and lengths of the edge and surface cracked specimens respectively. The load was then removed. Photomicrographs of the crack tips were taken using an optical microscope, giving a permanent record of the relative positions of the microhardness indentations and crack tips in order to determine any possible future environmental crack growth. The notch and fatigue precrack depths were measured using an image analyzing system, as outlined by Thomas et al [6].

Cantilever bend loading on both the edge and surface cracked specimens was used for the EAC tests. Crack tip loading was characterized using the linear elastic stress intensity factor, K. The maximum values imposed on the edge and surface crack specimens were 38 MPa√m and 31 MPa√m respectively.

Simulated ground water solution identified by TransCanada Pipelines Ltd. (TCPL) as NS4 was used as the electrolyte. Its composition was KCl-0.122 g/L, $NaHCO_3$ - 0.483g/L, anhydrous $CaCl_2$-0.093g/L and $MgSO_4.7H_2O$-0.131g/L. All the environmental tests were carried out in this electrolyte at a temperature of 25°C in an oxygen-free environment. To provide the anaerobic environment, a large polymethyl methacrylate (PMMA) box, which housed the steel specimen in a separate NS4 electrolyte chamber, was purged continuously with nitrogen. A controlled mixture of 5% carbon dioxide (CO_2) and nitrogen (N_2) was bubbled through the solution to maintain the pH at 6.5 to 7.0, which was monitored using a combination pH-saturated calomel reference electrode (SCE). It also was used to measure the corrosion potential (CP) and platinum potential (Pt). Small changes in pH were controlled by decreasing or increasing the flow of the CO_2/N_2 gas mixture. For long-term tests (~1 year), the pH was also measured periodically with litmus paper. The open-circuit potential for all the tests varied from ~0.75 V_{SCE} to -0.70 V_{SCE} and the Pt potential varied between -0.85 V_{SCE} and -0.35 V_{SCE} indicating that an anaerobic atmosphere was maintained throughout the test.

Cyclic loading was used in all the EAC tests with the stress ratios (R=minimum/maximum load) being controlled between 0 and 0.95. Loads were applied by a lever system and measured using a load cell. Examination of natural gas main pipeline data [7] revealed that most pressure fluctuations were in the range of R = 0.7 to 0.99. For pipeline inspection, pressure testing or other rare occasions, the stress ratio would drop to zero before returning to operating conditions. The present tests were carried out at frequencies of 5.8×10^{-2} Hz (5000 cycles per day) to 5.8×10^{-6} Hz (1 cycle every 2 days). Tests were carried out for periods lasting from 6 to 378 days.

Upon completion of each environmental test, the specimens were removed from the test rig and examined under the microscope. The final crack size was compared to the initial fatigue crack depth or length to compute crack growth. Following these observations, the specimens were immersed in liquid nitrogen for at least 10 min. and the cracks broken open under impact. The crack faces were cleaned ultrasonically for about 1 min in a solution containing 8.8 g 2-butyne-1.4-diol., 30 mL hydrochloric acid (HCl), and 100 mL distilled water. The cleaned surfaces were then examined in detail using the SEM to determine the fracture morphology and measure the extent of TGSCC.

3 Results and discussion

3.1 Fracture morphology

In general, transgranular fracture of the ferrite grains was observed following exposure in the NS4 electrolyte. As the exposure time increased, all cracks experienced secondary corrosion causing widening. The opposing crack flanks developed a mismatch, and the tip tended to become blunt and round. The degree to which these effects took place depended upon the test time and loading conditions. Based on measured crack widths at the microhardness indentation level, half crack depth and three-quarter crack depth, the average secondary corrosion rate was about $8.0\ (\pm 1.0) \times 10^{-10}$ mm/s, whereas the primary crack tip growth rate of about $2\ (\pm 1.0) \times 10^{-8}$ mm/s was much faster for K_{max} values in the range of 30 to 38 MPa√m.

Fracture surfaces consisting of transgranular facets mixed with microvoids (termed quasi-clearage) were observed on those specimens cycled at low frequencies and high R ratios. This fracture morphology was representative of TGSCC as opposed to corrosion fatigue which displayed ill-defined striations on specimens cycled at high frequencies and low R ratios. The influence of frequency (Hz) and R ratio can be summarized by the following equation:

$$R = 0.233\ log\ f + 1.19 \qquad (1)$$

When the left hand side of eqn. (1) exceeds the right, TGSCC takes place. For example, TGSCC was observed on specimens cycled at the lower frequencies of 4.6×10^{-4} Hz (40 cycles/day) and 4.6×10^{-3} Hz (400 cycles/day) with R ratios of 0.5 and 0.82, respectively. However, TGSCC was also observed after cycling at a faster frequency of 5.8×10^{-2} Hz (5,000 cycles/day), albeit at the higher R ratios of 0.90 and 0.95. For those specimens cycled at higher frequencies, particularly with a low R ratio of 0.5, the fracture surface features were flat and striated, typical of corrosion fatigue.

Many horizontal grooves were observed on the fracture surface. These were the original locations of MnS inclusions. It was anticipated that dissolution in the

NS4 solution produced a more acidic local environment, resulting in the formation of brittle striations in the immediate vicinity. The corrosion product, iron sulphide (FeS), is soluble in the local crack tip environment and would supply hydrogen sulphide (H_2S) to enhance crack growth [9]. Using higher temperatures, Kuniya et al [10], studied the effects of MnS inclusion dissolution on SCC in carbon steels in pure water. Flat transgranular regions associated with SCC were found next to MnS inclusions, indicating hydrogen assisted cracking (HAC) [11]. This morphology was similar to the surface features observed next to partly dissolved MnS particles in the present tests.

3.2 Superposition model

A superposition model has been applied to describe the observed the present EAC growth behaviour [5]. The cyclic (mechanical) component is added to the environmental (time dependent) component. This linear superposition model [5] is given by:

$$\left(\frac{da}{dN}\right)_T = \left(\frac{da}{dN}\right)_{cyclic} + \left(\frac{da}{dt}\right)_{SCC} \times \frac{1}{f} \quad (2)$$

where $(da/dN)_T$ is the total EAC growth rate (mm/cycle), $(da/dN)_{cyclic}$ is the cyclic crack growth rate in air (mm/cycle), $(da/dt)_{SCC}$ is the time dependent stress corrosion crack growth rate (mm/s) and f is the frequency (Hz). The cyclic component is given by:

$$(da/dN)_{cyclic} = C\Delta K^n \quad (3)$$

For an R-ratio of 0.5, the value of each constant in eqn.(3) is $C=2.92 \times 10^{-10}$ and $n = 3.97$ [12]. Using this data and applying a correction for R ratio according to the ASME Boiler and Pressure Vessel Code, Section XI [13] the crack growth rates in air at R ratios of 0.82 and 0.9 were predicted according to:

$$\left(\frac{da}{dN}\right)_{cyclic} = (2.03 \times 10^{-11} + 5.4 \times 10^{-10} R)\Delta K^{3.97} \quad (4)$$

Within experimental error, both edge and surface cracked specimens produced similar EAC results for the same testing conditions. For a given ΔK, the growth rate increased as the frequency decreased and, as expected, the crack growth rate per cycle decreased with ΔK for the same frequency.

The specimens cycled at the high R ratios of 0.82 and 0.9 and the low frequencies of 4.6×10^{-4} Hz (40 cycles/day) and 4.6×10^{-3} Hz (400 cycles/day) recorded crack growth rates higher than in air for the corresponding conditions. However, for those cycled at a frequency of 5.8×10^{-2} Hz (5,000 cycles/day) and an R ratio of 0.5, the crack growth rates in the NS4 solution fell below the air data. This suggested that a delay occurred in initiating an EAC crack from the fatigue precrack or that bifurcation occurred during early growth period resulting in lower growth rates.

In general, the superposition model described the experimental data extremely well. For higher ΔKs, the data converged to the cyclic component. At lower ΔKs, the data tended to a constant crack growth rate, independent of ΔK, and corresponded to the SCC growth rate divided by the frequency, given in eqn.(2). The SCC time dependent component was 2.94×10^{-9} mm/s.

EAC growth rate predictions for other combinations of stress ratios and frequencies can be made using the superposition model for this type of steel exposed to the dilute near-neutral pH solution. If only one cycle R=0 were imposed every four days (2.89×10^{-6} Hz), the predicted crack growth rate would be 1.37×10^{-3} mm/cycle. For comparison, when two surface cracked specimens were tested at the same frequency of 2.89×10^{-6} Hz and R ratio of 0 for 40 days in the NS4 solution, the crack growth rates averaged 8.0×10^{-4} mm/cycle in the first specimen and 1.63×10^{-3} mm/cycle in the second, giving an overall average of 1.22×10^{-3} mm/cycle, thereby showing good agreement between the observed and predicted crack growth rates.

3.3 Remaining life prediction

In order to apply the superposition model to predict the remaining life of a pipeline containing a known crack, threshold values of 5 MPa√m were assumed for both the cyclic (ΔK) and SCC (K_{max}) parts. This is reasonable since Marsh and Gerberich [14] showed that the threshold stress intensity factor range for a 400 MPa yield strength steel cycled at R=0 in hydrogen was 4 MPa√m. This environment is significant since hydrogen assisted cracking (HAC) is regarded as the operative SCC mechanism in near-neutral solutions [11]. As stated earlier, it is anticipated that atomic hydrogen, resulting from the cathodic reaction, diffuses into the steel and accumulates at regions of high stress triaxiality, lowering the cohesive strength ahead of a crack tip. Crack advance then occurs by tearing the weakened region, producing a quasi-cleavage fracture surface, as observed by the authors [5, 15].

Together with the superposition model, the threshold value can be applied to describe the growth behaviour of representative single cracks. Plumtree and Lambert [16] reported an average surface crack length and depth of 0.37 mm and 1.54 mm respectively, in a typical SCC colony on a 10 mm thick, 500 MPa (SMYS) natural gas pipeline. The deepest crack observed was 0.86 mm and the longest length was 3 mm. The nominal applied stress was 360 MPa, which corresponded to 72% SMYS. Using an R-ratio of 0.5, this resulted in initial stress intensity factor ranges of 4.3 MPa√m at the surface and 5.6 MPa√m at the deepest point. The maximum stress intensity factors were twice these values. With this information, a fracture mechanics model based on the superposition concept was developed for single crack growth in the pipeline [17]. Simulations were carried out until the crack reached 80% (8mm) wall thickness, which was regarded as failure. For a frequency of 4.6×10^{-4} Hz (40cycles/day), R ratio of 0.5 and SCC component of 2.94×10^{-9} mm/s, the average crack would be expected to grow to 80% thickness in 32.2 years. Using the same approach, the longest crack with a depth of 0.86 mm and length of 3.0 mm, would have a life of 26.2 years. However, it must be pointed out that most pressure fluctuations (99.7% on main transmission lines, 99.9% on lateral lines) were observed [7] to be within the range of R=0.7 to 0.99, as opposed to R=0.5, used in this example.

Bainbridge et al [18] analysed thirteen actual gas pipeline pressure histories. The data was processed using the rainflow counting technique to determine the stress ranges. This allowed EAC growth rates to be estimated by applying the superposition model. Simulations were performed for single semi-elliptical surface cracks in a 10 mm thick pipeline steel. Although the pipes for which the pressure histories were available had a variety of wall thicknesses, a standard thickness was used to provide the best basis for comparison of the effect of pressure history. By simulation, the crack was grown from an assumed initial depth of either 0.5 mm or 5 mm. The latter was used to assess how fatigue would affect the later stages of crack growth. In all cases, the final crack was again 8 mm deep. The crack aspect ratio (crack depth divided by the surface half-length) was 0.1 throughout the simulations, consistent with field observations. This constant value allowed pressure history effects to be isolated.

The expected life due to time dependent growth only, based on the constant crack growth rate over the remaining ligament (7.5 or 3 mm, depending on the initial crack depth) was considered, The time required to grow a crack from 0.5 mm to 8 mm deep would be 80.9 years, corresponding to a time dependent growth rate of 0.093 mm/year or 2.94×10^{-9} mm/sec, derived from the present experimental program. This is in agreement with the growth rates in the field which ranged from 3×10^{-9} mm/s to 2×10^{-8} mm/s. When the various pressure histories were considered, the total life was reduced from 80.9 yrs to a range of 68.7 to 80.8 yrs for the smaller initial crack size. For the larger initial 5 mm crack size, the life would be reduced from 32.4 years to within 21.3 to 32.3 years.

These are most interesting results, but it has to be stated that TGSCC has been found mainly in the crack colonies. A more practical approach would be to determine the growth in these colonies by considering the extent of interaction between neighbouring cracks. Plumtree and Lambert [16] considered a simple crack interaction model, based on examination of a limited series of solutions for pairs of surface cracks. In this case, crack coalescence was considered based on the simple proximity model. The results gave reasonable predictions of the growth behaviour of a crack colony. Coalescence dominated crack shielding. The predicted life of about 25 years was less than the worst single crack case, above.

4 Conclusions

1. Environmentally assisted cracking (EAC) in the form of transgranular stress corrosion cracking (TGSCC) has been observed in a ferritic-pearlitic steel used for natural gas pipelines (API X60) immersed in a dilute near-neutral pH solution at open-circuit- potential.

2. Quasi-clearage was observed on the fracture surface under the combined conditions of low cyclic frequencies and high R ratios, whereas corrosion fatigue was identified at low R ratios and high frequencies. A quantitative relationship between cyclic frequency and R is given that describes the conditions under which the two fracture mechanisms are likely to develop.

3. Application of a superposition model to express growth behaviour caused by different frequencies and loading conditions gave good agreement between observed and predicted rates.

4. Using limited operating data to analyze single surface crack growth behaviour formed a good base for predicting the remaining life of ferritic-pearlitic pipeline steels. However, determining the interactive crack growth within crack clusters would be more appropriate.

Acknowledgements

The authors acknowledge the financial assistance of TransCanada Pipelines Ltd and the Natural Sciences and Engineering Research Council of Canada. The technical assistance of S. J. Burany, M. Thomas, B. Williams and X.-Y. Zhang, University of Waterloo, is greatly appreciated. The authors thank Marlene Dolson and Jian Zou for typing the manuscript.

References

[1] Parkins, R.N., Blanchard, W.K. & Delanty, B.S., Transgranular stress corrosion cracking of high pressure pipelines in contact with solutions of near-neutral-pH, *Corrosion*, **50 (5)**, pp. 394 – 410, 1994.

[2] Parkins, R.N. & Beavers, T.A., Some effects of strain rate on transgranular stress corrosion cracking of ferritic steels in dilute near-neutral-pH solutions, *Corrosion*, **59 (3)**, pp. 258-273, 2003.

[3] Parkins, R.N., The controlling parameters in stress corrosion cracking, *Proc. 5th Symp. Pipeline Research*, AGA: Arlington VA, Catalog No. L30174, U-1, 1974.

[4] Parkins, R.N., *Stress Corrosion Cracking in Buried Pipelines*, TransCanada Pipelines Ltd., Calgary, Alberta, Report 76, 1990.

[5] Zhang, X-Y., Lambert, S.B., Sutherby, R. & Plumtree, A., Transgranular stress corrosion cracking of X-60 pipeline steel in simulated ground water, *Corrosion*, **55 (3)**, pp 297 – 305, 1999.

[6] Thomas M.A., Lambert, S.B., Sutherby, R. & Plumtree, A., Crack growth in API X-60 pipeline steel in NS4 solution, *Northern Area Western Conference*, Calgary, NACE: Houston, pp 1-7, 1999.

[7] Van Bovan, G., Sutherby, R. & King. F., Characterizing pressure fluctuations on buried pipeline in terms relevant to stress corrosion cracking, *Proc. 4th Int. Pipeline Conf.*, Calgary, ASME: New York, paper 27149, 2002.

[8] Hanninen, H., Torronen, K., Kemppainen, K. & Salonen, S., On the mechanisms of environmental sensitive cyclic crack growth of nuclear, reactor pressure vessel steels, *Corrosion Science*, **23(6)**, pp 663-679, 1983.

[9] Van Der Sluys, W.A., & Emanuelson, R.H., 3rd *Int. Symp. Environmental Degradation of Materials in Nuclear Power Systems – Water Reactors*, eds C.J. Theus & J.R. Weeks, The Metallurgical Society: Warrendale PA, pp 277 – 282, 1989.

[10] Kuniya, J., Anzai, H. & Masaoka, I., Effect of MnS inclusions on stress corrosion cracking in low-alloy steels, *Corrosion*, **48(5)**, pp 419-425, 1992.

[11] Gabetta, G., Transgranular stress corrosion cracking in low-alloy steels in diluted solutions, *Corrosion*, **53(7)**, pp 516-524, 1997.

[12] Ferraro, D, Personal communication, 1996, Department of Mechanical and Mechatronics Engineering, University of Waterloo, On., Canada.

[13] Rules for In-Service Inspection of Nuclear Power Plant Components, ASME Boiler and Pressure Vessel Code, Section XI, Div.1, ASME, NY., 1994.

[14] Marsh, P.G. & Gerbereich, W.W., Prediction of hydrogen-induced fatigue thresholds Δk_h^H in iron based materials, *Fatigue 93*, Vol. III, eds. J-P Bailon and J. I. Dickson, EMAS: Cradley Heath, UK, p1899, 1993.

[15] Ahmed, T.M., Lambert, S. B., Sutherby, R. & Plumtree, A. Cyclic crack growth rates of X-60 pipeline steel in a neutral dilute solution, *Corrosion*, **53(7)**, pp 581-590, 1997.

[16] Plumtree, A. & Lambert, S.B., Environmental crack growth in stressed pipeline steel, *Proc. Symposium on Environmentally Induced Cracking of Metals and Alloys*, eds. M. Elboujdaini, E. Ghali, W. Zheng, Canadian Inst. Met: Montreal, pp 7-15, 2000.

[17] Wang, X. & Lambert, S.B., Stress intensity factors for low aspect ratio semi-elliptical surface cracks in finite-thickness plates subjected to nonuniform stresses, *Eng. Fract. Mechanics*, **51(4)**, pp 517-532, 1995.

[18] Bainbridge, D., Lambert, S.B., Jack, T.R., Sutherby, R. & Plumtree A., Influence of pipeline operation pressure on SCC growth, *Eurocorr* in *Oil and Gas 2001, Corrosion Production*, Assoc. Ital. Met: Milan, Paper 240, 2001.

On the SCC behaviour of austenitic stainless steels in boiling saturated magnesium chloride solution

O. M. Alyousif[1] & R. Nishimura[2]
[1]*Mechanical Engineering Department, Kuwait University, Safat, Kuwait*
[2]*Department of Applied Materials Science, Osaka Prefecture University, Osaka, Japan*

Abstract

The stress corrosion cracking (SCC) of three austenitic stainless steels was investigated as a function of test temperature and applied stress in boiling saturated magnesium chloride solutions ($MgCl_2$) using a constant load method. Both type 304 and type 316 exhibited similar corrosion elongation curves, while the corrosion elongation curve of type 310 was different. The relationship between the time to failure and a reciprocal test temperature fell in two straight lines on a semi-logarithmic scale as well as the relationship between the steady-state elongation rate and a reciprocal test temperature. These regions were considered to correspond to a SCC-dominated region or Hydrogen Embrittlement (HE)-dominated region. The same behaviour was observed for applied stress dependence for these stainless steels.

The relationship between the time to failure versus steady state elongation rate on a logarithmic scale became a straight line, whereas the slopes of the line for the stainless steels were different for each other. It was found that the linearity of the relationship can be used to predict the time to failure for the stainless steels in the corrosive environment. The transition time to time to failure ratio values had a different constant value for each region. From the results obtained, it may be suggested that the cracking mechanism for type 304 and type 316 was different from that for type 310. This would be explained by whether or not a formation of martensite takes place.

Keywords: austenitic stainless steels, magnesium chloride solution, steady-state elongation rate, stress corrosion cracking, hydrogen embrittlement.

1 Introduction

The stress corrosion cracking behaviour of austenitic stainless steels in chloride and other corrosive solutions has been extensively investigated using various methods [1-7]. Stress corrosion cracking in the broad sense for austenitic stainless steels is caused by either stress corrosion cracking in the narrow sense (denoted as SCC hereafter) such as active path dissolution [1-2] and film-rupture [3-4], or hydrogen embrittlement (HE) [5-7]. Austenitic stainless steels such as type 304 and 316 steels, but not type 310 steel, may undergo phase transformation from $\gamma \rightarrow \alpha'$ martensite due to applied stress or hydrogen charging [8-9] and α'-martensite is considered to be directly related to brittle fracture [10]. However, the role of martensite in SCC and HE is still not fully identified. It was found that austenitic stainless steels suffered from cracking failure by two different mechanisms of SCC and HE depending upon test temperature [11].

In this study, the failure characteristics of type 304, 316 and 310 austenitic stainless steels were examined under constant loads in boiling saturated magnesium chloride solutions once with variable test temperature and constant applied stress and then with variable applied stress and at constant test temperature to evaluate the role of martensite in determining the cracking mechanism of austenitic stainless steels.

Table 1: Chemical compositions (wt%) and mechanical properties of the austenitic stainless steels used.

	C	Si	Mn	P	Ni	Cr	Mo	σ_{Yield} (MPa)	$\sigma_{Tensile}$ (MPa)
SS304	0.06	0.35	0.96	0.027	8.13	18.20	-	276	691
SS310	0.05	0.84	1.27	0.016	19.30	24.76	-	275	575
SS316	0.06	0.70	0.96	0.031	10.15	16.98	2.22	323	636

2 Experimental

The specimens used were the commercial type 304, 316 and 310 austenitic stainless steels whose chemical compositions (wt%) are shown in Table 1. The geometry for stress corrosion cracking experiments is as follows: the gauge length is 25.6 mm, the width 5 mm and the thickness 1 mm. The specimens were solution-annealed at 1373 K for 3.6 ks under an argon atmosphere and then water-quenched. Prior to experiments, the solution-annealed specimens were polished to 1000 grit emery paper, degreased with acetone in an ultrasonic cleaner and washed with distilled water. After the pretreatment, the specimens were immediately set into a stress corrosion cracking cell. Stress corrosion cracking tests were conducted in boiling saturated magnesium chloride solutions, the boiling temperatures of which were changed by the change in the concentration of magnesium chloride. For the experiments with variable test temperature and constant stress ($\sigma = 300$ MPa), the test temperature range was between 408 K and 428 K. For the variable applied stress and constant test temperature, the applied stresses were in the range from 0 to 500 MPa and the test temperatures used were 428 K (type 316), 416 K (type 304 and 310) and 408 K (type 304 and 316), where cracking failure was observed under constant applied stress condition [11]. All experiments were carried out under an open circuit condition.

3 Results

3.1 Corrosion elongation curve

Figure 1 shows the corrosion elongation curves for type 304 at 408 K and type 310 at 416 K under a constant applied stress condition (σ = 350 MPa) in boiling saturated magnesium chloride solutions, where the corrosion elongation curve of type 304 at 408 K was similar to that of type 316 at 408 K. From these curves the three parameters were obtained for each specimen: time to failure (t_f), steady-state elongation rate (l_{ss}) for the straight part of the corrosion elongation curve and transition time to time to failure ratio (t_{ss}/t_f), where the transition time (t_{ss}) is the time when the elongation curve begins to deviate from the linear increase. For type 304 at 408 K, t_f was considerably shorter than for type 310 at 416 K and t_{ss} was close to t_f. Thus the t_{ss}/t_f value was close to unity. For type 310 at 416 K, t_{ss} was approximately half the time to failure.

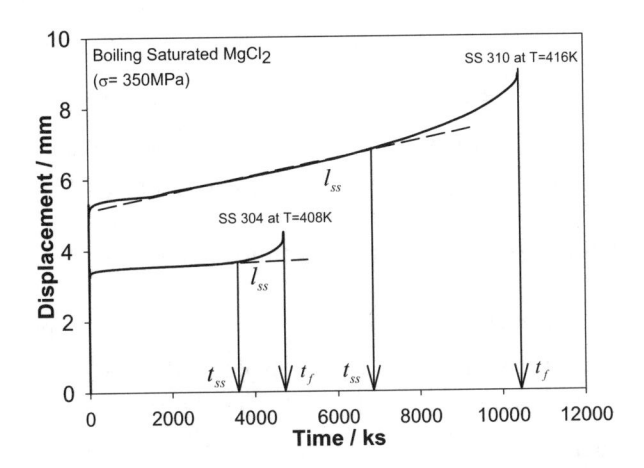

Figure 1: Corrosion elongation curves for type 304 steel at 408 K and type 310 steel at 416 K under a constant applied stress condition (σ = 350 MPa) in boiling saturated MgCl$_2$ solutions.

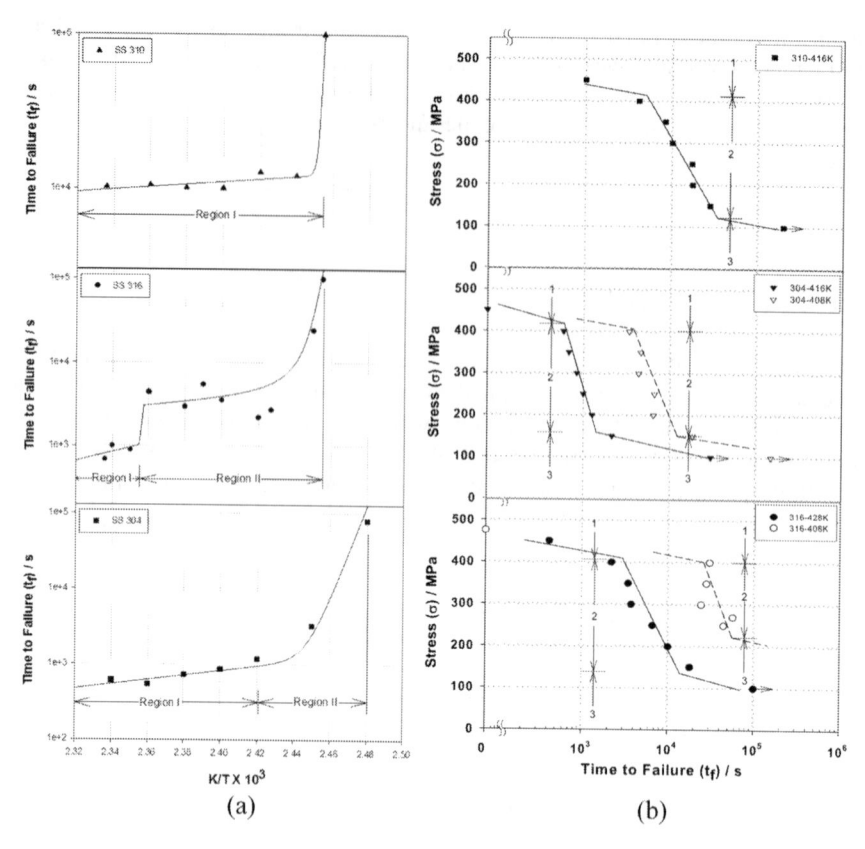

Figure 2: The logarithms of (t_f) versus the reciprocal of the test temperature (1/T) (a) and Applied stress (σ) versus t_f (b) for type 310, 304 and 316 steels in boiling saturated $MgCl_2$ solutions.

3.2 Effects of test temperature and applied stress on t_f

Figure 2 (a) depicts the logarithm of t_f versus the reciprocal of the test temperature for the three stainless steel types used in the experiments. For type 304, the relationship between t_f and 1/T falls in a straight line, shown as region-I, until a test temperature of about 413K (1/T = 2.42×10^{-3}), below which the relationship deviates from the linearity. The region from 413K (1/T = 2.42×10^{-3}) to a threshold test temperature, approximately 403K (1/T = 2.48×10^{-3}), below which little fracture takes place within a laboratory time scale, was called region-II. For type 316, the parameter t_f was also grouped into two regions; region-I at test temperatures above 424K (1/T = 2.36×10^{-3}) and region-II with test temperatures below 424K (1/T = 2.36×10^{-3}) up to a threshold test temperature of about 408K ($1/T = 2.45\times10^{-3}$). The t_f values of type 310 showed slight sensitivity to test temperature. The t_f values for type 310 were grouped in a straight line which constitutes region-I until a threshold test temperature of about 408K ($1/T = 2.45\times10^{-3}$) without region-II. Therefore, in Figure 1, the elongation curve for type 310 at 416K ($1/T = 2.40\times10^{-3}$) falls in region-I, while that for type 304 at 408K ($1/T = 2.45\times10^{-3}$) falls in region-II.

Figure 2 (b) depicts applied stress σ versus logarithm of t_f for the three stainless steels at three test temperatures used. All relationships can be divided into three regions shown by Arabic numerals 1-3 in Figure 3; region 1 is the stress-dominated failure region, region 2 the stress corrosion cracking SCC-dominated failure region and region 3 the corrosion-dominated failure region. In region 2, t_f values for type 304 at 416 K and for type 316 at 428 K are shorter than those at 408 K and t_f for type 304 is shorter than that for type 316. In addition, the applied stress range in region 2 for type 316 at 408 K is narrower than that for type 304 at 408 K; the maximum applied stress in region 2 is the same for both steels, but the minimum applied stress for type 316 is higher than that for type 304. At higher temperatures, the applied stress range in region 2 is almost the same for these stainless steels. It was also found that the behaviour of the steady state elongation corresponded to those in Figure 2 and the value of t_{ss}/t_f was different depending upon region; 0.57 in region I and 0.63 in region II.

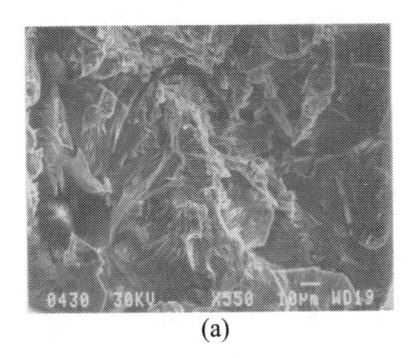

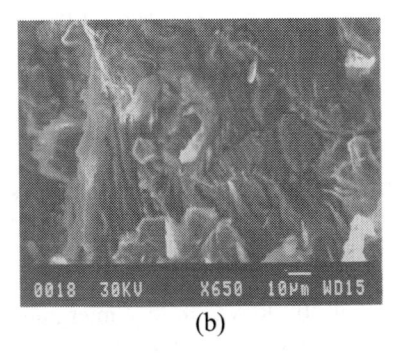

Figure 3: Fracture appearances for type 304 steel at 416 K and σ = 300 MPa; X550 (a) and for type 310 steel at 416 K and σ = 300 MPa; X650 (b) in boiling saturated MgCl$_2$ solutions.

3.3 Fracture appearance

Figure 3 shows the fracture appearances of type 304 (a) and 310 (b) in region 2 at 416 K. The fracture mode for type 304 and 310 is predominantly transgranular. The transgranular fracture was observed over the whole test temperatures in region I and the whole applied stresses in region 2. Type 316 in region 2 at 428 K exhibited the same transgranular fracture appearance as those in Figure 6. On the other hand, as shown in Figure 4, the fracture appearance in region 2 for type 316 at 408 K and all test temperatures in region II (408 K–423 K) is a mixture of intergranular and transgranular modes (a) at a low applied stress (300 MPa) and predominantly intergranular mode (b) at a high applied stress (400 MPa). The ratio of intergranular appearance to transgranular one increased with increasing applied stress in region 2 and finally the fracture mode became completely intergranular. The fracture appearances for type 304 at 408 K and at all temperatures in region II were mostly intergranular and almost the same as those for type 316. For type 310, the fracture mode was transgranular over the whole applied stresses and at all test temperatures in region 2.

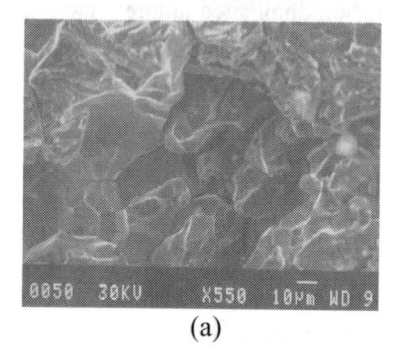

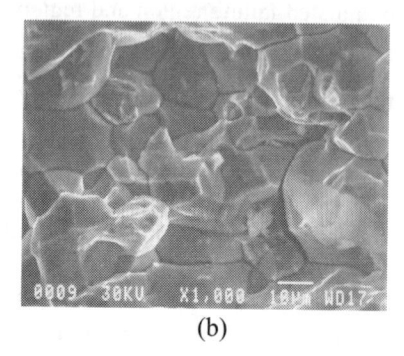

(a)   (b)

Figure 4: Fracture appearances for type 316 steel at 408 K and $\sigma = 300$ MPa; X550 (a) and for type 316 steel at 408 K and $\sigma = 400$ MPa; X1000 (b) in a boiling saturated $MgCl_2$ solution.

4 Discussion

4.1 Characteristics of stress corrosion cracking

In the applied stress region where the SCC-dominated failure occurred, the fracture mode for type 304 and 316 at 416 and 428 K was transgranular, while that at 408 K was mostly intergranular, in particular at higher applied stresses. In other words, for type 304 and 316 in boiling magnesium chloride solutions, the fracture mode was transgranular at higher temperatures and intergranular at the lower temperature. For type 310 only transgranular fracture was found as far as stress corrosion cracking occurred in any temperatures in boiling magnesium chloride solutions [11]. The minimum applied stress to induce intergranular cracking for type 316 was significantly higher than that for type 304. In general, transgranular cracking was characterized by a higher rate of elongation in both SCC-dominated and stress-dominated regions and shorter time to failure. By contrast the failure in intergranular cracking was slow and most of the time to failure was in the steady state elongation region indicating the small mechanical elongation until failure. Consequently, intergranular cracking for these austenitic stainless steels in the boiling magnesium chloride solution revealed more brittle nature in comparison with transgranular cracking.

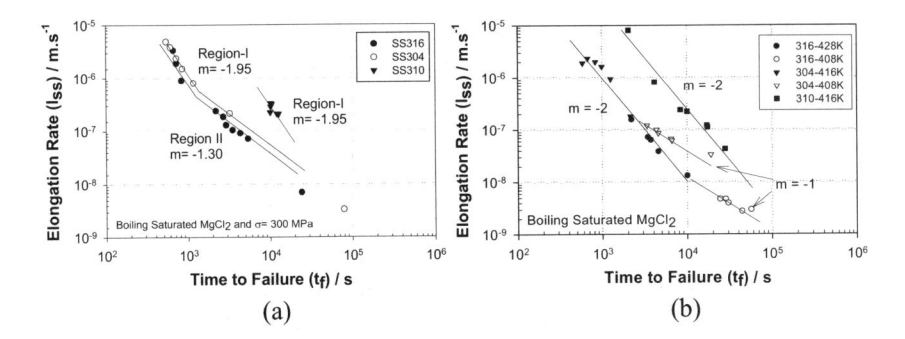

Figure 5: The relation between time to failure (t_f) and the logarithms of (l_{ss}), where m is the slope of the straight line. (a) for constant stress and variable test temperature and (b) for constant test temperature and variable stress.

4.2 A parameter for predicting time to failure

As shown in Figure 5 (a), the log-t_f versus log-l_{ss} relationship for types 304 and 316 obtained as a function of test temperature became two straight lines with the slopes of m = -1.95 and m = -1.30, where the former value was in region-I in Figure 2 (a) and the latter in region-II in Figure 2 (a). In the case of type 310, the log-t_f versus log-l_{ss} relation for type 310 became a single straight line with slope m = -1.95. Figure 5 (b) shows the relationships between the logarithm of l_{ss} and logarithm of t_f in region 2 in Figure 2 (b) for all stainless steels. The log l_{ss} versus log t_f curves in region 2 for types 304 and 316 become two straight lines with a slope of m = -2 at 418 and 428 K and with a slope of m = -1 at 408 K, and type 310 also shows a straight line with m = -2. Thus, the empirical equations for the stainless steels are as follows:

For type 304 and 316 at higher temperatures and type 310 steel in region 2 at constant stress or constant temperature conditions

$$\log l_{ss} = -2 \log t_f + C_1 \tag{1}$$

For type 304 and 316 in region 2

$$\log l_{ss} = m \log t_f + C_2 \tag{2}$$

where m = -1.3 for constant stress condition and m = -1.0 for constant temperature condition. C_1 and C_2 are constants and depend on material. It can be seen in Figure 8 that type 304 at 416 K and type 316 at 428 K show the same C_1

264 Simulation of Electrochemical Processes II

value, although type 310 steel shows different C_1 value, and that the C_2 value of type 304 at 408 K is different from that of type 316.

In addition to these characteristics, it can be said that steady-state elongation rate (l_{ss}) is a useful parameter for predicting the time to failure (t_f) independent of applied stress even in boiling saturated magnesium chloride solutions as well as in hydrochloric and sulphuric acid solutions [13-14], because l_{ss} can be obtained at a time within 10–20% of t_f from the corrosion elongation curve.

4.3 A qualitative explanation of fracture mechanism

Type 304 is a metastable austenite and is highly susceptible to martensite formation [16]. Type 316 is also a metastable austenite and is moderately susceptible to martensite formation [17]. On the other hand, type 310 is known to be less susceptible to martensite formation [18]. For metastable austenitic stainless steels like types 304 and 316, the strain-induced formation of martensite tends to take place along grain boundaries and particularly is facilitated by hydrogen entry [19-20]. The presence of martensite is known to induce hydrogen embrittlement (HE) because of very high hydrogen diffusivity coefficient and very small hydrogen content compared to those of the austenite [17]. Thus, martensite formed at grain boundaries is apt to be responsible for HE of intergranular mode, and types 304 and 316 have higher susceptibility to martensite-induced HE of intergranular mode, but type 310 has little susceptibility to martensite-induced HE of intergranular mode.

On the other hand, transgranular cracking is caused by propagation of cracks nucleated at slip steps and is not related to martensite formed at grain boundaries. Thus stress corrosion cracking in the narrow sense of the word (SCC) includes such transgranular cracking.

Types 304 and 316 suffered intergranular cracking at a lower temperature (408 K) and transgranular cracking at higher temperatures (428 K for type 316 and 416 K for type 304). In determining the cracking mode a competition between dissolution of material at slip steps inducing SCC and hydrogen entry inducing HE is the decisive factor as well as the formation of martensite. The hydrogen entry is determined by the difference between hydrogen absorption and hydrogen escape. Transgranular SCC for types 304 and 316 was caused by propagation of cracks nucleated at slip steps, not at martensite at grain boundaries [21]. Therefore, the cracking mode would be determined by the competition between the dissolution rate at slip steps and the hydrogen entry rate at grain boundaries with martensite. As test temperature increases, the dissolution rate increases. On the other hand, both hydrogen absorption rate and hydrogen escape rate increase with increasing test temperature, whereas the amount of hydrogen entry decreases, because the hydrogen escape rate becomes superior to the hydrogen entry rate with increasing temperature. This means that the amount of hydrogen entry decreases with increasing test temperature [22]. In addition, it is considered that the amount of martensite decreases with increasing test temperature, which suggests that the hydrogen entry rate decreases with increasing test temperature [23]. Thus, the dissolution rate becomes higher than the hydrogen entry rate at higher temperatures. This will result in transgranular SCC at higher temperatures.

On the other hand, it is known that the amount of martensite increases with strain or applied stress [17]. At 408 K for types 304 and 316, the fracture mode changed from a mixed transgranular and intergranular mode to a complete intergranular mode with increasing applied stress. If cracking failure at 408 K is caused by martensite-induced HE, the increase in intergranular mode with increasing applied stress is in agreement with an increase in the amount of martensite with increasing applied stress. Similarly, the minimum applied stress to induce intergranular cracking for type 316 steel was significantly higher than that for type 304 in agreement with the fact that type 316 is less susceptible to stress-induced martensite formation than type 304. Furthermore, type 304 was more susceptible to intergranular cracking in comparison with type 316 in agreement with the fact that type 304 has the higher susceptibility to strain-induced martensite formation in comparison with type 316.

Consequently, intergranular cracking observed at the lower temperature, that is, 408 K for types 304 and 316 is due to grain boundary martensite-induced hydrogen embrittlement (HE) and transgranular cracking for types 304 and 316 at higher temperatures at 416 and 428 K and for 310 is ascribed to stress corrosion cracking in the narrow sense of the word (SCC).

5 Conclusions

The stress corrosion cracking behaviour of three austenite stainless steels with different susceptibilities to stress-induced martensite formation was examined in boiling saturated magnesium chloride solutions with different boiling temperatures under constant load condition. The following conclusions can be drawn:

(1) The cracking mechanism for type 304 was transgranular at a higher test temperature more than 413K and intergranular at a lower test temperature less than 412K. The cracking mechanism for type 316 was transgranular at a higher test temperature more than 424K and a mixture of transgranular and intergranular cracking at a lower test temperature less than 424K. The cracking mechanism for type 310 was transgranular at test temperatures higher than 408K and no intergranular cracking behaviour was observed.

(2) The relationships between log l_{ss} and - log t_f for types 304, 316 and 310 became a good straight line. Their slopes depended upon fracture mode; -2 for SCC and -1 or -1.30 for HE.

(3) It was estimated that HE occurred by martensite formed along the grain boundaries with hydrogen entry, while SCC took place by a propagating crack nucleated at slip step by dissolution.

References

[1] T.P. Hoar & J.C. Scully, *J. of Electrochemistry*, **111**, pp. 348-352, 1964.
[2] R. Nishimura & Y. Maeda, *Corrosion Science*, **46**, pp. 343-355, 2004.
[3] T. Nakayama & M. Takano, *Corrosion*, **42**, pp. 10-14, 1986.
[4] M.B. Whiteman & A.R. Troiano, *Corrosion*, **21**, pp. 53-56, 1965.
[5] M.L. Holzworth, *Corrosion*, **25**, pp. 107-115, 1969.
[6] P.R. Rhodes, *Corrosion*, **25**, pp. 462-467, 1969.
[7] H. Hanninen & T. Hakkarainen, *Metallurgical Transactions A*, **10A**, pp. 1196-1199, 1979.
[8] N. Narita, C.J. Altstetter & H.K. Birnbaum, *Metallurgical Transactions A*, **13A**, pp. 1355-1365, 1982.
[9] J.C. Scully, The role of hydrogen in stress corrosion cracking, eds. A.W. Thompson & I.M. Bernstein, *Effect of Hydrogen on Behavior of Materials*, AIME, pp.129, 1975.
[10] O. M. Alyousif & R. Nishimura, *Corrosion Science*, **48**, pp. 4283-4293, 2006.
[11] R. Nishimura & H. Sulaiman, Stress corrosion cracking of sensitized type 316 austenitic stainless steel in pure sulfuric acid solution, *12th Int. Corrosion Congress: Corrosion control for low cost reliability*, pp. 4325, 1993.
[12] R. Nishimura & K. Kudo, *Corrosion*, **45**, pp. 308-316, 1989.
[13] R. Nishimura, *Corrosion Science*, **34**, pp. 1859-1868, 1993.
[14] R. Nishimura, *Corrosion Science*, **34**, pp. 1463-1473, 1993.
[15] C.L. Briant, *Metallurgical Transactions A*, **10A**, pp. 181-189, 1979.
[16] X. Sun, J. Xu & Y. Li, *Acta Metallurgica*, **37**, pp. 2171-2176, 1989.
[17] S.S. Birley & D. Tromans, *Corrosion*, **27**, pp. 63-71, 1971.
[18] T-P. Perng & C.J. Altstetter, *Acta Metallurgica*, **34**, pp. 1771-1781, 1986.
[19] H. Hanninen & T. Hakarainen, *Corrosion*, **36**, pp. 47-51, 1980.
[20] R. Nishimura, *Corrosion*, **46**, pp. 311-318, 1990.
[21] J.K. Tien, Diffusion and the dislocation sweeping mechanism, *Effect of Hydrogen on Behavior of Materials*, eds. A.W. Thompson & I.M. Bernstein, AIME, pp.309, 1975.
[22] R.E. Reed-Hill, *Physical Metallurgy Principles*, second ed., Brooks/Cole Engineering Division, pp. 656, 1973.

Electrochemical behaviour and corrosion sensitivity of prestressed steel in cement grout

E. Blactot[1], C. Brunet-Vogel[1], F. Farcas[1], L. Gaillet[1], I. Mabille[2], T. Chaussadent[1] & E. Sutter[2]
[1]*Laboratoire Central des Ponts et Chaussées Paris, France*
[2]*Ecole Nationale Supérieure de Chimie Paris, France*

Abstract

The durability of prestressed concrete bridges primarily depends on the steel wire cables. Several disorders in these bridges are well known, such as fracture of these cables by stress corrosion cracking due to water penetration in the prestressing ducts. This water can contain various aggressive constituents with respect to corrosion (like chlorides). Its penetration inside the duct is due to the presence of sealing defects of the structure or in the concrete (like cracks). The second category of defects relates to the cement grout injection, which protects cables. This occurs due to a degradation of the grout in contact with water (problems of bleeding and segregation) leading to brittle fractures of steels. The first step of the study consisted in obtaining a corrosive liquid typical of a segregated cement paste on which electrochemical tests were done. Electrochemical tests in synthetic solutions defined from a bibliographical study and analysis of real cases of segregated cement grout were also realized. High strength and ordinary steel specimens were used and all the tests were mainly done in deaerated conditions. The main results can lead one to consider that, in oxygen free conditions, steel specimens surface can be either in active or passive state. The susceptibility of steels to stress corrosion cracking was determined by the means of models applied to potentiodynamic polarisation.
Keywords: prestressed steel, corrosion, segregation, synthetic solution.

1 Introduction

In prestressed concrete structures, cement grout is injected in high-density polyethylene ducts containing the steel cables with the objective to ensure a protection against corrosion [1,2]. Nevertheless, under some particular conditions, this protection is not effective and could be sometimes the main initiator of steel passivation breakdown. Indeed, if the cement grout injection is badly carried out, it can exist bleeding, segregation, and even void zones [3,4]. These defects, far to be subjective, are an important concern for the concrete structure managers [5]. Solutions to avoid these types of defects exist but are used since relatively little time. So, for existing structures, cement grout heterogeneity could involve steel cable ruptures, which are the result of a stress corrosion phenomenon. Until now, the mechanisms of stress corrosion cracking (SCC) applied to this civil engineering case are not well known. The aim of this study is to evaluate the various parameters related on both the used materials and the environmental conditions, which can take part in these mechanisms.

2 Origins of the corrosion initiation

Protection conferred by cementitious materials to steel is well known. A passive layer is formed on the steel surface due to the high alkalinity, the quality of passive film however being able to be deteriorated by the presence of harmful species (chlorides, carbon dioxide). Within the framework of the present study, the characteristic of the corrosion phenomenon lies in the fact that the steel cables are under mechanical stress and that the surrounding medium is characterized by a bad cement grout, but also by the absence of oxygen in the ducts [6]. Heterogeneity of cement grout involves the presence of two different mediums: an healthy cement grout zone and a segregated cement grout zone, this last zone having the particularity to be strongly alkaline and to contain sulphate ions. Some of these zones are favourable to hydrogen formation (cathodic reaction), which can penetrate in the metal and weaken it. A stress corrosion cracking assisted by hydrogen can then occur and lead to the cable failure. Some other zones can be characterised by steel dissolution (anodic reaction) with loss of section.

Table 1: Chemical compositions of the studied steels (%).

	C	Mn	Si	P	S
Prestressing steel	0.81	0.68	0.23	0.01	< 0.01
Reinforced concrete steel	0.04	0.44	0.12	0.02	0.04

3 Materials and experimental techniques

3.1 Materials

Two commercial grade steel were used: a steel for prestressed concrete structure and an ordinary steel for reinforced concrete whose ultimate tensile strengths are respectively 1860 and 350 MPa. Their compositions are given in Table 1.

They are both ferrito-pearlitic steels and the main difference between them is the carbon amount and microstructure. Prestressing steel has a fully-oriented microstructure related to the manufacture process (cold-drawing).

Before experiments, the surface preparation of the steels specimens was carried out using silicon carbide paper up to grade 1200. Specimens of prestressing steel were also studied without surface preparation as it is in real structures. In these conditions, the surface is covered by adherent products used to facilitate cold-drawing operation. In all cases, after polishing or not, the specimens were washed with distilled water, degreased with acetone and dried.

The studied mediums were defined in order to represent the characteristics of the segregated phase (pH close to 14 and presence of sulphate ions). The behaviour of steel specimens was thus studied in three types of mediums: solution of NaOH (pH range between 12 and 14), solutions of NaOH and Na_2SO_4 (same pH) and segregated mediums (diluted or not) obtained by cement hydration under some conditions (additive to modify rheological characteristics of cement paste before hardening). The obtained segregated phase was composed of a liquid phase whose principal species are sulphate and hydroxyl ions and a white solid phase whose principal components are ettringite and portlandite. The compositions of the studied mediums are given in Table 2.

Within the framework of this study, all the solutions were deaerated before and during the tests by nitrogen bubbling.

Table 2: Compositions of the studied solutions.

Studied environments	SO_4^{2-} concentration (mol/L)	Measured pH
NaOH	0	13.6
NaOH + 0.15 g/L de Na_2SO_4	1.10^{-3}	13.7
NaOH + 15 g/L de Na_2SO_4	0.1	13.6
NaOH + 150 g/L de Na_2SO_4	1	13.6
Segregated medium	$1.85.10^{-3}$	13.7
Diluted segregated medium	$1.85.10^{-3}$	13.2

3.2 Experimental techniques and models

In order to determine if steel is sensitive or not to stress corrosion in a considering environment, several methods can be used:
- Tests with an applied stress in the considering environment and by following steel evolution (mechanical and/or electrochemical properties) with time until its possible rupture.
- Tests without mechanical stress by plotting potential curves and applying predictive methods allowing one to determine the sensitivity of the steel to stress corrosion [7,8]. This last test type was considered within the present study.

The steel failure by stress corrosion cracking (SCC) is related to the coexistence of anodic and cathodic processes, which locally lead respectively to the metal dissolution and the hydrogen penetration into the material makes.

The stress corrosion range in the anodic zone reveals the domain in which is likely to occur a reaction of localised dissolution on the steel surface. This phenomenon corresponds to the anodic mode of stress corrosion. In the majority cases of SCC, anodic dissolution is not the only cause but it is a necessary factor [9]. On the other hand, for cathodic potentials, when water dissociation becomes possible, a release of hydrogen can occur [9]. Thus, this range of potentials presents the risk of hydrogen embrittlement or hydrogen induced stress corrosion cracking (HISCC). The determination of the range of critical potential for which SCC occurs can be done using the potential curves. This exploitation makes it possible to confirm the role played by the active/passive transition in the cracking mechanism. To evaluate the sensitivity to SCC of the studied steels, potential curves were plotted at three different scanning rates: 0.25; 20 and 83.3 mV/s. These various speeds allow one to apply two predictive methods (that of Parkins [7] and that of Fang and Staehle [8]) based on the current intensity measurements according to the potential scanning and then on calculations.

Parkins [7] defined a method being able to predict SCC susceptibility. This method leads to determine the range of potential in which this susceptibility could exist. Method developed by Parkins is based on potential curve plotted with high and low potential scanning rates.

According to Parkins, the conditions involving a stress corrosion phenomenon are met if the following points are respected:
- In the case of high potential scanning curve, current densities must be higher than 1 mA/cm²
- The current density difference, at a given potential, between the high and low potential scanning rates, ΔI, divided by the current density corresponding to the slow scanning rate must exceed 1000.

$$(J_{high} - J_{low}) / J_{low} > 1000$$

However, various researcher works [10,11] showed that a threshold of 100 was enough to determine if the steel is sensitive or not to stress corrosion.

Interest of this method resides in the fact that it allows to identify, on the potential curve, the active/passive transitions for a studied steel in a considered environment and consequently the potential range where the sensitivity of steel to stress corrosion is maximum.

Fang and Staehle's method [8] is based on the determination of a "stress corrosion cracking" parameter, noted P_{SCC}. This parameter is defined as a ratio between current densities obtained at high and low potential scanning rates for a given potential:

$$P_{scc}(E) = (J(E)_{20\ mV/s})^2 / J(E)_{0.2\ mV/s}$$

According to these authors, it is thus possible to correlate the value of the parameter P_{SCC} with the susceptibility of steel to SCC: More the value of P_{SCC} is high, more the susceptibility of steel to SCC is important.

4 Results and interpretation

4.1 Description of electrochemical conditions

A preliminary study was realized to analyze electrochemical conditions of steel, when it is in contact with two different mediums: one defined by healthy cement paste and the other by a segregated state. The formation of these two mediums was naturally obtained by using cement (CEM I in the present study) and a suitable plasticizing additive. The measured potentials show the formation of an electrochemical cell in the considered cases ($\Delta V \approx 150$ mV for the first two months). Consequently, an electrochemical phenomenon exists and is able to involve corrosion in the long term. The output intensity decrease with time varying from about µA (10 to 29) to nA (70) after few months.

4.2 Open circuit potential

The open circuit potential measurements were realized in the different mediums. The obtained values depend on several parameters, and notably on surface state of the studied steel. More the value of the potential is negative, more the surface is active in term of corrosion.

Two mains groups of values were observed:
- -300 mV$_{sce}$ < E_{cor} < -500 mV$_{sce}$, passive surface state
- -600 mV$_{sce}$ < E_{cor} < -1200 mV$_{sce}$, active surface state

These two groups of corrosion potential values and the corresponding steel surface activity are given using potential curves (see 4.3).

The corrosion potentials values thus inform about the surface quality of studied steel. The steel will be more sensitive to corrosion phenomenon if its surface state corresponds to an active state.

The various corrosion potential measurements led to the same results distributed in a random way in the two groups of values, and this whatever the nature of steel and the composition of the solution. It can be noted that the corrosion potentials do not depend on the concentration of sulphate ions.

4.3 Determination of SCC susceptibility

Segregated zones of a badly cast cement paste are very alkaline and contain high amount of sulphate ions. For a better understanding of sulphate amounts and pH influences, potential curves at several potential scanning rates in real segregated mediums or synthetic solutions (Table 2) were plotted and analyzed.

4.3.1 Influence of sulphate ions
Tests were conducted to determine which could be the sulphate action on the electrochemical behaviour of the steel (for prestressing applications) according to the sulphate concentration at pH $\approx$ 13.6 (Figure 1).

When potential curves are plotted with high potential scanning rates, two anodic peaks are observed, one being located around -1000 mV$_{SCE}$, which could correspond to an intergranular stress corrosion and another around -750 mV$_{SCE}$ rather related to transgranular cracking according to [12].

Differences concerning the peak intensity can be considered as negligible, although there is a discrepancy of potentials. It can be admitted that in the case of a deaerated medium, at a high alkaline pH, sulphate ions have little influence on active/passive transition, and that sulphate concentration in the medium play a negligible role.

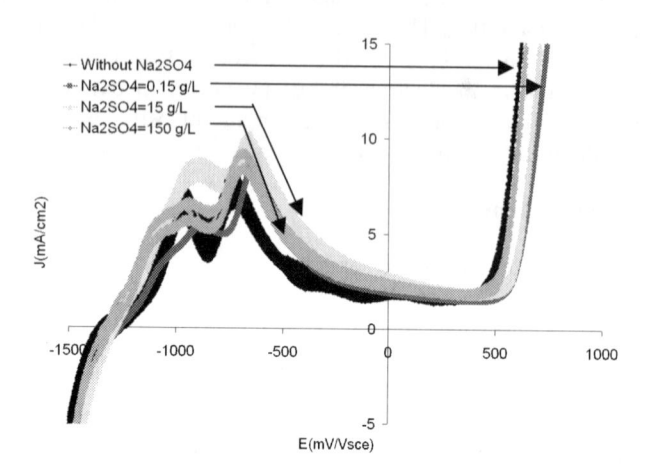

Figure 1: Influence of sulphate content on the potential curves – Raw prestressing steel (Scanning rate = 83.3 mV/s).

4.3.2 Influence of pH

As it can be seen on Figure 2, which collected potential curves at different pH in different mediums, it seems that the behaviour of prestressing steel strongly depends on the pH:
- When the steel is in contact with a solution whose pH is about 13.6, two distinct anodic peaks are observed:
    - The first one is at a potential of -1000 mV$_{sce}$
    - The second one at a potential of -750 mV$_{sce}$
- When the steel is in contact with a solution whose pH is about 12.9:
    - Only one peak is observed, much less intense than at pH 13.6, locating at a potential around -520 mV$_{sce}$,
    - Peaks are followed by a pseudo passivation stage extending up to +500 mV$_{SCE}$ at pH 13.6. At pH 12.9 and in a diluted segregated medium, the length of this stage is more important, the transpassivation potential locating at +800 mV$_{SCE}$. The increase of current in the transpassivation zone can be assigned to the oxidation of water.
- When the prestressing steel is in a diluted segregated medium at pH = 13.2, its behaviour is similar to that of the prestressing steel at pH = 12.9.

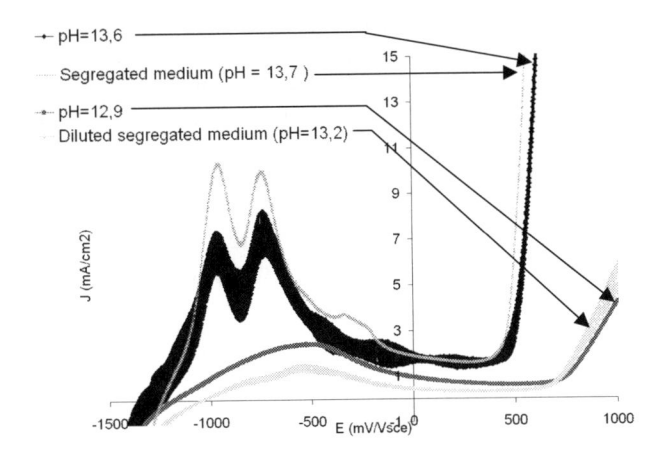

Figure 2: Influence of pH on the potential curves Raw prestressing steel (Scanning rate = 83.3 mV/s).

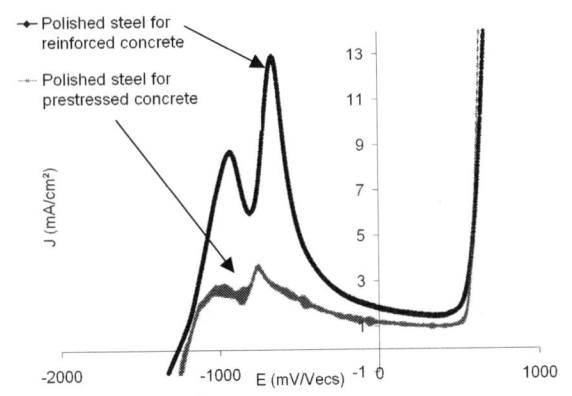

Figure 3: Electrochemical behaviour of polished prestressing steel and reinforced cement steel (pH = 13.6 – Scanning rate = 83.3 mV/s).

These observations confirm the important influence of the pH. In the case of low alkalinity (pH = 12.9), corrosion risks are lower: the oxidation peak is less intense, which correspond to a best aptitude to passivity. It can be observed the same electrochemical behaviour for steel in diluted segregated medium (pH = 13.2). An increase of pH to 13.7 leads to a more important sensitivity of steel to SCC, considering that potential curves at low potential scanning rate are almost identical for the different pH studied.

4.3.3 Influence of steel characteristics

To analyze the influence of the different steels (polished reinforced concrete steel and polished prestressing steel), comparison of potential curves in high alkaline medium is presented in Figure 3.

Heights of the peaks indicate that polished prestressing steel presents a better aptitude to passivation than the polished reinforced concrete steel. This fact can be explained by carbon amount (studies showed that low carbon steel is very sensitive to corrosion in high alkalinity medium [13]) that leads to different microstructure.

The non-polished prestressing steel presents at its surface a black layer, which is composed of residual products used for cold drawing. This layer seems to be partially destroyed (presence of strongly active and passive zones) during cathodic polarisation, which involves, as a consequence, an important surface reactivity (see Figure 4).

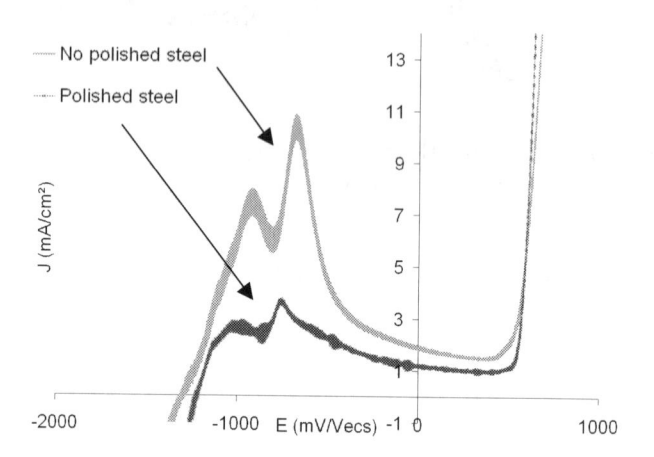

Figure 4: Electrochemical behaviour of polished or not prestressing steel (pH = 13.6 – Scanning rate = 83.3 mV/s).

Table 3: Determination of the susceptibility of steels to SCC using predictive methods.

Steel	Parkins' method	Fang et Staehle's method (P_{scc})
For concrete reinforcement	159	34
For prestressing concrete (polished surface)	49	5
For prestressing concrete (non-polished surface)	499	625

These observations are supplemented by calculations based on the previously described methods of Parkins and Fang and Staehle. Results of calculations reported in Table 3 give some information for the susceptibility of the studied steels in a NaOH solution at pH 13.6. It can be noted that the low carbon steel is susceptible to SCC, which is in compliance with the literature [14]. The non-polished prestressing steel is very susceptible to SCC in a high alkalinity medium, this susceptibility being in connection with the presence of a specific layer on the surface. The mechanisms involved in such behaviour and their real influences on SCC are not known for the moment.

5 Conclusions and prospects

Prestressing steel in contact with segregated cement grout is subjected to SCC phenomena.

The tendency of prestressing steel to SCC is evaluated by analyzing the impact of potential scanning rates on potential curves (Parkins' and Fang and Staehle's methods). The range of potential for which steel is subject to SCC is thus determined (between -800 and -600 mV$_{sce}$).

The experiments were carried out without mechanical stress and aimed to identify the influence on steel SCC susceptibility.

It can be concluded that:
- Sulphate content has a poor influence on SCC in theses conditions (pH > 12).
- High pH can generate important anodic peaks in the zone of active/passive potential.

In addition, the study relating to the type of steel makes it possible to confirm the best resistance of the polished prestressed concrete steel to SCC in high alkalinity medium compared to the reinforced concrete steel. Moreover, it can be considered that a specific layer present on prestressing steel surface due to cold drawing process breaks up and thus generates a high susceptibility to corrosion.

The whole of these tests were carried out without mechanical stress, further works are in progress to analyse the influence of stress on SCC mechanisms by using the same types of solutions.

References

[1] FIP, French technique for prestressed concrete structures, *XIe congress*, Hambourg, 1990.
[2] Ministère de l'aménagement du territoire, de l'équipement, du logement et du tourisme, Directive provisoire sur les injections des gaines des ouvrages en béton précontraint, mars 1973 (in French).
[3] US Department of transportation, Performance of Grouts for Post-Tensioned Bridge Structures, *Publication No FHWA-RD-92-095*, December 1993.
[4] Ganz, H.R., Vildaer, S. Grouting of post- tensioning tendons. VSL International LTD, May 2002.
[5] FIP commission on prestressing materials and systems, *Collection of monographs on premature fracture of prestressing reinforcement used in prestressed concrete structures or products and for other purposes*, second edition, march 1990.
[6] Nürnberger, U. Corrosion induced failures of prestressing steel. *Otto-Graf-Journal,* **13**, pp. 9-25, 2002.

[7] Parkins, R.N. Predictive approaches to stress corrosion cracking failure *Corrosion Science* **20**, pp.147-166, 1980.
[8] Fang, Z. & Staehle, R. W., Effects of the valence of sulfur on passivation of alloys 600, 690, and 800 at 25°C and 95°C. *Corrosion* **55 (4)**, pp. 355-379, 1999.
[9] Desjardin, D., Oltra R. (1990) *Corrosion sous contrainte : Phénoménologie et mécanismes*. Les éditions de Physique – Bombannes, Paris, 1990 (in French)
[10] Alonso, M.C. & Andrade, M.C. The electrochemical behaviour of steel reinforcements in Na_2CO_3 and $NaHCO_3$ solutions in relation to stress corrosion cracking. *Corrosion Science,* **29 (9)**, pp 1129-1139, 1989.
[11] Alonso, M.C., Procter, R.P.M., Andrade, C. & Saenz de Santa Maria, M., Susceptibility to stress corrosion cracking of a prestressing steel in $NaHCO_3$ solutions, *Corrosion Science* **34 (6)**, pp 961-973, 1993.
[12] Parkins, R.N. & Zhou, S., The stress corrosion cracking of C – Mn steel in $CO_2 - HCO^-_3 - CO_3^{2-}$ solutions II: Electrochemical and other data, *Corrosion Science* **39(1)**, pp 175-191, 1997.
[13] Saxena, A., Singh Raman, R.K. & Muddle, B.C., Slow strain rate testing for monitoring cracking of mild steels for vessels and pipes for processing using caustic solutions. *Pressure Vessels and Piping*, **83**, pp 399-404, 2006.
[14] Rihan, R.O. & Nesic, S., Erosion – Corrosion of mild steel in hot caustic – Part I: NaOH solution. *Corrosion Science* 48 **(9)**, pp 2633-2659, 2006.

Sub-Modelling Boundary Element approach for stress concentration in samples exposed to pitting corrosion

A. Peratta & C. Brebbia
Wessex Institute of Technology, UK

Abstract

The paper presents a Sub-Modelling - Boundary Element Method (BEM) for computing SCFs in samples exposed to pitting corrosion. The goal of this technique is to provide a detailed mapping of the stress concentration factor (SCF) including the effects of deep pits of several microns in a sample of a few centimetres size, provided that the topographical data of the sample is known in detail. The methodology is summarised as follows. First, the original geometrical data representing the top surface of the sample is used for building a hierarchical sequence of defeatured levels by means of a moving average approach. Second, an iterative three-dimensional BEM solving scheme is launched in order to explore and solve the sample data at a defeatured level. The calculation resolution level can be selected by the user. When the iterative solving scheme ends, an iterative assembly scheme starts in which the results of SCF and the geometrical error of the model are collocated at sampling points corresponding to the resolution selected by the user. Finally, and repeatedly if necessary, a new solving iteration sequence may be launched with increased resolution, in the regions of the sample data that require more accuracy, such as zones of high SCF, large geometrical error, or steep gradient of any of them. The process continues until convergence of results is achieved. The tool has been successfully tested in several real case scenarios, and the results obtained compare well with experimental findings.
Keywords: pitting corrosion, stress concentration, Boundary Element Method, Sub-Modelling.

1 Introduction

One of the most damaging and prevalent types of corrosive attack most likely to breach a metallic structure under stress is localized corrosion [1].

Pitting is a form of corrosion in metals characterised by its extremely local damaging effect.

Pitting corrosion is characterised by having the anodic and cathode areas, in close vicinity to each other. The former has poor access to oxygen in the electrolyte, while the latter appears in areas in contact with high concentration of oxygen. Hence, pitting can be regarded as a special case of galvanic corrosion, in which the anode and cathode are extremely close to each other. With time, the corrosion area tends to dig into the un-corroded metal, and further increase the local lack of oxygen in the anode.

Pitting corrosion usually leads to the creation of small and deep holes of sub-millimetric diameter in the exposed metallic surface. It is supposed that gravity effects contribute to the process. The structural damage induced by pitting is difficult to appreciate by simple visual inspection. Moreover, when the metallic structure is subject to mechanical stress, the holes induced by pitting can amplify the local stress concentration in factors ranging from two to five. Finally, pits originate cracks which will end up damaging the structure.

It is well known that pitting can also be initiated by surface defects and heterogeneities in the metal. Either a scratch or a local change in composition can cause pitting. In addition, homogeneous surfaces and metals characterised by uniform corrosion such as mild steels are less prone to pitting. In opposition, materials more resistant to uniform corrosion such as stainless steels, nickel and aluminium alloys are more susceptible to pitting.

The study of SCFs induced by pitting is motivated by the lack of low cost/efficient-ratio methods available nowadays for detecting and predicting the damage in industrial environments.

There have been a number of reviews of theories and mechanisms for describing pitting and crevice corrosion [1-3]. A thorough revision of the theoretical conceptual models for describing pitting corrosion has been done in ref. [4]. Detailed modelling of fatigue crack corrosion and crack initiation due to pitting has been studied with the Finite Element Method (FEM) in [5,6]. The need to retain in service expensive and large maritime structures beyond their lifetime expectations has triggered a significant number of research activities towards the prediction and risk assessment, where probabilistic approaches have recently proven quite useful [7].

The aim of this paper is to present a computer simulation tool that has been developed in order to assess the SCF of metallic samples exposed to pitting for which exact topography is known in detail, i.e. by scanning microscopy techniques. The characteristic size of a sample may be of few centimetres while the characteristic length of the dangerous pits in the sample might be of few tens of microns or microns. The conceptual model for the numerical test is illustrated in Fig. 1.

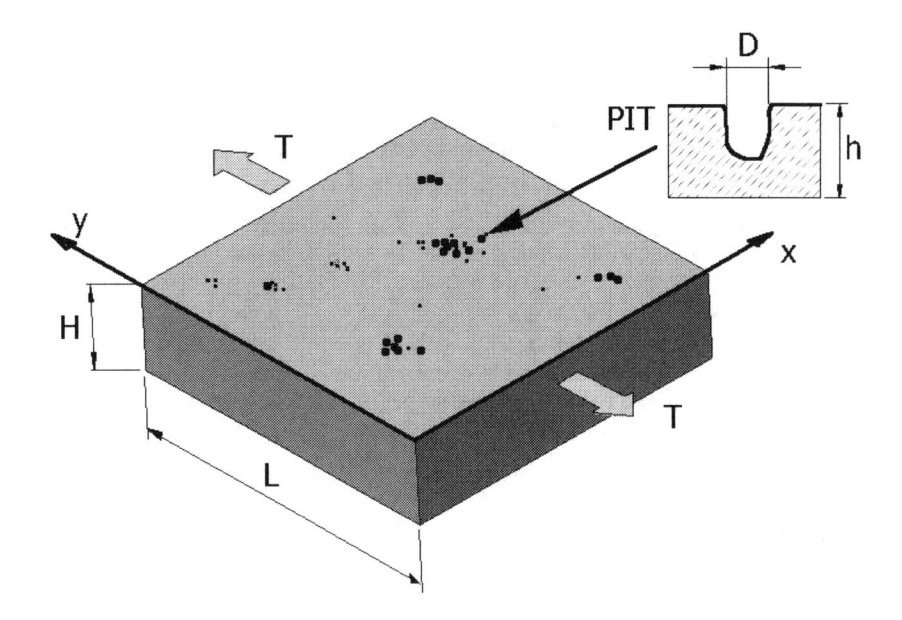

Simulation of Electrochemical Processes II 279

Figure 1: Conceptual model for the numerical test showing a sample with pitting defects exposed to lateral unitary stress along y direction.

The sample of characteristic size L and thickness H is constrained to lateral unitary stress T in y direction. The top surface of the sample may include a significant number of pits with characteristic diameter D and depth h.

The stress analysis is conducted by means of BEASY Mechanical Design module [8], a three dimensional computer code based on the Boundary Element Method (BEM) [9]. The key advantages of the BEM are first that it is based on the fundamental solution of the leading differential equation, and second that it does not need domain discretisation, i.e. in three dimensional models the computational mesh is required in the boundary only.

One of the major complications in BEM is that it yields fully populated unsymmetric linear systems of equations, and when the number of degrees of freedom becomes too large the calculation becomes prohibitive. Roughly speaking, a few tens of thousands degrees of freedom (i.e. 50000) may require a storage of few gigabytes (10 GBytes) in full format and could take a couple of days to run in a Pentium IV PC with 3GHz using a standard direct Gaussian solver. Noteworthy, the accuracy achieved with 50000 degrees of freedom is usually superior than the one obtained with other standard methods such as FEM, distributing the same amount of degrees of freedom in volume.

To illustrate the motivation of this approach, consider a problem in which a sample of L ~ 100mm with a significantly large number of pits of average size D~10μm randomly distributed is laterally loaded with constant stress. The challenge is to determine the stress concentration field defined as:

$$SCF = \frac{S_{yy}}{T} \quad (1)$$

In order to take into account the pits with a reasonable accuracy, it is necessary to use at least a 2 µm sampling resolution for the top surface. Henceforth, the top surface alone will produce a mesh of with 2.5×10^9 nodes. This figure yields a prohibitively large number of degrees of freedom, even unpractical for a supercomputer. On the other hand, it can be demonstrated that the stress field is a very local effect. If pits are separated by more than 5D, they can be considered as uncorrelated. This behaviour helps to decompose the large problem into smaller ones by means of a sub-modelling approach. A possible sub-modelling approach would be to solve 1000000 sub-models of 0.1mm by 0.1mm, covering the whole surface, where each one of them accounts for 2500 degrees of freedom. Models of few thousand degrees of freedom can be solved in a matter of seconds in an average PC. Then, roughly speaking, with a small cluster of four to eight PC's the complete problem can be solved within few days.

The method above outlined can be substantially improved if we start solving the problem in a rather defeatured geometry with few sub-models, and successively refined sub-modelling stages are applied in an iterative way in combination an adaptive mesh refinement that improves the resolution only in the regions where high SCFs are being discovered. This is the main idea implemented in the present work.

The paper is organised as follows. Section 2 outlines the technical implementation of the described approach. Section 3 provides basic guidelines on how to choose a safe sub-modelling approach. A case study is that briefly illustrates the applicability of the method is presented in section 4, and finally the corresponding conclusions are elaborated in Section 5.

2 Iterative solving scheme

Figure 2 shows the general sub-modelling approach with three calculation stages. Each rectangle represents a sub-model, characterised by certain space resolution and size. Both size and resolution of the sub-model should be chosen according to the results of the SCF obtained in the previous stage. The initial input consists of the original large sample with detailed geometrical information. In the first stage, the original sample is decomposed into N_1 sub-models with coarser resolution. The density of the hatch pattern represents the mesh resolution of the associated sub-model. Stage 2 (with N2 sub-models) performs another division to the previous iteration with increased space resolution. Stage 3 increases the number of sub-models but the mesh resolution is refined only in those sub-models which presented high concentration factor in the previous step. The number of stages may continue until the desired level of resolution and geometrical accuracy is achieved. At the end of the process, the pits with higher SCF will have been identified and solved. In some cases, the boundary conditions between sub-models must be readjusted iteratively, hence the method will iterate between two stages until the boundary conditions are properly adapted, same as in a standard multi-grid approach.

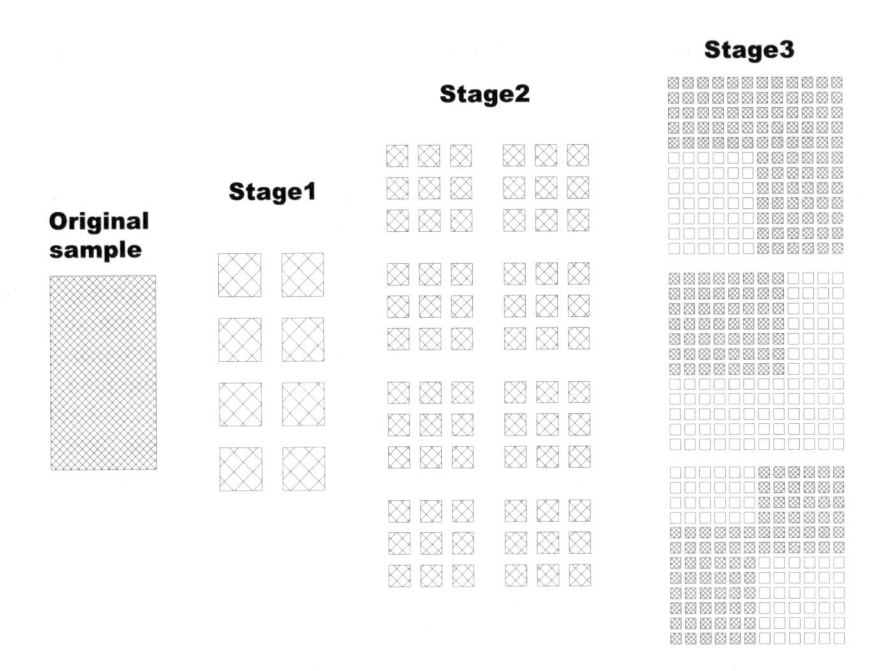

Simulation of Electrochemical Processes II   281

Figure 2:   Illustrative sketch showing different stages of the iterative sub-modelling approach. Each cell represents a sub-model, the density of the hatched area represents the space resolution of the computational grid.

The iterative solving scheme operates in the following way. An initial representation of the top surface, usually coarser than the original data, is used to automatically create a NurbSurface model by the geometry modeller BEASY-GiD [10]. Then, a parametric wireframe structure is connected to the created Nurb-surface in order to build a three-dimensional sub-model of the corroded surface. The remaining wireframe is then filled with flat surfaces to define a closed six-faced volume [8]. Next BEASY-GiD [8] creates a computational grid of the 3D model with the resolution and quality chosen either by the user or automatically by the software in a configuration file. Since BEASY is based on the Boundary Element Method, the computational grid is created only for the surface of the sub-model, thus avoiding meshing the volume. This represents an important advantage which helps to speed up model generation and further calculation time. Once the model is created by BEASY-GiD, the information is passed as input to the BEASY MECHANICAL DESIGN software which conducts the stress analysis. As output, BEASY provides the stress and strain tensors results in all points of the boundary mesh of the sub-model.

BEASY results of each sub-model are stored in a file for later assembly into the appropriate location of the overall result file.

The solving iterations stop once all the sub-models are completed. The above mentioned tasks are automatically performed in non-interactive mode, so as to avoid time consuming graphical interactions during the iteration over a large number of sub-models.

3 Sub-modelling

In each iteration stage a certain number of sub-models are solved after suitable space resolution and dimensions are chosen for them. A 3D mesh with boundary conditions is associated to each sub-model. The top surface of each sub-model undergoes three approximations. The first one happens when defeaturing the original sampled data with a moving average filter, the second one is introduced when representing the topography of the defeatured data with a rectangular grid and Nurb-Surfaces, and the third approximation occurs when meshing the surface approximated by Nurb-surfaces with a computational grid made of either linear (flat) or quadratic (curved) elements. The accuracy at each level of approximation can be adjusted according to the quality of results sought by the user.

We consider the original sample as a parallelepiped with L >> H. All sub-models in any stage will be also parallelepipeds of size $w \times w \times h$.

In order to decrease the computational burden the number of the degrees of freedom (NDOF) of the sub-models should be as small as possible. For a fixed space resolution, the NDOF is proportional to w and h. The window size w is directly affected by the local topography, while the depth can in principle be chosen arbitrarily. In order to determine a practical rule to determine the depth of the sub-model, a systematic numerical experiment has been performed. The results are summarised in Figure 3 which shows the scaled y-y component of the stress tensor S_{yy}, when a series of sub-models of different widths (1015μm by 1015 μm, and 507.5μm by 507.5 μm) and varying depth h are subjected to unitary stress. The original sample is 3.175 mm thick. It can be observed that S_{yy} converges to the same value when $h > 1,5\ w$. Finally, the minimum side w that the smallest sub-model could have should be at least twice the characteristic diameter of the surrounding features D, or the maximum depth fluctuation DZ. Then, the depth of the sub-model h should be chosen according to:

$$h = \min(H, 1.5w), \qquad (2)$$

where H is the thickness of the original sample, and w is the characteristic size of the submodel.

4 Results

This section shows the results of the sub-modelling technique in a test sample. The initial topography of the sample has been recorded with a grid spacing of 6.345 μm. Figure 5 shows the results of the SCF in two different sub-models, one containing the other. The space resolution in the larger sub-model is 50.76μm by 50.76 μm, the space resolution for the most inner sub-model is 6.345 μm by 6.345 μm. In this case, the small sub-model has captured a small pit responsible for a SCF of nearly 2.

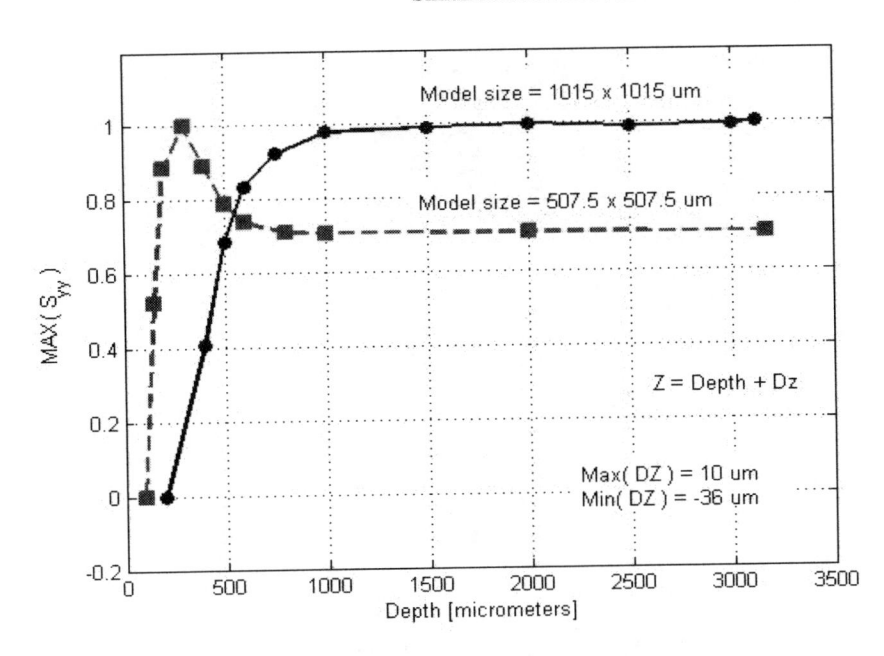

Figure 3: Normalised Y-Y component of the stress tensor for sub-models of different depths.

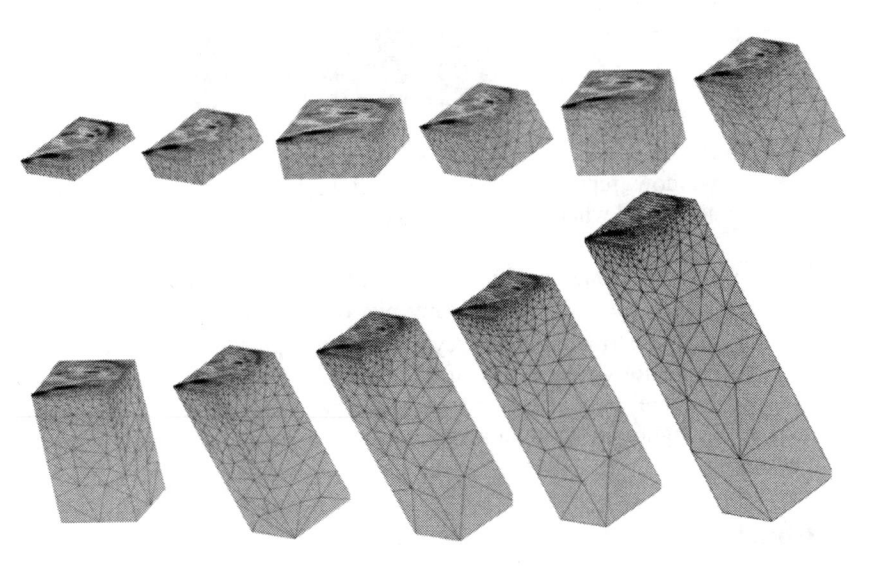

Figure 4: Sequence of sub-models with different depths used to determine a rule for defining the optimum size, for thick plates exposed to pitting corrosion. The greyscale in the top surface indicates S_{yy}.

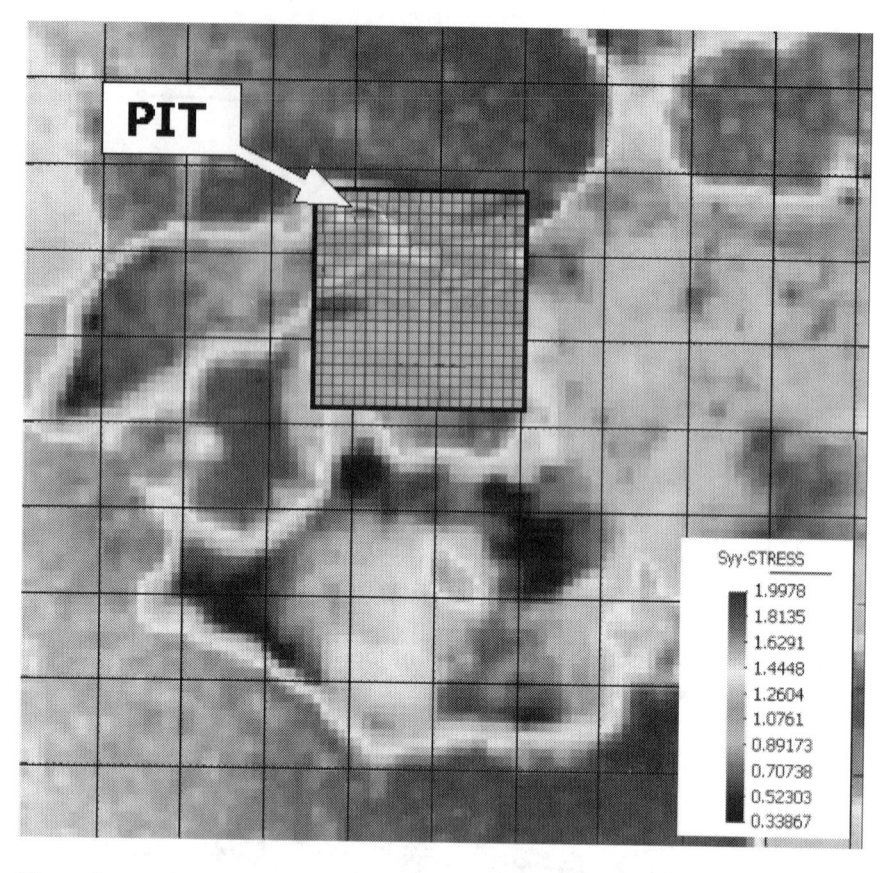

Figure 5: Pit detected during the sub-modelling iterations. The square window represents the region of high resolution covered by a submodel when scanning the whole sample.

Figure 6 shows some preliminary results obtained for a sample of $33mm \times 43mm \times 3.175mm$, in which many pits of different scales (typically 20 µm) are randomly distributed. The greyscale in the figure indicates SCF while the circles spot regions with peaks of SCF that significantly exceed the local average. Cracks are more likely to originate in these regions. The results obtained in this figure are in good agreement with experimental findings during fatigue tests.

5 Conclusions

A Sub-Modelling - Boundary Element approach for computing SCFs in samples exposed to pitting corrosion has been developed, and practical rules for defining the sub-model elaborated.

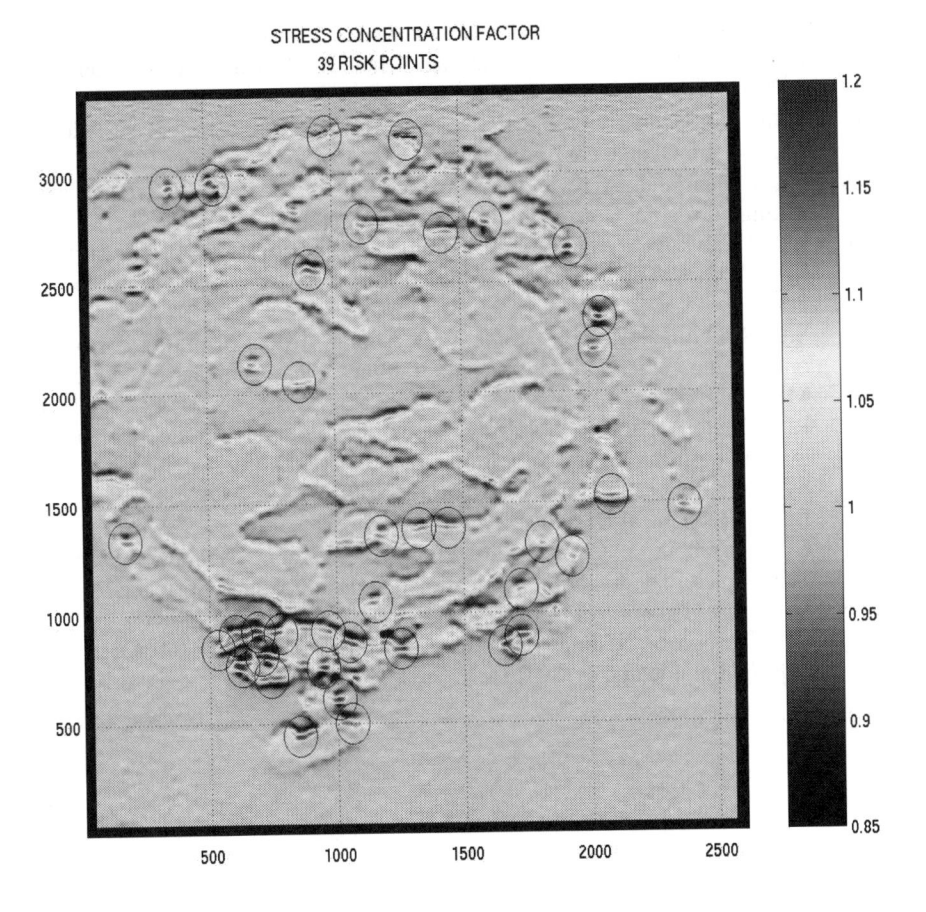

Figure 6.

The research behind this presentation is still being undertaken. The goal of representing a SCF mapping of a sampled surface to the most available level of accuracy has been achieved. The tool has been successfully tested in several real case scenarios, and the results obtained compare well with experimental findings.

Acknowledgements

The authors of this work would like to acknowledge the support provided by Dr John Baynham and Dr Robert Adey, and the use of BEASY MECHANICAL DESIGN software.

References

[1]   Kolotyrkin, Ya. M. Pitting corrosion of metals. Corrosion 19: 261t, 1963.

[2] Szklarska-Smialowska, Z. Localized Corrosion, Staehle, Brown, Kruger and Agrawal, eds., p. 312. National Assoc. of Corros. Engrs., Houston, 1974.
[3] Kruger, J. in Passivity and Its Breakdown on Iron and Iron Base Alloys, Staehle and Okada, eds., p. 91. Nat. Assoc. of Corros. Engrs., 1976.
[4] S. M. Sharland. A review of the theoretical modelling of crevice and pitting corrosion. Corrosion Science. 27 (3), pp 289-323, 1987.
[5] A. P. Jivkov. Evolution of fatigue crack corrosion from surface irregularities. Theoretical and Applied Fracture Mechanics. 40, pp 45-54, 2003.
[6] S. I. Rokhlin, J.-Y. Kim, H. Nagy, B. Zoofan. Effect of pitting corrosion on fatigue crack initiation and fatigue life. Engineering Fracture Mechanics 62 pp. 425-444, 1999.
[7] R. E. Melchers and R. J. Jeffrey, Probabilistic models for steel corrosion loss and pitting of marine infrastructure, Reliability Engineering and System Safety (2007), doi:10.1016/j.ress.2006.12.006.
[8] Beasy Mechanical Design software. www.beasy.com
[9] C. A. Brebbia, J. C. F. Telles, and L. C. Wrobel. Boundary Element Techniques. Theory and Applications in Engineering. Springer-Verlag, Berlin, Heidelberg, NY, Tokyo, 1984
[10] GID resources. Commercial pre and post-processor software. url: http://gid.cimne.upc.es

Corrosion of reinforcing steel in reinforced and prestressed concrete bridges

A.-K. M. Hussain & A. M. Rifai
*Civil Engineering Department, Engineering College,
Al-Marghab University, Libya*

Abstract

The corrosion of reinforcing steel has led to the premature deterioration of many concrete bridges in countries throughout the World, such as Libya, before their design life is attained. This has placed tremendous financial burden on such countries and local transportation organizations in their attempts to halt ongoing reinforcing steel corrosion in the existing structures that are still functional and to replace those structures that have already deteriorated to the point that it does not make any economic sense to keep on maintaining them. In addition, badly deteriorated bridges have considerable adverse effects on the nation's economic output and also place the safety of motorists at risk. In this study the effects of several factors on the corrosion behavior of steel have been studied. These factors are chloride ion concentrations and environmental variables such as temperature and relative humidity. The interactions of chloride ion concentration, temperature, and relative humidity on the corrosion behavior of steel are complex. In addition, the effect of the concrete mix components, such as water/cement ratio and proper selection of cement type, mineral admixture, and fine aggregate, coarse aggregate, and air content on the corrosion rate of reinforcing steel are studied.

Keywords: reinforcing steel corrosion, reinforced concrete bridges, corrosive environment, corrosion of reinforcing steel, corrosion of steel reinforcement in concrete.

1 Introduction

Reinforcing steel bars are placed in regions of low tension in concrete bridges.
Concrete has a tendency to crack eventually if proper curing is not observed. The resistance of mild steel bars to corrosion can not be significantly improved

by just modifying its composition, grade, or the level of stress. Therefore, to prevent corrosion of steel reinforcement in concrete located in corrosive environment either the conventional mild steel reinforcement must be coated with an effective and economical barrier to prevent contact with chloride, moisture, and oxygen, or reinforcement made of corrosion-resistant materials must be used as mentioned by Glasser and Sagoe-Crentsil [1], Hime [2], Baboian [3], Virmani and Clemena [4] and Thompson and Lamkard [5]. Among the above two options, application of a suitable coating on the mild steel reinforcement may be the most economical. The coated reinforcing steel bars must be resistant to damage during transport from a plant to a construction site, storage at construction site, and placement in the structure. It must also be durable in severe service environment and capable of maintaining its structural function throughout the service life of the structure, and be economical.

2 Corrosion control in new concrete constructions

Corrosion control of reinforcing steel bars in new concrete constructions, such as bridges, buildings exposed to the marine environments, and wastewater treatment tanks, require the use of a combination of different measures, such as: adequate depth of concrete cover, quality of concrete, corrosion inhibitors admixture and corrosion-resistant reinforcement. At least 50 mm of concrete cover reduces the ability of corrosive agents to penetrate the concrete and corrode the reinforcing steel bars. The most common cause of corrosion staining on concrete decks is poor placement of the reinforcing steel bars.

3 Quality of concrete

The quality of concrete is of utmost importance in determining the durability of reinforced concrete bridge members exposed to chlorides and subjected to intermittent wetting. Concrete contains a network of capillary channels that allows water and oxygen, both important to steel bars corrosion, to permeate into the concrete. Corrosion occurs because a salt solution, or water and carbon dioxide, penetrate the concrete. Lowering the porosity and the permeability of the concrete reduces the chances of contamination. At a low (W/C) ratio of 0.5 and good consolidation lead to low porosity in the concrete, which can lead to reduced permeability (hydraulic conductivity) coefficient of the concrete. Reduced permeability leads to reduction in the amount of chloride ions carried by water into the concrete. At a (W/C) ratio of 0.5, carbonation will penetrate 10 mm in ten years. A low (W/C) ratio also offers higher strength to the concrete, which would extend the time before stresses resulting from steel corrosion cause the concrete to crack. If the strength is low (20.7 MPa) carbonation can reach 25 mm in seven years. Strong concrete with a low (W/C) ratio will have a low permeability and be less susceptible to corrosion as mentioned by Gu *et al* [6]. The occurrence of corrosion of steel bars in concrete bridges can be avoided by investigation into how concrete material and mix variables affect the corrosion behavior of steel bars. The environmental factors of chloride concentration, temperature and humidity significantly affect the corrosion behavior of steel bars in concrete. A localized corrosion was noted at a location where the reinforcement had been exposed previously. In concrete with 0.50 (W/C) ratio, when galvanized bars are used, the corrosion rate was about 30% and the corresponding metal loss was 22%, in comparison to black steel. And, in the same concrete, when galvanized steel bars were used only in the top mat, the corrosion rate was twice of that observed when only black steel bars were used in both mats. Essentially, these results suggested that the use of galvanized steel bars would not provide extra benefit over black steel bars, if (W/C) ratio in the concrete is kept low as mentioned by McDonald et al [7, 8], Baboian [9], Thompson et al [10], Clear [11], Smith and Virmani [12], McDonald et al [13], Dagher and Kulendran [14] and Hussain et al [15].

4 Adequate concrete cover and chloride content

Chloride ion is the main corrosive agent. When chlorides penetrate concrete from external sources, such as de-icing salts and seawater, carbon steel bars corrode and rust forms; occupying a volume about three to seven times that of the original steel resulting in the surrounding concrete cracking as mentioned by Pettersson [16] and Virmani [17]. If a concrete structure is to be durable in a corrosive environment, it is necessary to provide an adequate depth of concrete cover over the first layer of reinforcing steel bars so that it would not be easy for chloride ions to reach the steel bars.

5 Case study – the model of the transient diffusion of chloride through a porous concrete

The adequate depth of concrete cover over the first layer of reinforcing steel bars can be determined by application of Fick's Second Law of diffusion, which adequately models the transient diffusion of chloride through a porous material, such as concrete. A closed – form solution of Fick's Second Law (a second-order differential equation) can be expressed as

$$C_{(x,t)} = Co \{ 1 - \text{erf}(y) \} \tag{1}$$

where

$$y = \frac{x}{2.\sqrt{D_c.t}} \tag{2}$$

and

$C_{(x,t)}$ = Measured chloride concentration at the desired depth x, for purposes of service life estimates, assumed to equal the chloride threshold, i.e. the chloride content that will initiate corrosion, (p.p.m.).

D_c = Chloride molecular diffusion coefficient (mm²/year). It is a function of the concrete permeability, environmental factors, and the presence of cracks.
x = Depth of reinforcing steel bars (mm).
C_0 = Constant mean surface chloride concentration measured at 12 mm below the deck surface (p.p.m.).
t = Time to reach the corrosive chloride threshold, i.e. the time for the chloride to reach the depth of reinforcing steel bars, (year).

Eqn (1) can be rewritten as

$$\text{erf}(y) = 1 - \{ C_{(x,t)} / C_0 \} \qquad (3)$$

Also, Eq. (2) may be rewritten as

$$t = \frac{x^2}{(4y^2 \cdot D_c)} \qquad (4)$$

Eqn (3) is applicable to concrete structures where the chloride ions enter from one direction, such as concrete bridge decks and piers. Using relationships (3) and (4), the time t required to reach the corrosive chloride threshold may be computed for a certain concrete cover, x and a known chloride concentration $C_{(x,t)}$ which corresponds to the onset of steel reinforcement corrosion. Repeating this process for different values of x and plotting the time t versus the concrete cover x shows the general trend of the relationship between the two variables. From such a plot, one can determine the minimum concrete cover required to prevent the onset of steel corrosion before the desired service life of the structure is reached.

For a known chloride concentration $C_{(x,t)}$, the value of erf (y) may be calculated using Eq. (3). The corresponding value of y may be interpolated from the table "Error function values y for the argument of y" given in Appendix A of Hussain [18] or from other suitable references. Eq. (4) is used to calculate the time of threshold of chlorides corresponding to a certain concrete cover x.

In this work it has been found that the chloride molecular diffusion coefficient D_c (the rate at which chloride permeates through concrete) is 32.26 (mm²/year), the constant mean surface chloride concentration in the bridges C_0 was 8300 (p.p.m.) and the typical chloride threshold for uncoated steel bars is 700 (p.p.m.).

Using these data items, and starting with x = 50 mm, Eq. (3) may be used to find the value of erf(y) as:

$$\text{Erf}(y) = 1 - \frac{700}{8300} = 0.9157$$

Next, from the table mentioned above the corresponding value of y is interpolated to be 1.2228. Using Eq. (4) the time t is calculated as:

$$t = \frac{50^2}{4(1.2228)^2 \cdot (3226)} = 13$$

Repeating this process for values of x of 55 mm, 60 mm, 65 mm, 70 mm, 75 mm and 80 mm and plotting the results as in Figure 1, we notice the relationship is almost linear. If we assume that the service life of the structure is about 25 years, we may conclude from the Figure that the minimum concrete cover should be about 68 mm. Note that Figure 1 applies to uncoated steel bars only. Once Figure 1 has been constructed, it serves as a quick way to estimate the required concrete cover for a certain service life of the structure.

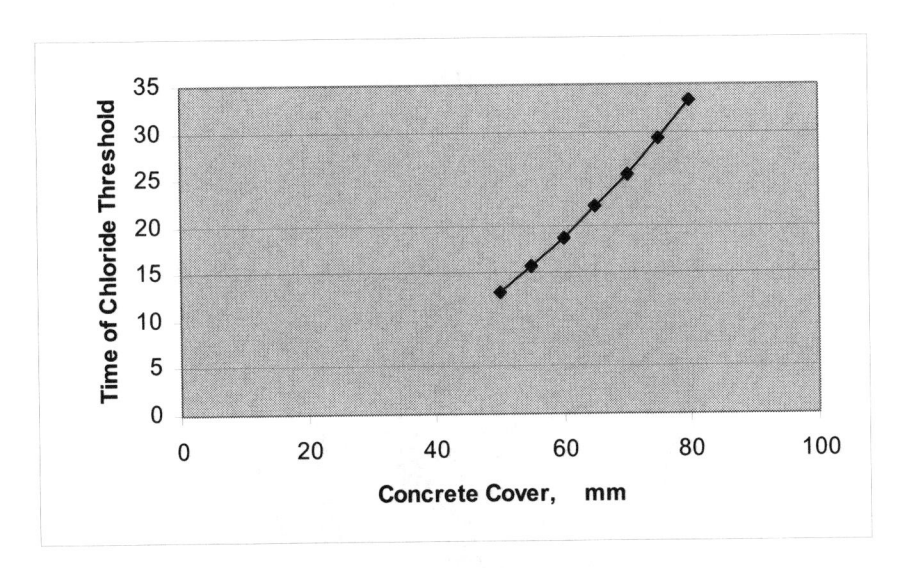

Figure 1: Relationship between concrete cover and time of corrosive chloride threshold.

However, even with adequate concrete cover, corrosion of reinforcement can still occur because, invariably, concrete will crack. In addition, presence of variances in the concrete cover and in the density of the placed concrete across a structure will eventually create corrosion micro-cells (consisting of cathodes and anodes), which drive steel corrosion. Therefore, other supplementary protective measures also need to be adopted in a new construction. Figure 2 provides a photograph of this damage. This explains why the problem of reinforcing steel bars corrosion is considerably more severe in the marine environment of Libya's northern cities, such as Al-Khoms, Tripoli and Musrata (because of high humidity) than in the other cities situated in the south. Figure 3 shows a photograph of corrosion-induced concrete spalling on a bridge piling.

6 Results

The following are the most important findings resulting from this study:
1. Chloride concentrations higher than 700 p.p.m can produce a wide range of corrosion behavior of reinforcing steel bars in concrete.
2. The interactions of chloride ion concentration and environmental variables, such as, temperature and relative humidity on the corrosion behavior of reinforcing steel bars are complex.
3. The corrosion rate of reinforcing steel bars can vary significantly based on the concrete mix components. The most significant beneficial effects were obtained from use of low (W/C) ratio and proper selection of cement type, mineral admixture and fine aggregate.

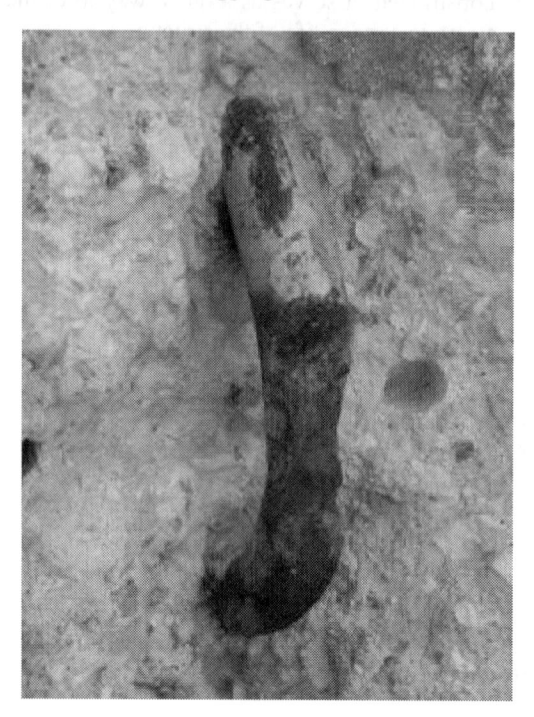

Figure 2: Corrosion of an exposed reinforcing bar.

4. Bridge utilizing epoxy powder-coated reinforcing steel bars revealed some problems, such as damage to the coating during transport and handling and the cracking of coating (in the bend areas) arising from the bending of bars at construction sites. To eliminate these problems, measures such as bending the steel bars before coating, increasing bar supports during shipping (to prevent abrasion between bars) and using padded bundling bands and nylon slings during loading and unloading were adopted.
5. A small exposed area of reinforcing steel bars would be susceptible to intense corrosion and require patching or repair of the coating damages.

This led to a decrease in the cost of epoxy-coated reinforcing steel bars and their wider use in bridges by many countries.
6. Some aspects, such as bar size, deformation pattern, grade of steel, and slab design used has effects on the occurrence of corrosion.
7. In general, the corroded reinforcing steel bars are placed in locations in the structures where concrete cover was relatively shallow, the chloride contents were high and there were cracks in the concrete.

Figure 3: Photograph of a cracked and spalled marine bridge piling.

7 Conclusions

The following conclusions can be made on the performance of reinforcing steel bars in the bridge decks:
1. Some cases of corrosion in bridge decks were centered at a construction joint.
2. Most cases of corrosion reinforcing steel bars were attributed to superficial corrosion that was already present on the reinforcing steel bars at the time of construction.
3. The use of quality concrete and adequate cover, proper finishing and curing of concrete, and proper manufacturing and handling of epoxy-

coated reinforcing steel bars are important in ensuring effective corrosion protection in concrete bridge decks.
4. Cracks in the concrete allow moisture and chloride an easy and direct access to the reinforcing steel bars as opposed to the normally slow diffusion process through sound concrete. The cracks probably allow the bars to remain wet longer than normal and they corrode.
5. Application of a suitable, stable organic coating, including epoxies (in both liquid and powder forms), on the reinforcing steel bars is the more economical solution to serve as a barrier for isolating the reinforcing steel bars from moisture, chloride ions, and oxygen, thereby preventing corrosion and keep many concrete bridges away from the premature deterioration.
6. To control the major causes of corrosion of reinforcing steel bars in concrete, the concrete must block the ability of salt to make contact with the steel bars and retard the progression of carbonation. To do this, the following ten ways should be considered: (1) -Reduce the permeability of the concrete. (2) - Ensure at least 50 mm of concrete cover the reinforcing steel bars. (3) - Coat the reinforcing steel bars with epoxy. (4) - Use a corrosion inhibitor in the concrete. (5) -Use cathodic protection. (6) - Keep the concrete dry. (7) - Use a blended cement in the concrete. (8) - Use a water-reducing agent; low-, mid-, or high-range. (9) - Cure the concrete properly. (10) - Use other special treatments.

References

[1] F. G. Glasser and K. K. Sagoe-Crentsil, "Steel in Concrete: Part II – Electron Microscopy Analysis", *Magazine of Concrete Research,* Vol. 41, No. 149, 1989, pp. 213-220.

[2] W. G. Hime, "The Corrosion of Steel—Random thoughts and Wishful Thinking". *Concrete International,* Vol.15, No.10, 1993, pp. 54-57.

[3] R. Baboian, "Environmental Aspect of Automotive Corrosion Control", *Materials Performance,* Vol.32, No.5, 1993, pp. 46-49.

[4] Y. Paul Virmani and Gerardo G. Clemena, 1998, Corrosion Protection-Concrete Bridges, Report No.FHWA-RD-98-088, Turner-Fairbank Highway Research Center, 6300 Georgetown Pike, McLean, Virginia 22102 - 2296, Office of Engineering Research and Development Federal Highway Administration, 6300 Georgetown Pike McLean, Virginia 22102 -2296.

[5] N. G. Thompson and D. R. Lankard, "Improved Concrete for Corrosion Resistance". Report No. FHWA-RD-96-207, Federal Highway Administration, Washington, D. C., 1995.

[6] P. Gu, Y. Fu, P. Xie, and J. J. Beaudoin, "Effect of Uneven Porosity in Cement Paste and Mortar on Reinforcing Steel Corrosion", Cement and Concrete Research, Vol. 24, No. 6, pp.1055-1064, 1994.

[7] D. B. McDonald, D. W. Pfeifer, and G. T. Blake, The Corrosion Performance of Inorganic-, Ceramic- and Metallic-Clad Reinforcing Bars and Solid Metallic Reinforcing Bars in Accelerated Screening Tests. Report No. FHWA-RD-96-085, Federal Highway Admin., Washington, D.C.,1996.

[8] D. B. McDonald, M. R. Sherman, D. W. Pfeifer, and Y. P. Virmani, "Stainless Steel Reinforcing as Corrosion Protection", Concrete International, Vol. 17, No. 5, pp. 65-70, 1995.

[9] R. Baboian, "Synergistic Effects of Acid Deposition and Road Salts on Corrosion", in Corrosion Forms and Control for Infrastmcture, STP 1137, V. Chaker, ed., Amer. Soc. for Test. and Materials, Philadelphia, Pa., 1992.

[10] N. G. Thompson, K. M. Lawson, D. R. Lankard, and Y. P. Virmani, "Effect of Concrete Mix Components on Corrosion of Steel in Concrete", presented at Corrosion/96, Denver, Colorado, 1996.

[11] K. C. Clear, "Effectiveness of Epoxy-Coated Reinforcing Steel", Final Report submitted to Canadian Strategic Highway Research Program, Toronto, Canada, 1992.

[12] J. L. Smith and Y. P. Virmani. "Performance of Epoxy Coated Rebars in Bridge Decks". Report No. FHWA-RD-96-092, Federal Highway Administration, Washington, D. C., 1996.

[13] D. B. McDonald, M. R. Sherman, and D. W. Pfeifer, "The Performance of Bendable and Nonbendable Organic Coatings for Reinforcing Bars in Solution and Cathodic Debonding Tests: Phase II Screening Tests", Report No. FHWA-RD-96-021, Washington, D. C., 1996.

[14] H. J. Dagher, and S. Kulendran, "Finite Element Modeling of Corrosion Damage in Concrete". *ACI Struc. Jour.*, Vol. 89, No. 6, 1992, pp. 699-708.

[15] S. E. Hussain, A. Rasheeduzzafar, A. Al-Mussallam, and A. S. Al-Gahtani, "Factors Affecting Threshold Chloride for Reinforcement Corrosion in Concrete". *Cement & Conc. Research,* Vol.25, No.7, 1995, pp.1543-1555.

[16] K. Pettersson, "Chloride Threshold Value and the Corrosion Rate in Reinforced Concrete". *Ceme.& Conc. Res.,* Vol. 20, 1994, pp.461-470..

[17] Y. P. Virmani, "Corrosion Protection Systems for Construction and Rehabilitation of Salt-Contaminated Reinforced Concrete bridge members", Proceeding of the International Conference on Repair of Concrete Structures—From Theory to Practice in a Marine Environment, Svolvaer, Norway, 1997, pp. 107-122.

[18] A-Kh. M. Hussain, "The Laboratory Determination of Longitudinal and Lateral Dispersion Coefficients in Unidirectional Groundwater Flow". A Thesis submitted to the University of Newcastle upon Tyne for the Degree of Doctor of Philosophy, Department of Civil Engineering, University of Newcastle upon Tyne, England, U.K., 1981, Appendix A, PP.275.

Section 6
Detection and monitoring of corrosion

Corrosion detection by multi-step genetic algorithm

K. Amaya[1], M. Ridha[1] & S. Aoki[2]
[1] *Tokyo Institute of Technology, Ookayama, Meguro, Tokyo, Japan*
[2] *Toyo University, Kujirai, Kawagoe, Saitama, Japan*

Abstract

A multi-step genetic algorithm is developed to eliminate the limitations of the inverse analysis method for detecting the corrosion profile on the steel matrices from a small number of potential data on the concrete structure. The corrosion profile represents the the number, locations and shapes of plural corrosion parts on the steel matrices. It was modeled into a binary string which is defined by discretizing the steel matrices into a suitable number of segments using a certain resolution. Each segment is encoded by one binary bit. A tree structure is employed in the multi-step genetic algorithm and the examination is localized for root and each branch. By this approach, we can avoid a long binary string that is required to encode the segments when a large examination area is discretized using a required resolution. Hence, for each step of examination, the standard genetic algorithm can be locally applied and performed efficiently. We demonstrate the effectiveness of the proposed method using an example of numerical simulation.

1 Introduction

Detection of reinforcement corrosion in the early stage is important to reduce maintenance cost and increase the durability of the concrete structure [1].

The boundary element inverse analysis [2] has been introduced to estimate the locations and sizes of plural corrosion in the concrete structure by modeling them into a set of unknown parameters. The inverse analysis was carried out by minimizing the cost-function using a down-hill simplex method. However, a such method required to predetermine the number and shapes of plural corroded parts. In addition, some difficulties were occured related with an appropriate initial guess of the set of unknown parameters.

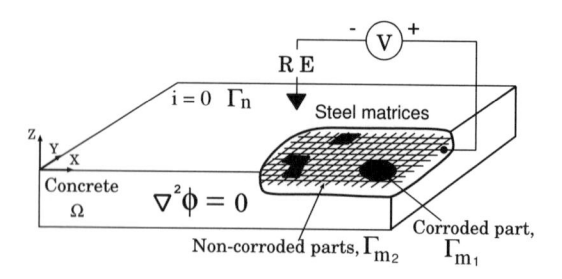

Figure 1: Model of concrete domain for corrosion analysis.

The purpose of this research is to develop a multi-step genetic algorithm (GA) for eliminating the above limitations. In this method, A tree structure is employed and the examination was carried out separately for the root and the branches. For each step of examination, the corrosion profile is modeled into a binary string which is defined by discretizing the steel matrices of the examination area into a suitable number of segment using a certain resolution.

2 Numerical analysis for corrosion detection

Let us consider that the concrete domain (Ω) is surrounded by the surface of the concrete (Γ_n) and the surface of the steels matrices (Γ_m). The steel matrices are divided into corroded and non-corroded parts.

In the previous work [2], the Laplace Equation in Equation (1) has been used to describe the potential in the Ω. The related boundary conditions (BCs) are given in Equation (2) through Equation (3).

$$\nabla^2 \phi = 0 \quad in \ \Omega \qquad (1)$$

$$i(\equiv \kappa \frac{\partial \phi}{\partial n}) = 0 \quad on \ \Gamma_n \qquad (2)$$

$$-\phi_m = \begin{cases} f_{m_1}(i) & on \ \Gamma_{m_1} \\ \\ f_{m_2}(i) & on \ \Gamma_{m_2} \end{cases} \qquad (3)$$

If all of the BCs are prescribed, the direct BEM can be used to solve the Laplace's equation in Equation (1). Hence, the potential (ϕ) and current density (i) on the whole concrete domain can be determined.

In actual cases, however, the BCs in Equation (3) are unknown. Therefore, an inverse analysis is necessary in order to detect the corrosion profile that represents the number, location and shapes of the plural corrosion parts on the steel matrices in the concrete structure.

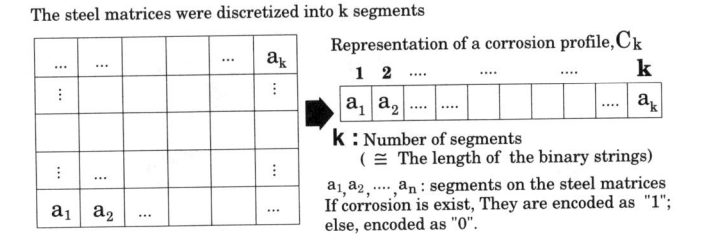

Figure 2: Representation of the corrosion profile.

To simplify the inverse problem, the polarization curves of corroded and non-corroded parts of the steel matrices are prescribed. The inverse analysis is carried out by minimizing a cost-function, $\varepsilon(C_k)$, in Equation (4).

$$\varepsilon(C_k) = \sum_{l=1}^{N} \left(\frac{\phi(x_l, y_l; C_k) - \bar{\phi}(x_l, y_l)}{\bar{\phi}_{max}} \right)^2 \tag{4}$$

where C_k is the corrosion profile on the steel matrices which is represented by a binary string. The x_l, y_l are the location of potential measurements on the structure in x and y directions. The N is the number of measurements, the ϕ and $\bar{\phi}$ are the calculated and measured potential values, respectively. The calculated values of potential were obtained by solving the Laplace's equation in Equation (1) using BEM.

3 The multi-step genetic algorithm

To define the binary string of the C_k, the steel matrices is discretized into a suitable number of segments using a certain resolution as given in Figure 2. The number of segment indicates the length of the binary string. The corroded and non-corroded segments are encoded by a binary bit "1" and "0", respectively.

When a large area of the steel matrices is discretized using a required resolution, however, a very long binary string is necessary to represent all of the segments. Therefore, a long calculation time is required in carrying out the standard GA.

A tree structure, shown in Figure 3, is employed in the multi-step GA to eliminate the above difficulty. In this method, the examination is performed separately for the root and each branch. Hence, for each step of examination, the standard GA can be applied locally and performed efficiently.

In the first step, to define the binary strings of the C_k, a low resolution (coarse segment) is used to discretize the entire steel matrices into a suitable number of segments.

The standard GA is carried out by employing some potential data which are spread out uniformly on the surface of the structure. The $\varepsilon(C_k)$ in Equation (4) is used to measure the fitness of each binary string as a candidate of the true C_k. Two conditions were used to terminate the iteration process: (1) the $\varepsilon(C_k)$, is smaller

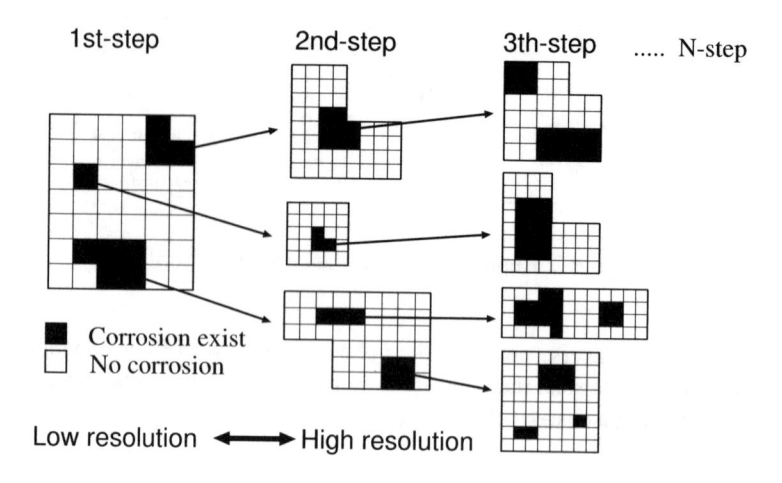

Figure 3: Concept of the multi-step GA.

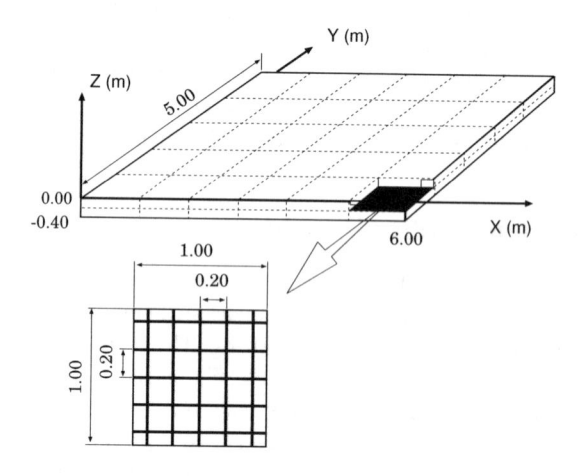

Figure 4: Model of the reinforced concrete specimen and detail of the steel matrices.

than ε_o, or (2) the number of generations in the GA is reached. From this step, the corrosion is localized and one or more corroded parts are obtained.

In the second and next steps, the examination is carried out separately for each corroded part. The local steel matrices is remeshed into a certain segment using a higher resolution. The procedures of the GA is the same as the previous step but they may used a smaller size of the segments, a different length of the binary string and a different number of populations and generations.

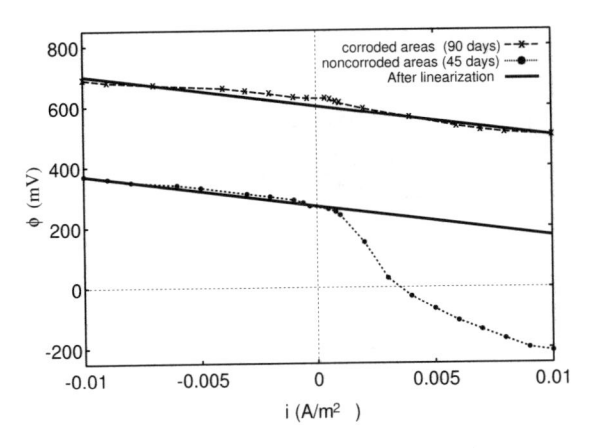

Figure 5: Two linear and parallel polarization curves for corroded and non-corroded parts of the steel matrices. These curves were approximated from the literature [5].

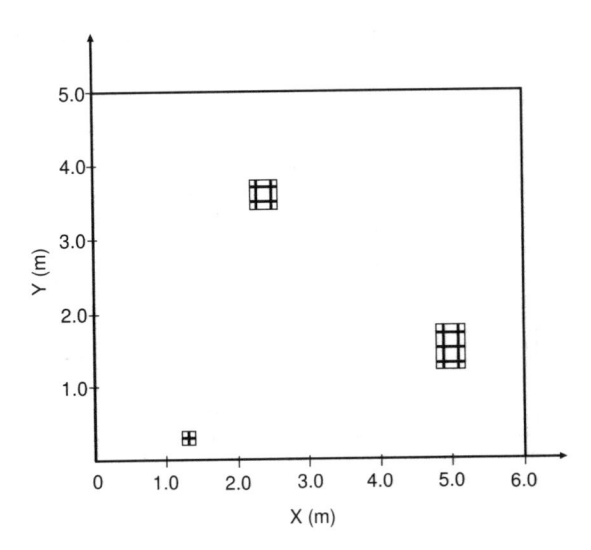

Figure 6: The profile of corrosion area on the steel matrices that were used in direct BEM.

The final step is carried out when the area of examination become suitable to be discretized using the high resolution. Once this condition is achieved, then the standard GA can be performed to accurately detect the corrosion profile on the steel matrices.

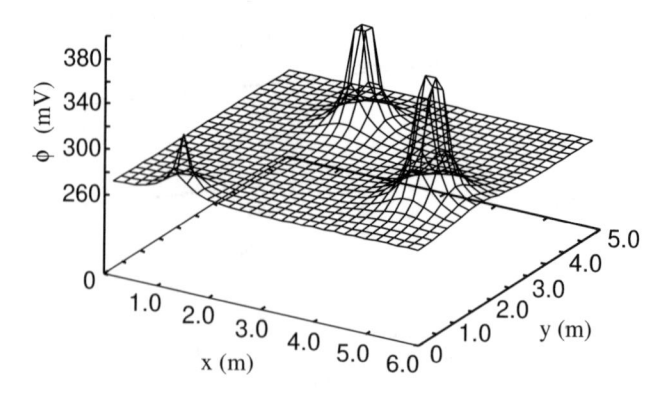

Figure 7: Potential distribution on the surface of the concrete specimen (result of the direct BEM calculation).

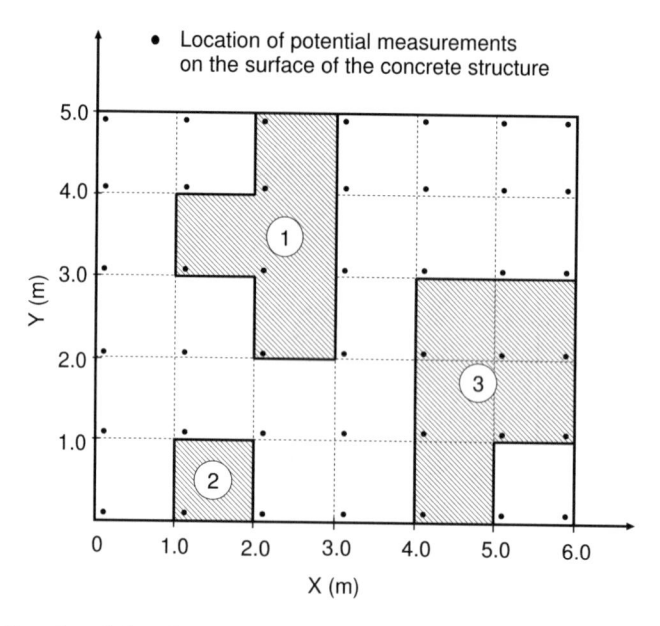

Figure 8: Result of the first-step of the multi-step genetic algorithm and the locations of potential measurements on the surface of the concrete structure.

4 Numerical simulation and discussion

An example using simulation data was used to verify the effectiveness of the proposed multi-step genetic algorithm. A preliminary numerical simulation was carried out to obtain the potential values on the surface of the concrete structure.

Binary string = 28 bits   Binary string = 12 bits

Binary string = 29 bits

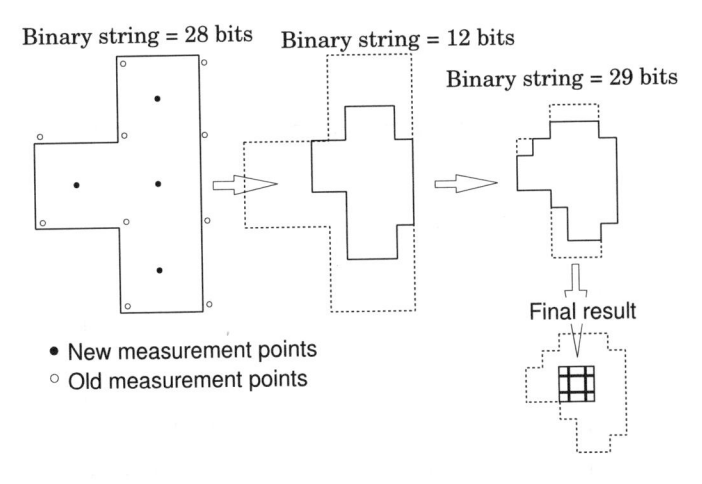

Final result

- New measurement points
- Old measurement points

Figure 9: The multi-step genetic algorithm for detecting the first corroded part.

By giving all of the boundary conditions in Equation (2) and Equation (3), the Laplace's equation in Equation (1) was solved directly using BEM.

For carrying out the multi-step genetic algorithm, a small number of potentials were chosen from the obtained potential values. These potential values were rounded off into three digits in order to simulate modeling and measurement errors and used as the measured potential data.

4.1 Model of the reinforced concrete structure

We considered a model of reinforced concrete specimen shown in Figure 4, where the electrical conductivity (κ) of the concrete was 7.10^{-3} $1/\Omega.m$, the steel matrices were cast in the concrete in the depth of 0.07 m, the diameter of the steels were 0.01 m, and the distance between two steels in the concrete specimen was 0.2 m in both of x and y directions.

The steel matrices were divided into corroded and non-corroded parts. Two polarization curves of the steel in concrete which were immersed in a 10% NaCl solution for 90 days ($E_{corr} = -0.635$ Volt vs. SCE.) and 45 days ($E_{corr} = -0.270$ Volt vs. SCE.), that were reported by Wheat et al. [5], were employed to represent the polarization curves of the corroded and non-corroded parts of the steel matrices, respectively.

To reduce the computational time, we linearized the polarization curves as shown by solid-line in Figure 5. The formula of the linear curves for the corroded and non-corroded parts were $\phi_a = -10.i + 0.600$ Volt and $\phi_c = -10.i + 0.270$ Volt, where i was the current density A/m^2. It noted that when the original non-linear curve is used, the proposed method also can be carried out normally, but the calculation time for each iteration of BEM calculation became increase.

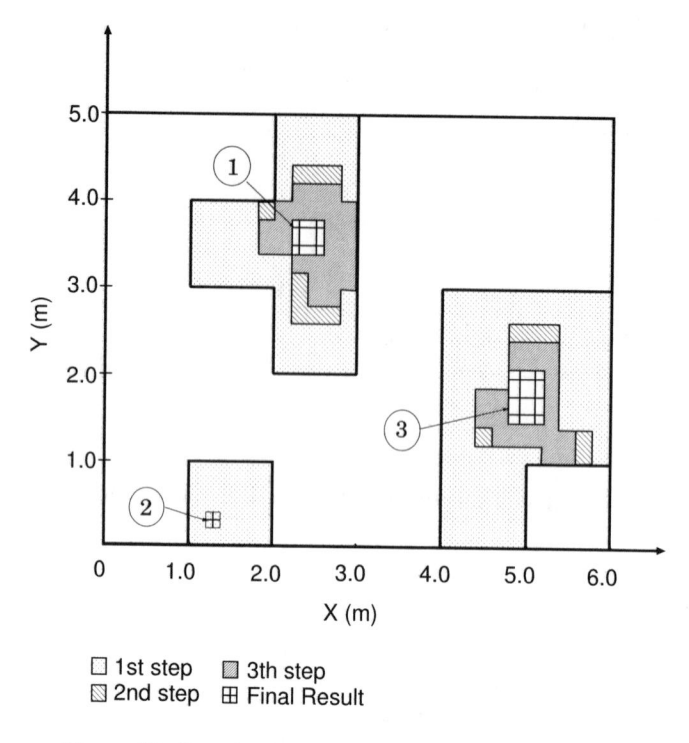

Figure 10: Result of the multi-step genetic algorithm.

4.2 Direct boundary element calculation

Three corroded parts were assumed on the steel matrices in the concrete specimen. Their locations and shapes were given in Figure 6.

In the BEM calculation, the surface of the steel matrices was discretized into 1500 constant elements, while the surface of the concrete, except for the top surface, were discretized into 970 constant elements. The type of the elements for the concrete and steel were the rectangular element (size 0.2×0.2 m) and the pipe element (length 0.2 m and diameter 0.01 m), respectively. The potential value at any point on the top surface of the concrete specimen were calculated by using the method of images [7].

Figure 7 shows the potential distribution on the surface of the specimen as a result of the direct BEM calculation. In order to simulate modeling and measurement errors in carrying out the multi-step genetic algorithm, however, only about 50 values of the potentials were chosen and rounded off into three digits. These potential values were used as the measurement data of potential on the surface of the concrete specimen.

4.3 Result of the multi-step genetic algorithm

In order to carry out the multi-step genetic algorithm for solving the inverse problem using the simulation data, firstly, let us assumed that the number, locations and shapes of corrosion on the steel matrices were unknown and only a small number of potential data on the surface of the concrete specimen were available. The electrical conductivity (κ) of the concrete and the polarization curves of corroded and non-corroded steels were assumed to be the same as in the previous sections.

A multi-step genetic algorithms was employed to obtain the corrosion profile (C_k) on the steel matrices in the concrete structure. The C_k was encoded by a binary string. For each step of examination, we used the genetic algorithm driver which was developed by D. L. Carroll [6].

In the first step of examination, to define the length of the binary string of the C_k, the steel matrices were discretized using a low resolution (coarse mesh) into 30 segments. Each segment consisted of 50 steel-elements.

Forty-two simulation data of potential, which uniformly spread out in the entire surface of the concrete structure, were employed as the measurement values of potential on the surface of the concrete structure. The distance between two locations of measurement was one meter.

The genetic algorithm was carried out by setting the number of populations of 50 and the number of generations of 100. In addition, the tolerance (ε_o) was 10^{-5}. The result of the first-step of the multi-step genetic algorithm was given in Figure 8. It can be seen that the corrosion was localized into three parts on the steel matrices.

In the second and next steps, the examination was localized for each corroded parts. The procedures of the genetic algorithm were the same as in the first step, but it may uses a smaller size of segment, different length of the binary strings, some new measurement data and/or different number of populations and generations. Figure 9 shows result of the multi-step genetic algorithm for detecting the the first corroded part. In examining the corrosion of this part, three new measurement data were added. The accurate result of the corrosion profile was obtained after performing the forth-step of the standard genetic algorithm.

Similar procedures were also carried out for detecting the second and third corroded parts that were obtained in the previous step. The result of the multi-steps genetic algorithm for identifying the corrosion on the steel matrices was summarized in Figure 10. It can be seen that the second and third corrosion parts were accurately detected after carrying out two and four steps of the standard genetic algorithm, respectively.

For the given example, the complete examination process consisted of 8 steps of the genetic algorithm. The calculation time for the complete problem to convergence was about 7 hours (using VT-Alpha 6-667, 667 MHz, OS:Tru64 Unix, memory 2G). However, Since each corroded part is examined separately, then we can employ parallel computation to reduce the calculation time for the complete problem to convergence.

5 Conclusions

A multi-step GA was developed to eliminate the difficulties of the inverse method for detecting the corrosion profile from a small number of potential data. The corrosion profile represents the number, locations and shapes of corrosion on the steel matrices. It is encoded by a binary string which is defined by discretizing the steel matrices into a number of segments. A tree structure was employed and the examination was localized for the root and the branches. Therefore, the standard GA can be applied locally and performed efficiently. An example using simulation data was used to demonstrate the effectiveness of the proposed method.

References

[1] J. P. Broomfield: "Corrosion of Steel in Concrete", London: E & FN Spon (1998).
[2] M. Ridha, K. Amaya & S. Aoki, Proc.12th Compt. Mech. Conf., JSME, 461-462 (1999).
[3] S. Aoki, K. Amaya & M. Miyasaka: Boundary Element Analysis on Corrosion Problems, Shokabo (1998).
[4] D. E. Goldberg, "Genetic Algorithms in Search, Optimization and Machine Learning", Addison- Wesley (1989).
[5] H. G. Wheat, Z. Eliezer, Corrosion, **41**, 11, 640-645 (1985).
[6] D. L. Carroll, "FORTRAN Genetic Algorithm Driver", http://www.staff.uiuc.edu/~carroll/ga.html, (1998).
[7] C. A. Brebia, The Boundary Element Technique in Engineering, (London: Newnes-Butterworths, 1980), p.102.

A reactive transport model for evaluating the long-term performance of stainless steels in concrete

M. Boulfiza
Department of Civil Engineering, University of Saskatchewan, Canada

Abstract

One significant challenge facing the world today is the decay of infrastructure. A major component of infrastructure decay is the degradation of bridges and highways due to corrosion of the embedded reinforcing steel. This problem has prompted the development of a variety of alternative strategies for increasing the service life of reinforced concrete structures exposed to harsh environments. Over the last two decades, many advanced metallic and non-metallic materials have been developed for withstanding severe corrosion typically encountered in concrete bridge decks. However, adoption of these materials by industry has fallen far short of initial expectations, due in a large extent to a lack of tools to evaluate their long term in-service performance. At present, very little reliable information is available for evaluating the corrosion performance of advanced reinforcement materials available in the market for bridge decks over the entire service life of the structure. Based on first principles, a comprehensive reactive transport model is proposed in this study for predicting the occurrence of general and localized corrosion in concrete bridge decks. Preliminary results show excellent agreement with experimental data under various environmental conditions.

Keywords: corrosion, stainless steel, concrete, bridge deck, reactive transport, modelling, service life.

1 Introduction

The cost of protecting and restoring concrete structures worldwide is a large and growing problem (>$100 billion/yr). A major component of infrastructure decay is the degradation of bridges and highways due to corrosion of the embedded reinforcing steel. This problem has prompted the development of a variety of alternative strategies for increasing the service life of reinforced concrete structures exposed to harsh environments. However, adoption of these materials by industry has fallen far short of initial expectations, due in large extent to a lack of tools to evaluate their long term in-service performance. At present, no technically sound procedure is available for evaluating the corrosion performance of advanced reinforcement materials available in the market for bridge decks over the service life of the structure.

Reinforcing steel in concrete is initially surrounded by a highly alkaline (pH > 13) pore solution. At this high pH, a thin protective layer of ferric oxide is formed on the surface of the steel. Disruption of the protective layer of ferric oxide takes place when the concentration of chloride ions in the vicinity of the rebars exceeds a critical threshold value. Based on first principles, a comprehensive model is proposed in this study for predicting the occurrence of localized corrosion which is known to govern the corrosion behavior of most reinforcing materials in chloride laden environments. Pitting Corrosion (localized) is assessed by calculating the corrosion and repassivation potentials as a function of pore water chemistry and exposure conditions for two stainless steels (304 and 316) and compared with carbon steel. The ingress of chloride ions and the amount of the various aqueous species available in concrete along the depth of concrete cover are calculated using a Pitzer-based reactive transport model which considers all possible chemical interactions with the concrete solid phases and pore water solution.

2 Mathematical model

The ability to understand, and, therefore, to predict the onset and subsequent evolution of corrosion in concrete requires the knowledge of: flow and transport processes; chemical interactions between dissolved constituents in the pore solution and the solid phases (concrete and steel reinforcement); as well as coupling between the transport and the chemical processes. Flow and mass transport in the presence of chemical reactions, usually known as reactive transport plays a key role in a variety of problems that may face the bridge engineer who deals with the design and management of structures. Reactive transport processes range from relatively simple approaches, such as solving the transport equation with an appropriate retardation term, to the simulation of complex, multispecies models that couple transport with different types of chemical reactions. The model proposed in this study belongs to this latter class.

A reactive transport model for chlorides ingress and carbonation in concrete has been extended to include corrosion modeling of high performance reinforcing steels. This model deals with transport of carbon dioxide and chloride ions in the presence of chemical reactions. A detailed description of the mathematical model at the core of the simulations in this study has been presented in earlier papers [1, 2] and only the salient features of this model will be reviewed here.

At the core of this model lies the diffusion-reaction equation, which can be written as

$$\nabla \cdot \phi D \nabla c_i + \phi S_i = \frac{\partial \phi c_i}{\partial t} \qquad (1)$$

in which c_i denotes the concentration of the ith species, S_i represents the source (or sink) term (i.e. the change in mass due to chemical reactions) of the ith species per unit volume of solution, D denotes the coefficient of molecular diffusion in concrete and ϕ denotes the porosity. In the above equation the mass of species are increased or destroyed in the source term by a number of mechanisms depending on the problem being considered [3]. These mechanisms may include both homogeneous and heterogeneous reactions. Homogeneous reactions, refer to processes that occur within the aqueous phase itself while other fluid phases present in the void space and the solid matrix are not involved. Heterogeneous reactions, on the other hand, refer to processes that involve interactions between components in the aqueous phase and constituents of the solid phase, or transfer across interphase boundaries within the void space. Examples include reactions between aqueous species and solid phases in concrete and reinforcement. Reactions of either type that contribute to the overall source term S_i may be described by using equilibrium or kinetic models [2, 3]. In equilibrium models the concentrations of the multiple species are simultaneously adjusted within the aqueous phase in accordance with equilibrium relationships and mass balance constraints. Such models are appropriate when the rates of reactions are relatively rapid (in comparison with other transport and transformation mechanisms). For example, equilibrium models are appropriate for the description of carbonation or chlorides binding using ion exchange theory. In other situations, when reaction rates are relatively slow, like in the case of corrosion, kinetic models become necessary.

The mass balance and mass action expressions are used to determine how the masses of different chemical elements in an aqueous system, represented as components, are distributed among various species that may be present. For corrosion problems, one needs to explicitly account for redox reactions which involve the exchange of valence electrons between reactants and products. An understanding of redox processes is essential for predicting the onset and evolution of corrosion because the different redox states exhibited by an element can behave very differently from one another. The rates of redox reactions involved in corrosion problems are notoriously slow. Typically, this is because redox reactions involve breaking relatively strong covalent bonds.

Thermodynamics suggests that systems that are in a non-equilibrium state will try to achieve thermodynamic equilibrium. Typical electrochemical reactions involved in corrosion tend to proceed slowly toward equilibrium, and concepts of chemical kinetics must be invoked to characterize how fast the reaction is moving towards equilibrium conditions. A way to characterize corrosion has been through polarization curves and the associated potentials and current densities [4, 5]. Although, most electrochemical reaction rates are usually represented by current densities, it is always possible to convert corrosion rates to mm/year using simple formulas. It is worth remembering that kinetic expressions are, generally, empirical in nature, and must be determined from experimental measurements. A model, based on the use of experimentally measured repassivation potentials and calculated corrosion potentials, is used here for predicting the long-term performance of various alloys in chloride-laden environments. If the corrosion potential is above the repassivation potential, then localized corrosion initiates.

A two-step approach is used for coupling of the source term in the mass balance equation with the chemical and electrochemical reaction equations governing the interaction of the pore water solution with the concrete solid phases as well as the reinforcement alloys. This approach involves the separate solution of the transport and chemical reaction equations over small times steps without iteration between the two.

3 The concrete environment

Cement paste is the product of hydration reaction of the constituents of cement. When cement is exposed to water, a series of reactions, called hydration of cement, occur between cement constituents and water. Both calcium silicates (C_3S and C_2S) combine with water to form gel-like calcium silicate hydrate (C-S-H) and calcium hydroxide (CH). A portion of the $Ca(OH)_2$ is consumed during the hydration of the other constituents of Portland cement, C_4AF and C_3A. When practically all gypsum has been consumed the hydration reactions of C_4AF and C_3A produce AFm phases [6]. The minerals of these AFm phases are anionic clay minerals, which have positively charged main layers. Anions incorporated between the layers balance the charge on the main layers. Water molecules are also incorporated into the interlayer. These phases crystallize as thin pseudo-hexagonal plates, composed of distorted portlandite-like octahedral layers, in which one third of the Ca^{2+} sites are occupied by Al^{3+} or Fe^{3+}. The substitution of Al^{3+} or Fe^{3+} for Ca^{2+} generates the net positive charge in the octahedral layers, and this substitution always occurs in a stoichiometric 2:1 ratio. In addition to anions, the inter-layer contains water molecules, and its water content varies with the type of anion. The inter layer anion can be easily exchanged.

Reinforcing steel in concrete is initially surrounded by a highly alkaline (pH > 13) pore solution [6]. At this high pH, a thin protective layer of ferric oxide is formed on the surface of the steel. Disruption of the protective layer of ferric oxide takes place when the concentration of chloride ions in the vicinity of the rebars exceeds a critical threshold value [7]. Not all of the chloride ions, which penetrate the concrete, remain free in the pore solution. Some of the ions get bound to the hydration products through a chemical reaction to form calcium chloroaluminate hydrate (Friedel's salt) while others get adsorbed to the various hydrates of cement. Thus, only a portion of the chloride ions remains free. It is this latter one that is responsible for causing damage to the concrete structures by corroding rebars. Table 1, provides a description of a typical concrete pore water solution.

Table 1: Ions in initial concrete pore water solution.

H_2O	53.613	mol/l
Na^{+1}	0.185	mol/l
K^{+1}	0.2665	mol/l
Ca^{+2}	1.25E-03	mol/l
SO_4^{-2}	0.028627	mol/l
$Al(OH)_4^{-1}$	8.00E-05	mol/l
OH^{-1}	0.42797	mol/l

The lack of a sound and comprehensive explanation of the mechanisms controlling changes in the pore solution chemistry due to chlorides binding and carbonation in cement-based materials has led to difficulties in predicting the effect of pore solution chemistry on the corrosion behaviour of rebars inside the concrete. Based on the concepts of ion-exchange theory, a recently developed mechanistic model has proven to be very effective in predicting the effects of temperature and carbonation on the evolution of the concrete pore water solution, especially the binding of free chlorides and release of bound chlorides [2]. A very brief description of the model is provided here.

Chemical binding of chlorides is the result of chemical reactions between chloride ions and certain cement hydration products. The final product of chemical binding of chloride ions is Friedel's salt ($3CaO.Al_2O_3.CaCl_2.10H_2O$), which is another AFm phase and its composition is $[Ca_2Al(OH)_6.2H_2O]^+.Cl^-$. The ion exchange mechanism, assumed in this study, describes chlorides binding as a replacement of hydroxide ions present in the interlayers of hydroxy AFm phase by chloride ions present in the pore solution. This ion exchange mechanism leads to the formation of Friedel's salt as expressed in the following reaction,

$$R^+ - OH^- + Cl^- \rightarrow R^+ - Cl^- + OH^- \quad (2)$$

where R is the principal layer of hydroxy AFm, $[Ca_2Al(OH)_6.2H_2O]^+$. The equilibrium constant for eqn (2) is given by:

$$K_{Cl/OH} = \frac{\{R-Cl\}\{OH^-\}}{\{R-OH\}\{Cl^-\}} \quad ; \text{ Braces in the equation denote activities.} \quad (3)$$

4 Corrosion performance of various alloys in bridge decks without overlays

The case study explored here involves three different reinforcing steels (carbon steel, stainless steel 304, and stainless steel 316) embedded in a typical bridge deck without overlays. Bridge decks without overlays use conventional concrete without any additional protection on the top surface of the deck (Figure 1). As shown in Figure 1, typical nominal dimensions for the steel reinforced deck include an overall height of 250 mm, 70 mm top cover and 50 mm bottom cover.

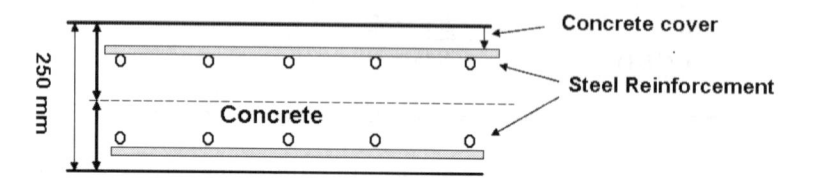

Figure 1: Schematic representation of a bridge deck without overlay.

Service life prediction has emerged, over the last few years, as a major task in the design of concrete structures. Although chloride ions in concrete do not directly cause severe damage to the concrete, they initiate and contribute to the corrosion of steel reinforcement in the structures, when the chlorides concentration at the surface of rebars reaches a threshold level. The formation of rust is associated with large volume expansion which may result in cracking, spalling, and delamination of the concrete cover and in the case of severe corrosion reduces the load-carrying capacity of structures due to a reduction of the cross-sectional area of reinforcement. In cold countries, use of de-icing salt on roads and bridges in winter causes premature deterioration of structures.

Corrosion of reinforcing steel in concrete is generally modeled in two stages. The first stage, known as the corrosion initiation time, represents the time period needed by chloride ions penetrating from the surface to reach a threshold concentration at the rebar level for corrosion to initiate. The second stage, known as active propagation time, is the time from the onset of corrosion until the end of the functional service life for concrete bridge decks due to delamination, spalling, deformation and excessive cracking of the concrete.

It is worth noticing that the numerical model used in this study calculates both the initiation period as well as the corrosion rates during the propagation stage without attempting to determine the time to cracking and spalling and their subsequent effects on the corrosion process. Time to cracking and spalling has often been used considered as the end of the useful life of the bridge deck, time at which concrete cover and the top reinforcement would typically need to be replaced. However, this should not be too hard to establish based on the computed corrosion rates. Indeed, having corrosion rates as a function of time, one could estimate the corrosion depth (loss of reinforcement cross section) and the associated pressure (generated by the change in volume of corrosion products) that would develop on the concrete cover as a function of time. Once a critical pressure is reached, cracking and spalling will occur. Table 1, provides a description of the alloys investigated in this study. Although a relatively long history of field performance is available for carbon steel, very little historical data about the field performance of high performance steels like the considered alloys is still lacking. Hence, the need for mechanistic approaches, such as the one proposed in this study, for predicting their long term performance under realistic field condition scenarios.

Table 2: Composition of the reinforcement alloys considered.

Alloy	Fe (wt%)	C (wt%)	Mn (wt%)	Cr (wt%)	Ni (wt%)	Mo (wt%)
Carbon Steel	98.50	0.50	1.00	-	-	-
Stainless 304	71.90	0.10	-	19.00	9.00	-
Stainless 316	66.90	0.10	-	17.00	13.00	3.00

Concrete quality and concrete cover thickness, two of the most widely used strategies for mitigating the effect of chlorides on the corrosion performance of the considered steels would be investigated here. The surface chloride concentration on bridge decks depends on the environment and amount of de-icing salt application on in winter. Chloride concentration values ranging from 1 to 19 kg/m^3 have been reported in the literature [1,6,7]. A surface chloride concentration of 13 kg/m^3, which corresponds to a relatively severe chloride environment, is used here for the numerical simulations.

4.1 Effect of concrete cover thickness

Three cover thicknesses (50 mm, 70 mm, 90 mm) are considered to investigate the potential benefit of using thicker concrete covers on the corrosion response of the various alloys considered. Figure 2a) shows the faster build-up of chlorides at the rebar location as the cover thickness is reduced. As can be seen in Figure 2b) and Figure 2c), only a marginal improvement in corrosion rates is achieved by using stainless steel 304 instead of carbon steel. However, a substantial improvement in pitting corrosion rate reduction is achieved by using stainless steel 316 instead of carbon steel. The change is so significant that a different scale than the one used with stainless 304 and carbon steel had to be used in order to be able to appreciate the evolution of corrosion rates in the case of stainless 316.

4.2 Effect of concrete quality

The effect of concrete quality as measured by the diffusion coefficient is investigated in this section. An effective chloride diffusion coefficient of 1.0×10^{-11} m^{2}s^{-1} is used to represent a low quality concrete, a diffusion coefficient of 5.0×10^{-12} m^{2}s^{-1} is used to represent a medium quality concrete, whereas

1.0×10^{-13} m^{2}s^{-1} is used to represent a high quality concrete. Figure 3(a) shows that concrete quality can have a tremendous impact in limiting the accumulation of chlorides at the top layer of steel reinforcement. A high quality concrete (Do = 1.0×10^{-13} m^{2}s^{-1}) would lead to a very slow build-up of chlorides at the reinforcement level, which in turn translates into lower pitting corrosion rates. As can be clearly seen from Figure 3(b) and Figure 3(c), only a marginal benefit would be achieved by using stainless 304 instead of carbon steel in conjunction with this mitigation strategy. Figure 3(d), on the other hand, shows that a substantial improvement would be achieved by using stainless 316 instead of stainless 304 or carbon steel.

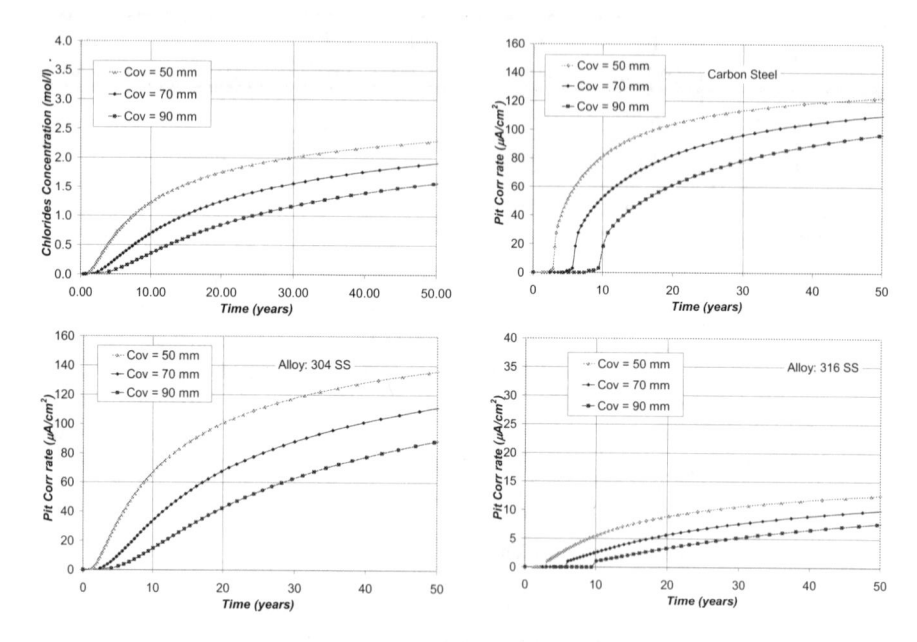

Figure 2: Effect of concrete cover thickness on (a) the chlorides accumulation at the top reinforcement location and (b), (c), (d) the pitting corrosion rates of the various steel reinforcements used.

5 Conclusions

The following conclusions can be drawn:

1. Based on first principles, a comprehensive reactive transport model is proposed in this study for predicting the ingress of chloride ions and subsequent occurrence of localized (pitting) corrosion in concrete bridge decks.

2. The model can predict both the corrosion initiation time and the corrosion rates during the propagation period prior to concrete cracking and spalling, which is usually considered the end of the useful life of a bridge deck since repairs would be required

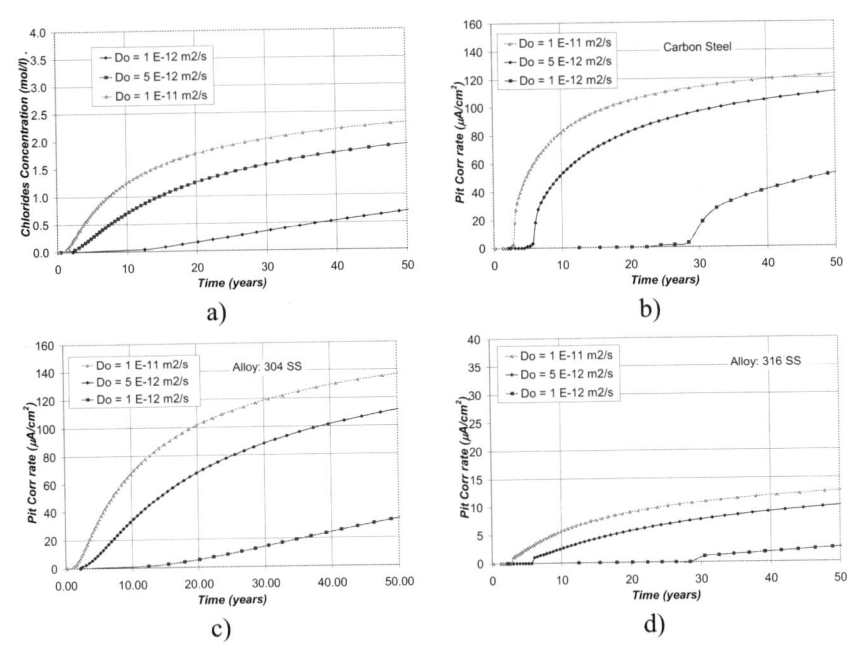

Figure 3: Effect of concrete quality on a) the chlorides accumulation at the top reinforcement location and b), c), d) the pitting corrosion rates of the various steel reinforcements used.

3. The relative performance of various corrosion mitigation design strategies has been investigated in a chloride-laden environment under constant chlorides ingress conditions for more than 50 years.

4. It has been shown that concrete quality, as measured by the apparent chlorides diffusion coefficient, has a more dramatic impact of chlorides ingress, and hence on corrosion rates, than the cover thickness.

5. Only a marginal corrosion rate reduction would be achieved by using stainless 304 instead of carbon steel whereas a substantial improvement would be achieved by using stainless 316 instead of stainless 304 or carbon steel.

Acknowledgements

The author would like to thank NSERC and ISIS Canada for their generous support.

References

[1] Munshi, S. & Boulfiza, M. Effect of Cations and Carbonation on Chlorides Binding in Concrete. *First International Structural Specialty Conference, ISSC-1*, Calgary, Alberta, Canada, May 23-26, 2006.

[2] Boulfiza, M. & Munshi, S. A reactive transport model for carbonation and chlorides ingress in concrete. *ISTC'06 International Symposium of Theoretical Chemistry*, Algiers, June 12-15, 2006.

[3] Lichtner, P. C., Steefel, C. I. & Oelkers, E. H.; (eds). *Reactive transport in porous media*. Reviews in Mineralogy, Vol. 34, Mineralogical Society of America.

[4] DeGiorgi, V.G. Corrosion basics and computer modeling. Naval Research Laboratory, Department of the Navy, Washington D. C. 203755000, U.S.A.

[5] Brebbia, C.A., DeGiorgi, V.G. & Adey, R.A., (eds) *Simulation of Electrochemical Processes*. ELECTROCOR 2005. First International Conference on the Simulation of Electrochemical Processes, 2 - 4 May 2005, Cadiz, Spain.

[6] Taylor, H.F.W., *Cement Chemistry*, 2nd edition, Thomas Telford, London, 1997.

[7] Broomfield, J.P. *Corrosion of Steel in Concrete: Understanding, Investigation and Repair*. Taylor & Francis; 2nd edition 2007.

Characteristic evaluation of corrosion of aluminum-based rigid conductor line

S. Mohri & Y. Sato
Railway Technical Research Institute, Japan

Abstract

Aluminum-based rigid conductor lines have been widely used for tunnel sections due to a cost-saving factor and their lightweight structure. Aluminum, however, is easily corroded, and it is therefore important to understand the corrosive characteristics of aluminum under the practical environment. The corrosion of the conductor lines inside the tunnel section is mainly due to groundwater seepage, containing alkali components or seawater. The authors carried out an accelerated corrosion test with an aqueous solution of sodium hydroxide to comprehend the corrosive characteristics with alkali components. Another problem is a corrosion that is likely to occur at areas where different materials contact each other. The authors also carried out saltwater spraying tests for the contact area of aluminum-base and bronze contact wire with various corrosion resistances to investigate the corrosion caused thereby. The results of these tests are described elsewhere in this paper, and the authors have indicated the predicted lifetime of aluminium-based conductor lines under the various environments.

Keywords: corrosion, rigid conductor line, aluminum, tunnel section, water seepage.

1 Introduction

Rigid conductor lines have been widely provided with tunnel sections or subway systems because the height of power collection equipment can be lowered. Aluminium-based rigid conductor is especially effective for power collection quality due to its lightweight structure. Aluminium however easily corroded [1], it is therefore important to understand the corrosive characteristics

320 Simulation of Electrochemical Processes II

of aluminium under the existing environment. The authors have carried out corrosion tests under various conditions, and predicted the lifetime of conductor lines. The results of these investigations are described elsewhere in this paper.

2 Characteristics of corrosion of aluminum base

2.1 Outlines of aluminum corrosion

The corrosion of aluminum is classified into uniform corrosion, galvanic corrosion and local corrosion. The aluminum is covered with oxide layer in the air. The uniform corrosion occurred when the layer is worn out. The layer is easily dissolved for the water whose pH is less than 4 (acid) or more than 8.5 (alkali) [2]. The galvanic corrosion is observed at areas where different metals contact [3]. The corrosion occurs according to the galvanic potential of the metals. The aluminum easily corroded because the potential of aluminum is less noble than that of copper, steel and other metallic materials. The local corrosion derived by the adhered ions, chloride, or sulfide for example. These ions cause pitting or filiform corrosion.

2.2 Characteristics of corrosion under the practical environment

Fig. 1 shows the cross section of aluminum-based rigid conductor line. Main aspects of the corrosion of the conductor line installed in tunnel sections are as follows:
(a) The uniform corrosion due to groundwater seepage containing alkali components
(b) The galvanic corrosion due to the contact between aluminum-base and bronze contact wire

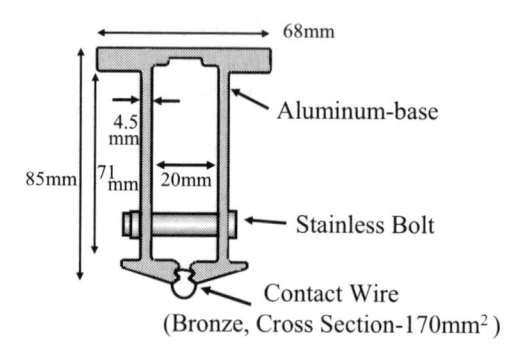

Figure 1: The Cross Section of Aluminum-based Rigid Conductor Line (Length of specimen for the test: 100 mm).

2.3 The tests for evaluating of corrosion

For evaluating these corrosions, the authors carried out corrosion tests as follows:

(a) Accelerated corrosion test by aqueous solution of sodium hydroxide was carried out for the investigation of the corrosion caused by alkaline water. The authors examined corrosion state and carried out tensile test of specimens.
Section 3.1 has described the results.
(b) The authors carried out tests to evaluate the effects of measure for galvanic corrosion. For example, tin-plating, anodic oxide layer and insulating of the contact area.
Section 3.2 has described the results.
(c) The authors carried out a saltwater spraying test to estimate the speed of corrosion at areas where aluminum-base and bronze wire contact with various corrosion resistances.
Section 3.3 has described the results.

3 Tests results

3.1 The corrosion tests with alkali water

3.1.1 The outline of the tests
Fig. 1 shows the aluminium-based rigid conductor lines immersed into aqueous solution of sodium hydroxide. The pH of the aqueous solution was 13. For a comparison with the test specimen, hard-drawn aluminum stranded conductors, whose cross section of 200 mm^2 was immersed together. The authors also carried out material fatigue tests after the specimens exposed for 123 hours.

3.1.2 The results of the tests
Prior to the evaluation of corrosion of the aluminum base, the authors measured the remains of aluminum in the stranded conductors by the eddy current wire diagnosis device as developed by our Institute. In comparison with the results of the conductors as installed in the existing railway system, it is possible to estimate corresponding test time to the existing system. Table 1 shows the summarized results. From the results, 123 hours under the test environment corresponds to 57 years under the practical environment.

Table 1: The comparisons of test hours with years under the practical environment.

Specimen	Test Hours (h)	Remains of Aluminum (%)	Estimated Years under the Practical Environment (y)
NEW	0.0	100	0.0
A	36.5	84	12.0
B	53.0	74	19.7
C	58.7	58	32.6
D	123.3	26	57.4

Table 2 shows the corrosion state of the specimens as classified in Table 1. The mass of the specimen D reduced by 14.1% in comparison with the specimen before exposed. The electric conductivities and modulus of elasticity are almost equal to the new specimen. The results indicate that the material properties themselves did not vary after the test completed.

Table 2: The corrosion state of the specimens.

Specimen	Mass of Unit Length (kg/m)	Mass Reduction Rate (%)	Electric Conductivity (%)	Module of Elasticity (GPa)
NEW	3.10	-	56.4	66.9
A	2.93	5.5	57.0	65.8
B	2.87	7.5	56.9	66.7
C	2.79	10.1	56.7	66.4
D	2.66	14.1	55.9	66.9

Table 3 shows tensile test results. The fracture strengths of the specimen reduced according to hours exposed. The tensile strengths however, hardly varied because the thickness of cross sections also reduced. Under the installed condition, the stresses for the aluminum base mainly caused by the weight of the aluminum base itself and contact wire. Therefore, the specimen D has a sufficient strength usable for the existing system. It means the aluminum base is available in terms of the strength for 57 years in practical application with no problem.

Table 3: The results of tensile test.

Specimen	Cross Section (mm^2)	Thickness of Cross Section (mm)	Fracture Strength (N)	Tensile Strength (MPa)
NEW	110.1	4.50	25795	234
A	100.9	4.23	23108	229
B	98.3	3.88	22375	228
C	93.4	3.77	21105	226
D	48.6	1.95	11968	246

3.2 The evaluation of measures for galvanic corrosion

3.2.1 The outline of the tests

The authors spayed saltwater to the specimens to evaluate the measures for galvanic corrosion as shown in Fig. 1, and measured the mass reductions of the specimens. The density of the saltwater was adjusted to 5.0%, and the temperature was to 50 degrees centigrade. Table 4 indicates the measures for the specimens [4].

3.2.2 The results of the tests
The mass reductions of the aluminum base throughout the test reached to 100 days as shown in Fig. 2. The specimen of the largest mass reduction is specimen A, and those of the smallest are specimen D and E. The results indicated that the most effective measures for galvanic corrosion are anodic oxidation and the insulation of contact area. Plating metals also have some effect [5]. The authors also carried out a grip test as shown in Fig. 3. The max gripping force between the aluminum base and the contact wire were obtained as shown in Fig. 4. The results indicate that the all specimen satisfied the specified value (5500 N) for gripping the contact wire.

3.2.3 The conclusions of the evaluation
The most effective measures against the galvanic corrosion are to insulate contact wire or oxidize the aluminium base. These measures however are unsuitable to reduce the circulation current, because keeping the base and the wire to the same electric potential is the most effective to reduce the current. Plating should be paid attention in view of both suppressing the corrosion and reducing the circulation current; therefore, the authors decided to adopt tinning to the existing system. Furthermore, all the specimens have a sufficient strength to grip contact wire until the test reached to 100 days. The largest amount of mass reduction for 100 days is 5.2 g as shown in Fig. 2 (specimen A at 100 days). Therefore, the authors assumed that the mass reduction of 5.2 g is the limitation for use in the existing system.

Table 4: The measures of the specimens.

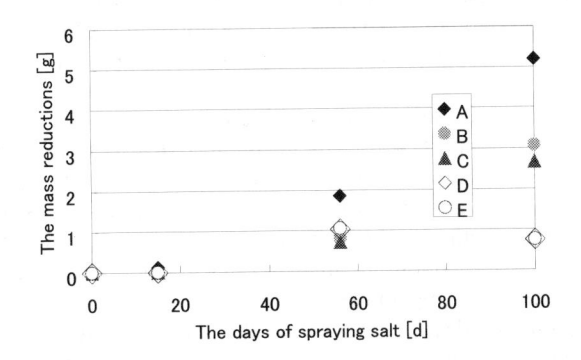

Measure	A	B	C	D	E
Contact wire (Bronze)	-	Tin - Plating	Zinc - Plating	-	-
Base (Aluminum)	-	-	-	Anodic Oxidation	Insulated

Figure 2: The mass reductions of the specimens.

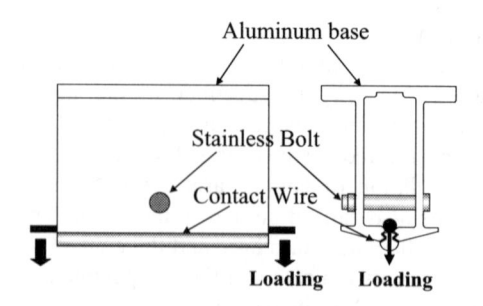

Figure 3: The outline of the grip test.

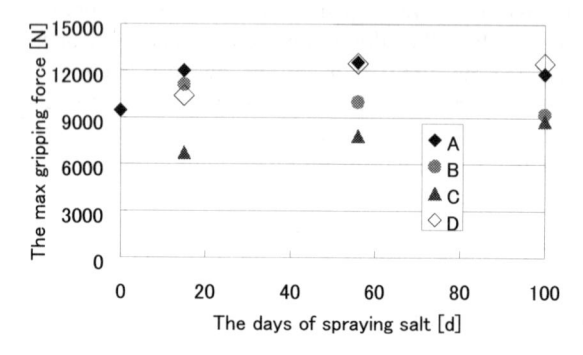

Figure 4: Max-gripping force between base and contact wire.

3.3 The estimation for the speed of the galvanic corrosion

3.3.1 The outline of the whole tests

The authors carried out to saltwater splaying tests to estimate the speed of the galvanic corrosion under the practical environment. The saltwater dropping tests, which simulated the practical environments, were also carried out. The authors decided the test conditions by the component analyses of groundwater seepage in tunnel sections. The mass reductions attempted to formulate by comparing the results of saltwater spraying test and dropping test.

3.3.2 The outline of the saltwater spraying tests

A cross section of contact wire will be reduced by wear. It will affect the corrosion speed. Another factor which affects the corrosion speed is diminishing of the plated tin. The authors carried out saltwater spraying test to investigate the effects of these factors to the corrosion speed. Fig. 1 shows the specimens and Table 5 shows the conditions of the contact wire. The density of the salt adjusted to 5%, and the temperature adjusted to 50 degrees centigrade. The term of the test is up to 100 days. Fig 5 shows the mass reductions of the aluminum base. The result shows that the mass reduction speed of the aluminum base with the worn out wire is less than that of the new wire. The result also shows that the plated tin is suppressing the mass reduction.

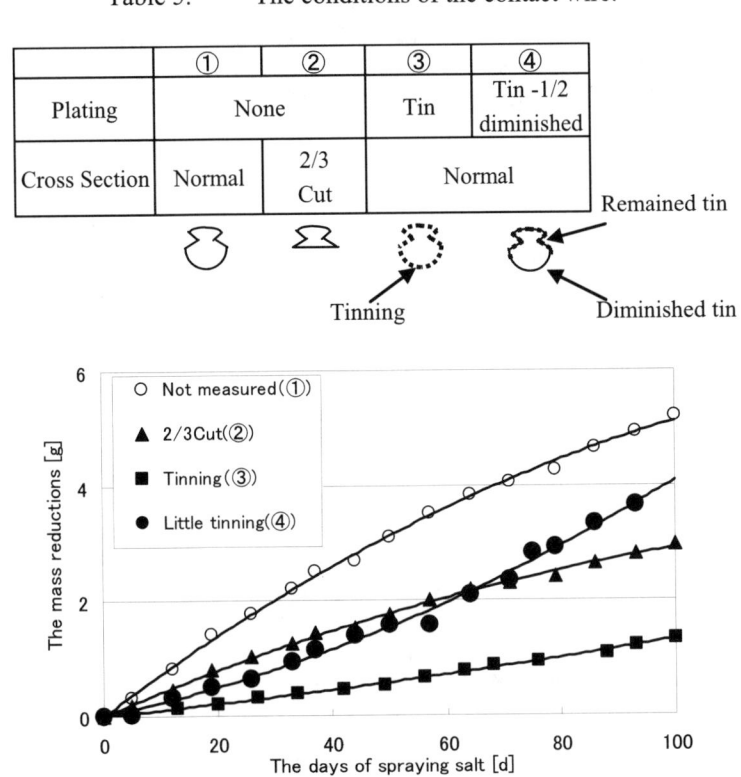

Table 5: The conditions of the contact wire.

	①	②	③	④
Plating	None		Tin	Tin -1/2 diminished
Cross Section	Normal	2/3 Cut	Normal	

Tinning   Diminished tin   Remained tin

Figure 5: The mass reductions of the aluminum bases under the various conditions of contact wires.

3.3.3 The outline of the saltwater dropping test

To simulate the corrosion under the practical environment, the saltwater dropping test was carried out. The analyses of the groundwater seepage in tunnel, the conditions of the saltwater dropping were decided as shown in Table 6. The formulation of the speed of the mass reduction was made using the results of the saltwater dropping test. The reduction of the aluminum base can be formulated by exponential function (as $y = ae^{-bt}$ for the specimen 1 and 2, as $y = ae^{bt}$ for the specimen 3 and 4). Table 7 shows the results of formulation.

Table 6: The conditions of the saltwater dropping tests.

Conditions of Water Seepage	Tunnel section near Sea		Tunnel Section near Mountain	
	Much Seepage	Little Seepage	Much seepage	Little Seepage
Densities	0.30% NaCl		0.0025% NaCl	
Conditions of Dropping	Every 10sec 0.1ml	Every 60sec 0.1ml	Every 10sec 0.1ml	Every 60sec 0.1ml

Table 7: The formulations of the speed of the corrosion.

Specimen	Tunnel Section near Sea		Tunnel Section near Mountain	
	Much Seepage	Little Seepage	Much Seepage	Little Seepage
1	a=3.7x10^{-3} b=2.0x10^{-4}	a=3.2x10^{-3} b=2.0x10^{-4}	a=1.1x10^{-3} b=6.0x10^{-4}	a=5.0x10^{-4} b=3.0x10^{-4}
2	a=2.1x10^{-3} b=2.0x10^{-4}	a=1.8x10^{-3} b=2.0x10^{-4}	a=6.2x10^{-4} b=6.0x10^{-4}	a=2.8x10^{-4} b=3.0x10^{-4}
3	a=5.0x10^{-4} b=2.0x10^{-4}	a=4.0x10^{-4} b=2.0x10^{-4}	a=2.0x10^{-4} b=6.0x10^{-5}	a=7.0x10^{-5} b=3.0x10^{-5}
4	a=1.1x10^{-3} b=2.0x10^{-4}	a=9.0x10^{-4} b=2.0x10^{-4}	a=3.0x10^{-4} b=7.0x10^{-5}	a=2.0x10^{-4} b=3.0x10^{-5}

Specimen 1 and 2
$$y = ae^{-bt}$$

Specimen 3 and 4
$$y = ae^{bt}$$

y: the Speed of Mass Reduction (g/d)
t : the Days in Use (d)

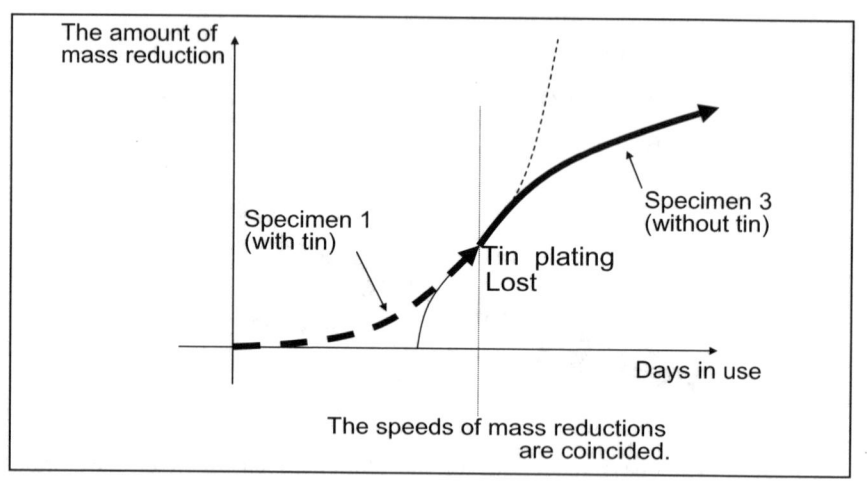

Figure 6: The way to calculate the term.

Table 8: The terms for the lost of plated tin.

Conditions of Water Seepage	Tunnel section near Sea		Tunnel Section near Mountain	
	Much Seepage	Little Seepage	Much Seepage	Little Seepage
Years	8.3	8.7	27.4	41.9

3.3.4 The prediction of the lifetime of the aluminum base

The plated tin will diminish gradually, so it is important to take into account of the effect of diminishing of plated tin for predicting the lifetime of aluminum base. The term for which the plated tin is lost is calculated in comparison with the specimen 1 and 3 indicated in Table 7. Fig. 6 shows the way of formulation about the term, and Table 8 shows the assumed terms.

On the other hand, the formulation of the mass reduction of the aluminum base was carried out. The factor a in the speed of the mass reduction (formulated as $y = ae^{-bt}$) is approximated from the results of specimen 1 (new specimen) and 2 (cut specimen) shown in Table 7. The factor a is also affected by diameter of contact wire. It is expressed as a linear function of used years. Fig. 7 shows the conditions of the formulation.

The calculations of the amount of the mass reductions are carried out from the investigations mentioned above. Fig. 8 shows the procedure of the calculations, and Fig. 9 shows the results of calculations of the mass reductions. As mentioned in Section 3.2.3, the mechanical strength for use of the aluminum base is guaranteed until the mass reduction reached to 5.2 g, and the authors assumed that this term is a lifetime of the aluminum base. Table 9 shows the predicted lifetimes of aluminum bases as obtained from Fig. 9.

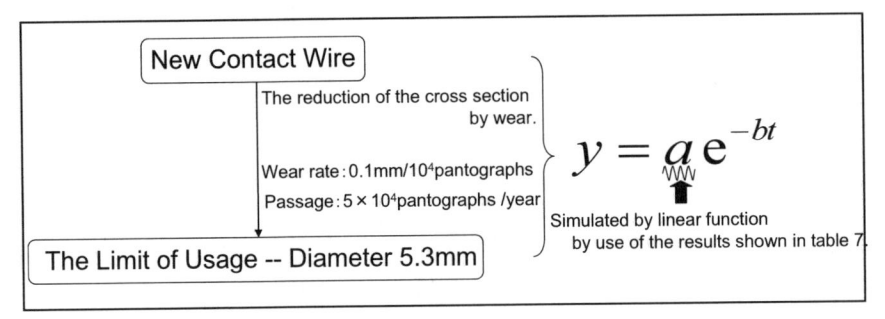

Figure 7: The conditions of the formulation.

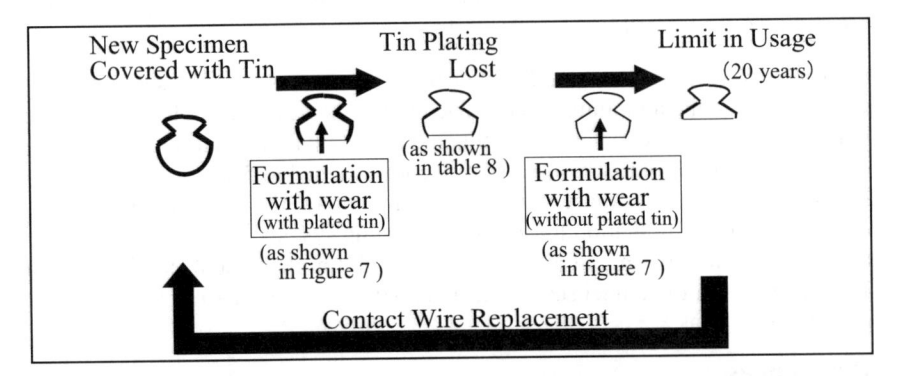

Figure 8: The procedure of the calculations of the mass reduction.

4 Conclusions

The authors carried out corrosion tests of the aluminum-based rigid conductor lines. The authors took into account two aspects of corrosion. One is caused by ground water seepage which contains the alkali materials. The other is galvanic corrosion at the contact area of the aluminum base and the contact wire. The conclusions of this paper are as follows:

(1) The material characteristics, such as tensile strength and others did not vary throughout the accelerated corrosion test with aqueous solution of sodium hydroxide. The test term was corresponded to 57 years in practical environment. The results mean that the aluminum base has no problem for strength until 57 years in use.

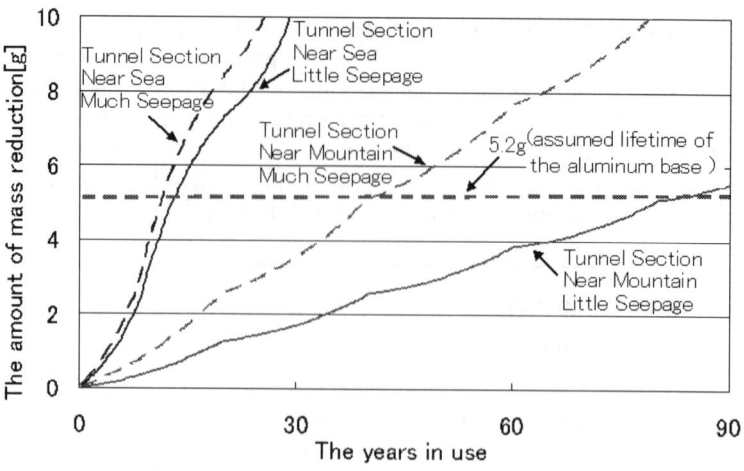

Figure 9: The amount of the mass reduction of aluminum base (calculated).

Table 9: The predicted lifetimes of aluminum bases.

Conditions of Seepage	Tunnel section near Sea		Tunnel Section near Mountain	
	Much Seepage	Little Seepage	Much Seepage	Little Seepage
Lifetime (y)	11	13	40	80

(2) The most effective measure for the galvanic corrosion was to insulate contact wire or oxidize the aluminium base. The plating, however, was also effective in both suppressing the corrosion and reducing the circulation of current.
(3) The aluminum base, whose mass reduction reached to 5.2 g, had a sufficient strength to grip the contact wire. This amount of the mass reduction was conceivable as the limitation for use in the existing system.
(4) The predicted lifetime of aluminum base with tin plated contact wire is from 11 to 13 years in the tunnel section near seashores, and from 40 to 80 years near mountainous areas.

References

[1] Samuel, L.C. & Kent, R.V.H., *Aluminum in Iron and Steel,* John Wiley & sons, Chapman & Hall: New York and London, pp.82 83, 1953.
[2] Wilson, C.L. & Oates, J.A., *Corrosion and the Maintenance engineer,* Hart Publishing Company: New York, pp. 68 69, 1968.
[3] Boris, H.T., Isidore, G. & Herman S.P., *Theory of Corrosion and Protection of Metals,* Macmillan Company & Collier-Macmillan limited: New York and London, pp. 165 193, 1966.
[4] Schmidt, D.P., Shaw B.A., Sikora, E. & Shaw, W.W., Corrosion Protection Assessment of Barrier Properties of Several Zinc-Containing Coating Systems on Steel in Artificial Seawater, *Corrosion*, 62(4), pp.323 339, 2006.
[5] Johnson, C.M., Tyrode E. & Leygraf, C., Atmospheric Corrosion of Zinc by Organic Constituents, *Journal of the Electrochemical Society*, 153(3), pp. 113 121, 2006.

Special weathering steel, a contribution to environmental protection

E. S. Ayllón
*Department of Materials Science and Technique
(Departamento Ciencia y Técnica de Materiales) – CITEFA,
V. Martelli, Argentina*

Abstract

The term weathering steels describes a class of low-alloy structural steels that develop an adherent protective rust layer, called patina, during exposure to the atmosphere. As a result of this, weathering steels, unlike ordinary plain carbon steels, do not require painting, which means important advantages: avoiding associated costs of initial painting, periodic repainting, containment and disposal of blasting debris during repainting.

The development of the protective patina depends on the steel microalloying elements which it contains, consequently, the corrosion process destroys a certain amount of steel in order to supply the microalloying elements. However, only part of all the Fe involved in this process gets into the patina. In fact, a percentage of Fe in the form of soluble corrosion products is swept away by the rainwater or dew, staining other materials with rough surfaces on the structures.

The completion of the protective patina takes at least several years.

In order to analyse the influence of non-metallic inclusions morphology and composition on the initial stages of the corrosive atmospheric attack, an experimental steel called PA (P: patina-A: acceleration) was designed.

A great quantity of non-metallic inclusions with spherical morphology, great chemical reactivity and very small size was chosen. Considering the structural behaviour, this morphology is advisable since it produces a greater isotropy of the mechanical properties.

Experimental samples of PA steel were tested in outdoors exposure producing, in only one year, a very effective passive patina. Consequently it can be observed that these tests clearly showed a vanishing effect of all staining contaminant products, which means an outstanding advantage towards the environmental protection.

1 Introduction

The development of the passivity depends on the steel microalloying elements in the patina, which leads to the destruction of a certain amount of steel to form the magnetite and patina main components (α, β, γ, and δ FeOOH). However, not all the Fe involved in this process gets into the patina. In fact, a percentage of Fe in the form of soluble corrosion products is swept away by the rainwater or dew, generally producing an unpleasant effect when staining other materials with rough surface on weathering steel structures.

This happens, for instance, in the concrete foundations of bridges, viaducts and high-tension towers. Since they are built with a porous material, the corrosion products are trapped.

Several solutions have been proposed to avoid product corrosion stains the formerly mentioned materials. Nevertheless, they did not meet he expectations.

In previous works, the detrimental effect of sulphur in the corrosion resistance of steels was highlighted. The non-metallic inclusions formed by this element and the manganese are plastically deformed at steel rolling temperatures, thus showing the inclusions and elongated morphology in the rolling direction, having the metal an anisotropy of the mechanical properties. This is quantified by tension tests, whose striction and strain results differ when the tests are carried out in the direction of the rolling or perpendicular to the already mentioned rolling. Charpy impact values change and the bending test on mandrel shows cracks when it is carried out in the direction of the MnS elongated inclusions chains.

There is a remarkable improvement when the morphology of the non-metallic inclusions is modified, thus substituting the elongated structure chains for a spherical spatial uniform distribution [1].

As it was reported in previous works, the nucleation of steel atmospheric corrosion is related to non-metallic inclusions, particularly to the MnS ones [2].

The change proposed to improve the mechanical properties outlines its possible influence on the nucleation of the atmospheric corrosion, on the patina formation and protective attributes.

Previous results verified by optical microscopy, electron scanning, and electron probe microanalysis showed that the addition of rare earths modifies steel behaviour towards atmospheric corrosion because of the different chemical composition of the non-metallic inclusions developed in the process.

Furthermore, and according to the Group IIIA element of the Periodic Table, a difference in the behaviour was also found, which was used as a non-metallic inclusions modifier. The Free Energy ($\Delta G°$) of the CeS formation is lower than the CeO's, coinciding with the fact that both of them were stable in the air at environmental temperature. The La thermodynamic values are higher than Ce's, and show that oxidation is benefited under the same environmental conditions.

A. Szummer and M. Janik-Czachor [3] point out that the presence of non-metallic inclusions, which were morphological modified by rare earths, is detrimental for corrosion resistance, and that the CeS attack nucleating effect is bigger than the MnS one, usually found in steels.

2 PA steel formulation

According to the above mentioned, the conclusion is that a formulation containing the necessary microalloying elements to produce the corrosion product film of a steel (Cu, Cr, and Si) and non-metallic inclusions – modified by rare earths additions (particularly Ce) – is considered to join both criteria: weatherability and better mechanical properties.

Following these criteria, it was planned a weathering steel experimental tapping with the addition of rare earths as modifiers of non-metallic inclusions from their morphological as well as from their chemical reactivity point of view.

A minimum C content was planned to provide the mechanical strength and ferritic attribute of structural steels. The percentage of the main microalloying elements, Cu and Cr, were fixed on practical basis. The available ferroalloys for steel production were considered as well as their use under the fusion conditions to be adopted. The recommendations given by specialised literature were used for the rare earths percentage and addition process to the ladle [4].

The percentage of Ni added – almost similar to the Cu's percentage – is the suggested one to neutralise the detrimental effect of the latter in the hot-rolled process, widely known as **hot shortness**. The vanadium – together with the aluminium – is a ferritic grain-refining agent. Due to its availability and easiness to operate, the rare earths silicide was adopted (Fe: 35%; Si:31%; and Rare Earths: 34%; with a Ce composition of: 50%; La:25%; Nd: 10%; and Pr:5%) in a percentage of 1.0 g/kg of melting steel as a means of containing these rare earths.

The chemical composition of steel – analysed in the hot-rolled product – in comparison with the one of the presently produced Cor Ten A and QST-52 non-alloyed steel is shown in Table 1.

Table 1: Chemical composition of hot-rolled PA steel, Cor Ten A steel and QST-52 non-alloyed steels used as a reference samples.

Steel	C	Si	Mn	P	S	Cu	Cr	Ni	Al	Sn	As	V
PA	0.15	0.41	1.01	0.086	0.105	0.46	1.0	0.38	0.021	0.01	0.013	0.006
Cor Ten A	0.10	0.36	0.34	0.093	0.018	0.31	0.60	0.12	-------	-----	0.005	0.008
QST-52	0.16	0.36	1.36	0.013	0.035	0.05	0.07	0.06	-------	-----	-------	-------

An induction furnace was used and 4 ingots of 60 kg net were tapped (without dead heads). The final temperature of the process after a careful bath deoxidisation by Mn, Al and Si addition was of 1673°C in the furnace and of 1620°C in the ladle. The temperature of this process was between 1590 and 1600°C.

After the tapping, the ingots were heated again at a rolling initial temperature of 1150°C. The final rolling temperature was between 950-1000°C. The final thickness was of 3.5 mm and the hardness was 283 H_B.

Upon completion of the rolling, the steel underwent a thermal treatment process to set the mechanical properties to the usually recommended ones under the Cor Ten A steel specifications (ASTM A-588 and A-242).
- Austenitised treatment: T = 870/880°C for 10 minutes.
- Water cooling at room temperature.
- Tempering at 750°C for 1 hour.
- Cooling inside furnace up to 400°C, then to the air.
- Final hardness of H_B = 160, equal to a resistance to rupture of 53 kg.mm^{-2} (52 MPa).

One of the ingots was cold-rolled. Its final thickness was 2 mm, obtained from 10 times in the rolling machine (at 0.15 each time, starting from a hot-rolled thickness of 3.5 mm.).

3 Open air exposure test

Long-lasting tests were carried out in the urban site of VILLA MARTELLI outdoor test station and in the rural site of IGUAZU station. Specimens were obtained from the hot-rolled material at 3.5 mm thickness and from the cold-rolled material at 2 mm thickness with 45 mm wide and 300 mm long. Holes with a diameter of 22 mm were made to set the samples in the same frame insulators, which were used for the previous tests [5] (Figure 1). The superficial preparation was the sanding up to Sa3 grade of SIS 05 59 00 STANDARD (1967) and weighing the samples. Three samples exposed at 3, 6, 9, 12, 18, 24, 30, 36, 42, 48, and 60 months were removed each time, and after measuring their electrochemical potential *in situ* [6] they were weighed (increased weight).

Figure 1: PA steel outdoor test in VILLA MARTELLI urban site.

Then, an acid pickling was carried out at 80°C in a HCl solution (50%) with urotropine (0.1%) as inhibitor of the steel attack. The samples were weighted at intervals of 10 minutes each until achieving almost linear weight variations between 30 and 50 minutes, according to DP/8407 ISO standard. The ΔW (g) value was determined by the intersection of the line obtained by linear regression by minimum squares with the vertical axis (Figure 2) [7].

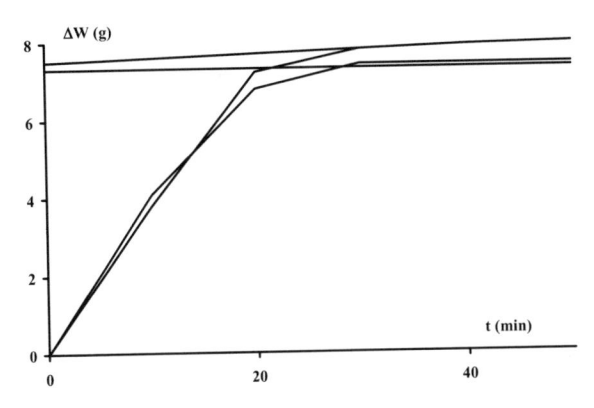

Figure 2: Weight loss based on pickling time, and pertinent linear regression.

Three hot-rolled PA steel specimens exposed for 5 years in VILLA MARTELLI station.

The results were graphically represented from (1).

$$\Delta W = k.t^n \qquad (1)$$

k and n constants were determined from experimental data by regression analysis; results are shown in Table 2.

Table 2: k and n constants of the linear regression.

Steel	Thickness		Urban Site	Subtropical Site
PA HR	3 mm	k	1.2177	0.8054
		n	0.1652	0.2354
PA CR	2 mm	k	1.9232	1.057
		n	0.3641	0.8232

HR: Hot-rolled   CR: Cold-rolled.

k coefficient represents the weight loss corresponding to 1 year of exposure. As regards the initial stage of the atmospheric corrosion, the comparison between both values reveals a greater aggressiveness in the urban site than in the rural site. n exponent values for hot-rolled PA steel are lower than Cor Ten A values and QST 52 non-weathering steel values, which were determined in previous tests [5] (Figures 3 and 4).

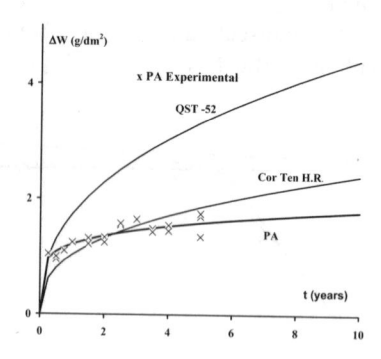

Figure 3: Time-based weight loss of hot-rolled PA steel, extrapolated at 10 years. Urban site.

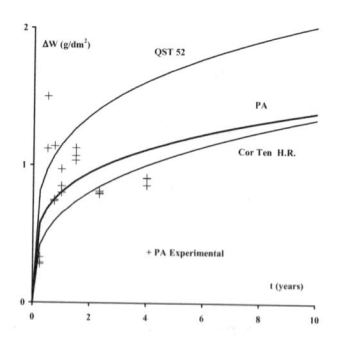

Figure 4: Time-based weight loss of cold-rolled PA steel, extrapolated at 10 years. Subtropical site.

Figures 5 to 8 show the ΔW (g) weight loss on the basis of the weight of the corrosion products that form the patina as well as the line of experimental data obtained by linear regression in VILLA MARTELLI urban site. In those same figures, we can see two lines passing through the coordinate origin, which correspond to two extreme cases of patina chemical composition. If all the Fe dissolved in the corrosive process had combined and formed FeOOH, the representative points would have met on the lowest slope line; if it had been fully combined forming magnetite, they would have met on the highest slope line with respect to the horizontal axis.

Since a mixture of the two compounds forms the patinas, the representative points should meet in the area between these two boundaries, [8].

The abscissa difference between the line resulting from linear regression of experimental data, and the lines pertaining to Fe_3O_4 and FeOOH gives us the extreme values of the corrosion products weight loss by rain wash. If we compare this experimental result with the results of the hot-rolled Cor Ten A steel and QST – 52 steel (Figures 7 and 8) [5], we will notice that there is a decrease of the corrosion products that are let loose among PA steel, Cor Ten A steel and non-weathering steel, when they are exposed to the open air.

The metal weight swept away by dew or rain-wash may be estimated as follows:

The abscissa difference between the line obtained by regression from ΔW (g) experimental data, and the one corresponding to the FeOOH line (W_c (g)) –as the most unfavourable case- represents the weight loss by ΔW_L wash:

$$\Delta W_L = W_c - W_P$$

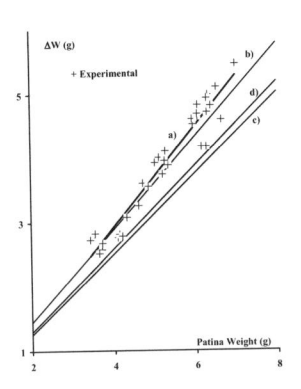

Figure 5: Hot-rolled PA steel in urban site; a) Regression; b) Fe_3O_4; c) FeOOH; d) 0.925 FeOOH + 0.075 Fe_3O_4.

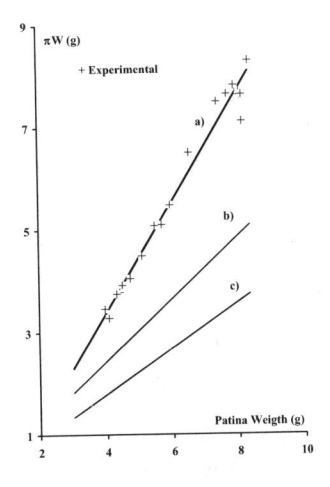

Figure 6: Cold-rolled PA steel in urban site; a) Regression; b) Fe_3O_4; c) FeOOH.

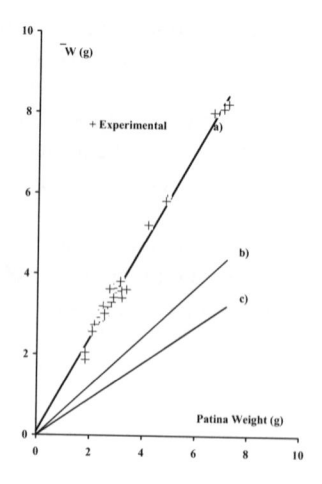

Figure 7: Hot-rolled Cor Ten A steel in urban site; a) Regression; b) Fe$_3$O$_4$; c) FeOOH.

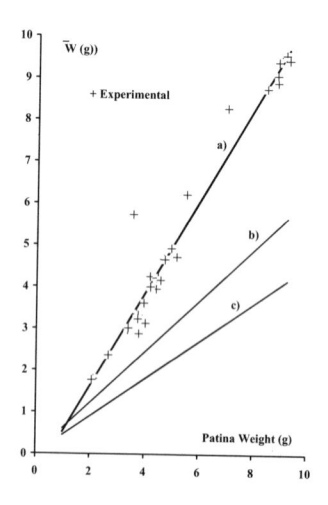

Figure 8: Hot-rolled QST – 52 steel in urban site; a) Regression; b) Fe$_3$O$_4$; c) FeOOH.

Considering as abscissas the previous ordinates and vice versa:

$$\Delta W_L = (b_c - b_P)\, \Delta W \qquad (2)$$

where: b_c is the angular coefficient with respect to the ΔW (g) vertical axis of the FeOOH line; b_p is the angular coefficient with respect to the ΔW (g) vertical axis of the line obtained by regression from experimental data; and ΔW (g) is the weight loss of the specimens.

According to (1)

$$\Delta W(\frac{g}{dm^2}) = \frac{\Delta W(g)}{A(dm^2)} = k.t^n$$

A (dm²) represents the specimen area exposed to the open air. k and n: Regression coefficients. They depend on the outdoor exposure site and the material.

Substituting (2)

$$\Delta W_L = (b_c - b_p).A.k.t^n = K.t^n \quad (3)$$

The values for the calculus are shown in Table 3, and results in Figure 9, where a) ΔW Cor Ten A; b) ΔW PA HR; c) ΔW_L Cor Ten A; d) ΔW_L PA HR

The variation of the weight loss ΔW_L with respect to time:

$$\frac{d}{dt}\Delta W_L = nKt^{(n-1)} \quad (4)$$

Table 3: Values for the graphic representation of (3) and (4).

Steel	k	n	A (dm²)	b_c	b_p	K	nK
Urban Site
PA. HR	1.2173	0.1652	2.87	1.5910	1.2834	1.0746	0.1775
PA. CR	1.9232	0.3641	3.01	1.5910	0.9522	3.6979	1.3664
Cor Ten A HR	1.2734	0.1901	3.00	1.5910	0.8598	0.8598	0.5310
Subtropical Site
PA HR	0.8054	0.2354	2.87	1.5910	1.0588	1.2302	0.2896
PA CR	1.057	0.8232	3.01	1.5910	1.1378	1.4419	1.1870

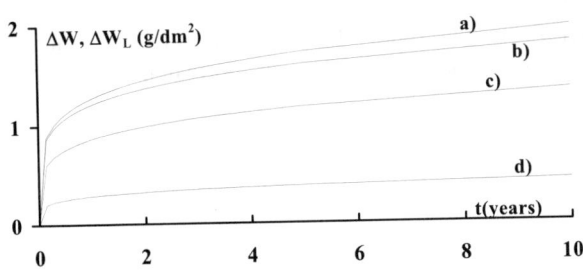

Figure 9: Total weight loss and the corresponding one to dew and/or rain-wash in urban site.

4 Conclusions

In order to obtain acceleration in the formation of the patina, the goal was to increase the intensity of the initial attack, first year of atmospheric exposition.

This aim was achieved through an increase of a non-metallic inclusions density with spherical morphology, great chemical reactivity and very small size of them.

Non-metallic inclusions modified by rare earths additions, particularly Ce, were considered to join: weatherability and better mechanical properties.

Experimental samples of PA steel were tested in two-outdoors exposure which produced, in only one year, a very effective passive patina. Consequently these tests clearly showed a vanishing effect of all staining contaminant products, meaning an outstanding advantage towards the environmental protection.

References

[1] K.G. Trubin and G.N. Oiks "Steel Making ". Mir Publishers. P. 553, (1974).
[2] E.S. Ayllón and B.M. Rosales, Revista Latinoamericana de Metalurgia y Materiales (Latin American Magazine on Metallurgy and Materials), 6, N°1 and 2, p.49, (1986).
[3] A. Szummer and M. Janik–Czachor, Werkstoffe und Korrosion, 33, p.150, (1982).
[4] R. Zenarruja, H: Moral, H. Spira, G. Waellkens, E.Ribera, A. Borras, L. Ferro and H. Reggiardo, Comité de Acería (Steel Mill Committee), IAS, Chapter 2, p.55, (1983).
[5] E.S. Ayllón, S.L. Granese and B.M. Rosales, Corrosion Reviews, 9, N° 3-4, p. 245 (1990).
[6] M. Pourbaix, Rapports Techniques, CEBELCOR, 109, R.T. 160 (1969).
[7] A. Fernandez, J.A. Navarro and B.M. Rosales, PROC, 4° Congreso Iberoamericano de Corrosión y Protección, (4° Lantin American Congress on Corrosion and Protection), Mar del Plata, Argentina, 1, p.143, October 25th-30th (1992).
[8] S. Hoerlé, F. Mazaudier, Ph. Dillman and G. Santarini, Advances in understanding atmospheric corrosion, Corrosion Sci., 45, p.1431, (2004).

Author Index

Abdurrahim A. 203
Adey R. 89, 123, 213
Aerts T. 193
Akid R. 203
Alyousif O. M. 257
Amaya K. 299
Aoki S. 299
Arurault L. 163
Ayllón E. S. 329

Barnard N. C. 3, 53
Baynham J. M. W. 89, 123, 213
Bes R. S. 163
Bisanovic S. 133
Blactot E. 267
Blanc J. 235
Bortels L. 103
Boulfiza M. 309
Brebbia C. 277
Breugelmans T. 173, 183
Brown S. G. R. 3, 53
Brunet-Vogel C. 267

Caire J. P. 63
Chaussadent T. 267
Chausse A. 13
Chen Z. Y. 33
Curtin T. 123

De Graeve I. 193
Deconinck J. 103, 153, 183, 193
DeGiorgi V. G. 113, 143, 225
Demilier L. 235
Depauw M. 183
Dorochenko A. 103
Durand C. 235

Farcas F. 267

Gaillet L. 267
Grolleau A. M. 235
Guibert A. 235

Hamburger J. 73
Hogan E. A. 113, 143
Hogan E. 235
Hubin A. 173, 183
Hussain A.-K. M. 287

Javidi A. 63

Keddie A. J. 225
Kelly R. G. 33
Krupa M. S. 113
Krupa M. 235

Lambert S. B. 247
Le Coz F. 163
LeDoux A. L. 113
Lemieux E. J. 113

Mabille I. 267
Mandin P. 73
McElman J. E. 113
Mendy H. 13
Mohri S. 319
Muharemovic A. 133
Murugaian V. 89

Nelissen G. 193
Nishimura R. 257

Peratta A. 23, 277
Picard G. 73
Pintelon R. 173
Plumtree A. 247
Pocock M. D. 225
Pongsaksawad W. 43
Powell A. 43
Purcar M. 103

Rannou C. 235
Ridha M. 299
Rifai A. M. 287
Roustan H. 73

Sato Y. 319
Stafiej J. 13
Sutter E. 267

Taleb A. 13
Terryn H. 193
Thomas J. F. 153
Tomasoni F. 153
Tourwé E. 173, 183
Turkovic I. 133

van Beeck J. 153
Van den Bossche B. 103
Van Parys H. 183
Vautrin-Ul C. 13

Williams D. 89
Wimmer S. A. 143
Wüthrich R. 73

Yildiz D. 153

WITPRESS

Computer Methods and Experimental Measurements for Surface Effects and Contact Mechanics VIII

Edited by: **J.T.M De HOSSON**, University of Groningen, Netherlands, **C.A. BREBBIA**, Wessex Institute of Technology, UK and **S-I NISHIDA**, Saga University, Japan.

The importance of contact and surface problems in modern engineering and their combined effects has led to the Eighth International Conference on Computer Methods and Experimental Measurements for Surface and Contact Mechanics. Nowadays the importance of contact and surface problems in many technological fields is well understood: they are complex and inherently non-linear due to their moving boundaries and the different properties of materials, particularly along the contact surfaces.

The use of surface treatments can reduce the cost of components and extend the life of the elements. Their effect is of particular importance in the case of surfaces undergoing contact, a problem which is addressed throughout the book. Topics featured: Surface Treatment; Surface problems in Contact Mechanics; Fracture Mechanics; Coupled analysis and experiments; Thin Coatings; Thick Coatings; Contact Mechanics; Material Surface in Contact; Applications and Case Studies.

WIT Transactions on Engineering Sciences, Vol 55
ISBN: 978-1-84564-073-6   2007
apx 400pp
£130.00/US$235.00/€195.00

Environmental Deterioration of Materials

Edited by: **A. MONCMANOVA**, Slovak Technical University, Slovakia

This book deals with the fundamental principles underlying the environmental degradation of widely used and economically important construction materials.

The invited contributions cover aspects such as the deterioration mechanisms of materials and metal corrosion, environmental pollutants, micro- and macro-climatic factors affecting degradation, the economic impact of damaging processes, and fundamental protection techniques for buildings, industrial and agricultural facilities, monuments, and culturally important objects. Basic details of ISO standards relating to the classification of atmospheric corrosivity and low corrosivity of indoor atmospheres are also included.

Designed for use by materials, corrosion, civil and environmental engineers, designers, architects and restoration staff, this book will also be a useful tool for managers from different industrial sectors and auditors of environmental management systems. It will also be a suitable complementary course book for university students in all of the above disciplines.

Series: Advances in Architecture, Vol 21
ISBN: 978-1-84564-032-3   2007 336pp
apx £98.00/US$188.00/€147.00

Find us at
http://www.witpress.com

Save 10% when you order from our encrypted ordering service on the web using your credit card.

WITPRESS

Structural Studies, Repairs and Maintenance of Heritage Architecture X

Edited by: **C.A. BREBBIA**, *Wessex Institute of Technology, UK*

The importance of architectural heritage for the historical identity of a region, town or nation is now widely recognized throughout the world. In order to take care of our heritage we need to look beyond borders and continents to benefit from the experience of others and to gain a better understanding of our cultural background.

Featuring contributions from the Tenth International Conference on Structural Studies, Repairs and Maintenance of Heritage Architecture, this book covers a broad spectrum of topics including: Heritage Architecture and Historical Aspects; Regional architecture; Structural Issues; Seismic Vulnerability Analysis of Historic Sites; Maintenance; Seismic Behaviour and Vibrations; Surveying and Monitoring; Material Characterization; Material Problems; Protection and Prevention; Simulation Modeling; Environmental Damage; Assessment and Retrofitting; Preservation and Prevention; Historical Dockyards, Shipyards and Buildings; Underwater Heritage; Surveying Techniques; Rivers, Lakes, and Canals Heritage; Site Protection, Oral Traditions and Stories.

WIT Transactions on The Built Environment, Vol 95

ISBN: 978-1-84564-085-9   2007
apx 900pp
apx £290.00/US$520.00/€435.00

Digital Architecture and Construction

Edited by: **A. ALI**, *University of Seoul, Korea* and **C. A. BREBBIA**, *Wessex Institute of Technology, UK*

Digital Architecture is a particularly dynamic field that is developing through the work of architecture schools, architects, software developers, researchers, technology, users, and society alike. Featuring papers from the First International Conference on Digital Architecture, this book will be of interest to professional and academic architects involved in the creation of new architectural forms, as well as those colleagues working in the development of new computer codes of engineers, including those working in structural, environmental, aerodynamic fields and others actively supporting advances in digital architecture. Expert contributions encompass topic areas such as: Database Management Systems for Design and Construction; Design Methods, Processes and Creativity; Digital Design, Representation and Visualization; Form and Fabric; Computer Integrated Construction and Manufacturing; Human-Machine Interaction; Connecting the Physical and the Virtual Worlds; Knowledge Based Design and Generative Systems; Linking Training, Research and Practice; Web Design Analysis; the Digital Studio; Urban Simulation; Virtual Architecture and Virtual Reality; Collaborative Design; Social Aspects.

WIT Transactions on The Built Environment, Vol 90

ISBN: 1-84564-047-0   2006   272p
£85.00/US$155.00/€127.50

All prices correct at time of going to press but subject to change.
WIT Press books are available through your bookseller or direct from the publisher.

WITPRESS

Simulation of Electrochemical Processes

Edited by: **C.A. BREBBIA**, Wessex Institute of Technology, UK,
V.G. DEGIORGI, Naval Research Laboratory, USA and **R.A. ADEY**, Wessex Institute of Technology, UK

This book contains most of the papers presented at the First International Conference on the Simulation of Electrochemical Processes held in Cadiz, Spain in May 2005. The meeting was organised by the Wessex Institute of Technology. The motivation for the meeting was to bring together researchers who have made significant developments in the area of Electrochemical modelling over recent years. Electrochemical processes are used by engineers to protect structures against corrosion, to apply coatings and paints, and as a manufacturing process. However, until recently, Engineers had to use experimental testing or frequent surveys to ensure the adequacy of a design as sophisticated prediction models were not available. The papers presented at this conference demonstrate the major advances that have been made in computational modelling to enable the most complex processes to be simulated. The papers in this book are divided into the following main topics: Modelling of Cathodic Protection Systems, Electrodeposition and Electroforming, Modelling Methodologies, Modelling Coatings. With chapters including Cathodic protection systems; Modelling methodologies and Modelling stress corrosion cracking and corrosion fatigue.

WIT Transactions on Engineering Sciences, Vol 48

ISBN: 1-84564-012-8 2005 264pp
£92.00/US$147.00/€138.00

e-Manufacturing

Fundamentals and Applications

Edited by: **K. CHENG**, Leeds Metropolitan University, UK

This book begins by presenting the concepts of and an engineering-oriented approach to e-manufacturing. Next the enabling technologies and implementation issues for e-manufacturing, including topics such as Java programming, database integration, client-server architecture, web-based 3D modelling and simulations and open computing and interaction design, are reviewed. There is then an exploration of application perspectives through a number of application systems developed by the authors based on their own front-end research and first-hand engineering practices. These include Internet based design support systems, mass customization, Java based control and condition monitoring, digital and virtual manufacturing systems, e-supply chain management and e-enterprise for supporting distributed manufacturing operations.

Designed for final year undergraduate elective courses on e-manufacturing and introductory courses on e-manufacturing at postgraduate level, this book can also be used as a textbook for teaching e-engineering in general. It will also provide a useful reference for design and manufacturing engineers, company managers, e-business/e-commerce developers and IT professionals and managers.

ISBN: 1-85312-998-4 2005 344pp
£133.00/US$213.00/€199.50

WITPress
*Ashurst Lodge, Ashurst, Southampton,
SO40 7AA, UK.
Tel: 44 (0) 238 029 3223
Fax: 44 (0) 238 029 2853
E-Mail: witpress@witpress.com*

WITPRESS

Computer Methods and Experimental Measurements for Surface Effects and Contact Mechanics VII

Edited by: **J.T.M. DE HOSSON**, University of Groningen, The Netherlands, **C.A. BREBBIA**, Wessex Institute of Technology, UK and **S-I. NISHIDA**, Saga University, Japan

The research activities in the field of surface engineering have been greatly driven by the realization that the surface is usually the most important part of any engineering component. The scientific research featured deals with fundamental and applied concepts of surface engineering, in particular focusing on the interplay between applied physics, materials science and engineering, computational mechanics and mechanical engineering. The book is devoted to fundamental and applied studies of four interconnected aspects: processing, microstructural features, functional performance as well as the design of an appropriate theoretical and predictive framework of protective surfaces.

This volume contains papers presented at the Seventh International Conference on Computational Methods and Experiments in Contact mechanics and Surface Treatment Effects.

WIT Transactions on Engineering Sciences, Vol 49

ISBN: 1-84564-022-5   2005   416pp
£145.00/US$250.00/€217.50

All prices correct at time of going to press but subject to change.
WIT Press books are available through your bookseller or direct from the publisher.

Surface Treatment VI

Computer Methods and Experimental Measurements for Surface Treatment Effects

Edited by: **C.A. BREBBIA**, Wessex Institute of Technology, UK, **J.T.M. de HOSSON**, University of Groningen, The Netherlands and **S.-I. NISHIDA**, Saga University, Japan

Papers from the sixth international conference on this topic. Highlighting fundamental and applied concepts in this interdisciplinary field, the contributions focus on the interplay between applied physics, materials science and engineering, computational mechanics and mechanical engineering.

WIT Transactions on Engineering Sciences, Vol 39

ISBN: 1-85312-962-3   2003   356pp
£133.00/US$205.00/€199.50

Find us at
http://www.witpress.com

Save 10% when you order from our encrypted ordering service on the web using your credit card.

WIT eLibrary

Home of the Transactions of the Wessex Institute, the WIT electronic-library provides the international scientific community with immediate and permanent access to individual papers presented at WIT conferences. Visitors to the WIT eLibrary can freely browse and search abstracts of all papers in the collection before progressing to download their full text.

Visit the WIT eLibrary at
http://library.witpress.com